Systems Analysis and Operations Research

Systems Analysis and Operations Research

Contributors

Qian Zhang, Zhongming Shen et al.

AURIS
Reference

www.aurisreference.com

Systems Analysis and Operations Research

Contributors: Qian Zhang, Zhongming Shen et al.

Published by Auris Reference Limited

www.aurisreference.com

United Kingdom

Printed in 2017 for Sale in the Indian Subcontinent

Systems Analysis and Operations Research

ISBN: 978-1-78154-710-6

British Library Cataloguing in Publication Data
A CIP record for this book is available from the British Library

Printed in the United Kingdom

Exclusively distributed by CBS Publishers & Distributors Pvt. Ltd.

Sales & Distribution Rights only for India, Pakistan, Bangladesh, Sri Lanka, Nepal and Bhutan.This book is not to be sold outside these territories.

Contents

List of Abbreviations

ACO	Ant Colony optimization
ACS	Adaptive current search
ANN	Artificial neural networks
BNPB	National Disaster Management Board
CDF	cumulative distribution function
CLP	Constraint Logic Programming
CO	Combinatorial optimization
CS	Chaos search
DEA	Data Envelopment Analysis
DES	Discrete event simulation
DMUs	Decision Making Units
DSCM	Demand shortage clearance model
DT	Disappointment Theory
EUT	Expected Utility Theory
FLGA	Fuzzy guided genetic algorithm
GA	Genetic Algorithm
GEUT	Generalized Expected Utility Theories
HSPs	Heat shock proteins
KKT	Karush-Kuhn-Tucker
LRP	Location-routing problem
MAS	Multi Agent systems
MCLP	Multiple Criteria Linear Programming
MCQP	Multiple Criteria Quadratic Programming
MEU	Maximum Expected Utility
MIP	Mixed integer programming
MST	Mathematical Systems Theory
NN	Neural Networks
ODP	Open Dimension Problem
PSO	Particle Swarm Optimization
RED	Region with Excess Demand
RT	Regret Theory
SMO	Sequential Minimal Optimization
STH	Systems Thinking Hierarchy
SVM	Support Vector Machine
TC	Termination criteria
TED	Total Excess Demand
TS	Tabu search
TSP	Travelling Salesman Problem

List of Contributors

Qian Zhang
School of Economics and Finance, HuaQiao University, Quanzhou, China

Zhongming Shen
School of Economics and Finance, HuaQiao University, Quanzhou, China

Xianji Zhang
BPS Global Group, Hong Kong, China

Asif Rahman
Department of Computer Engineering, Dalarna University, Borlänge, Sweden;
Department of Information Technology, Örebro University, Örebro, Sweden.

Siril Yella
Department of Computer Engineering, Dalarna University, Borlänge, Sweden;

Mark Dougherty
Department of Computer Engineering, Dalarna University, Borlänge, Sweden;

Nian Yan
College of Information Science and Technology, University of Nebraska at Omaha, Omaha, USA
Nebraska Furniture Mart, Berkshire Hathaway Company, Omaha, USA

Zhengxin Chen
College of Information Science and Technology, University of Nebraska at Omaha, Omaha, USA

Yong Shi
College of Information Science and Technology, University of Nebraska at Omaha, Omaha, USA
Research Center on Fictitious Economy and Data Sciences, Chinese Academy of Sciences, Beijing, China

Zhenyuan Wang
Department of Mathematics, University of Nebraska at Omaha, Omaha, USA

Guimin Huang
Guilin University of Electronic Technology, Guilin, China

Jaklin Tripto
Department of Science and Technology Education, Ben Gurion University of the Negev, Beer Sheva, Israel

Orit Ben-Zvi Assaraf
Department of Science and Technology Education, Ben Gurion University of the Negev, Beer Sheva, Israel

Miriam Amit
Department of Science and Technology Education, Ben Gurion University of the Negev, Beer Sheva, Israel

Oliver C. Ibe
Department of Electrical and Computer Engineering, University of Massachusetts, Lowell, USA

Edouard Kujawski
EJK Associates, Berkeley, USA

Irfan Ali,

Department of Statistics & Operations Research, Aligarh Muslim University, Aligarh, India

S. Suhaib Hasan

Department of Statistics & Operations Research, Aligarh Muslim University, Aligarh, India

Mohamed A. Shalaby
Mechanical Design and Production Department, Faculty of Engineering, Cairo University, Giza, Egypt

Mohamed Kashkoush
Mechanical Design and Production Department, Faculty of Engineering, Cairo University, Giza, Egypt

Novia Budi Parwanto
National Graduate Institute for Policy Studies (GRIPS), Tokyo, Japan
Institute of Statistics (STIS), Jakarta, Indonesia

Hozumi Morohosi
National Graduate Institute for Policy Studies (GRIPS), Tokyo, Japan

Tatsuo Oyama
National Graduate Institute for Policy Studies (GRIPS), Tokyo, Japan

Supaporn Suwannarongsri
Rattanakosin College for Sustainable Energy and Environment (RCSEE), Rajamangala University of Technology Rattanakosin, Nakhon Pathom, Thailand

Tika Bunnag
Rattanakosin College for Sustainable Energy and Environment (RCSEE), Rajamangala University of Technology Rattanakosin, Nakhon Pathom, Thailand

Waraporn Klinbun
Faculty of Engineering and Technology, Panyapiwat Institute of Management, Nonthaburi, Thailand

Zoltán Varga
Institute of Mathematics and Informatics, Szent István University, Godollo, Hungary

Manuel Gámez
Department of Mathematics, University of Almería, Almería, Spain

Inmaculada López
Department of Mathematics, University of Almería, Almería, Spain

Mazen Arafah
The Department of Industrial Engineering, Faculty of Engineering & Technology, The University of Jordan, Amman, Jordan

Tetsuo Morimoto
Department of Biomechanical Systems, Ehime University, Matsuyama, Japan

Md Parvez Islam
Department of Biomechanical Systems, Ehime University, Matsuyama, Japan

Kenji Hatou
Department of Biomechanical Systems, Ehime University, Matsuyama, Japan

Preface

Systems analysis is a problem solving technique that decomposes a system into its component pieces for the purpose of the studying how well those component parts work and interact to accomplish their purpose. Operational research is the use of advanced analytical techniques to improve decision making. The text *Systems Analysis and Operations Research* focuses on recent advances and application of system analysis and operations research. The effective method for location routing problem (LRP) using a hybrid algorithm based on clustering analysis and chaos search has been proposed in first chapter. Second chapter gives a review on recent developments and challenges associated with simulation and optimization techniques. In third chapter, we use a nonadditive measure to model data with interactions and propose new nonlinear classification models. The goal of fourth chapter is to characterize the concept map as a tool to test system thinking in the context of the human body. In fifth chapter, we consider vacation queueing systems with server timeouts. Sixth chapter proposes a portfolio allocation model with stochastic optimization as a flexible and transparent method for defining choice problems and determining hybrid solutions for critical systems with desirable properties such as diversification and robustness. In seventh chapter, the problem of optimum allocation of repairable and replaceable components in a system is formulated as a Bi-objective stochastic nonlinear programming problem. In eighth chapter, the common industrial problem "2-D Irregular Strip Packing Problem" is addressed, and a heuristic sequence-based approach is developed for solving it. In ninth chapter, we formulate a multi-commodity transshipment network flow optimization problem under various types of uncertain situations. The aim of tenth chapter is to present the optimization of energy resource management in a car factory by the adaptive current search (ACS)—one of the most efficient metaheuristic optimization search techniques. In eleventh chapter, we report on recent developments in the methodology and application areas of monitoring in complex ecological systems. In last chapter, we consider a number of critical success factors (CSFs) that are most commonly discussed in quality improvement projects.

Chapter 1

RESEARCH ON LOCATION ROUTING PROBLEM (LRP) BASED ON CHAOS SEARCH (CS) AND EMPIRICAL ANALYSIS

Qian Zhang[1], Zhongming Shen[1], Xianji Zhang[2]

[1]School of Economics and Finance, HuaQiao University, Quanzhou, China

[2]BPS Global Group, Hong Kong, China

ABSTRACT

Due to the problem complexity, simultaneous solution methods are limited. A hybrid algorithm is emphatically proposed for LRP. First, the customers are classified by clustering analysis with preference-fitting rules. Second, a chaos search (CS) algorithm for the optimal routes of LRP scheduling is presented in this paper. For the ergodicity and randomness of chaotic sequence, this CS architecture makes it possible to search the solution space easily, thus producing optimal solutions without local optimization. A case study using computer simulation showed that the CS system is simple and effective, which achieves significant improvement compared to a recent LRP with nonlinear constrained optimization solution. Lastly the pratical anlysis is presented relationship with regional logistics and its development in Fujian province.

INTRODUCTION

All companies that aim to be competitive on the market have to pay attention to their organization related to the entire supply chain. In particular, companies have to increase the efficiency of their logistics operation. It is essential that reasonable optimal systems of logistics distribution for enterprises with the development of e-commerce and integrated logistics. Location-routing problem (LRP) is an important branch of routing optimization in integrated logistics systems, which has been solving for every logistics distribution corporations.

The conceptual foundation of LRP studies dates back to Von Boventer (1961), Maranzana (1965), Webb (1968), Lawrence and Pengilly (1969), Christofids and Eilon (1969) and Higgins (1972) [1]. Although these earlier studies are far from capturing the total complexity of LRP, they first recognized the close interface between location and transportation decisions. Watson-Gandy and Dohrn（1973）may be some of the first authors credited to consider the multiple-drop nature of the vehicle routes within the location-transportation framework [2]. Many scholars have developed more efficient problem solving techniques using the concept of integrated logistic systems, which include optimal models and their algorithms. Due to its complexity, it is models and algorithms that are core for solving LRP, which is NP-hard. More and more scientists have been interested in these works. Exact algorithms and heuristics are given in some articles. Exact algorithms for LRP can be classified into four categories: direct tree search (branch-and-bound), dynamic programing, integer programming, and non-linear programming. Heuristic includes location-allocation-first, route-second, route-first, and location-allocation-second, savings/insertion, and improve/exchange [3].

Chaotic system is a common phenomenon exiting in nonlinear dynamical system with popular characteristic of dynamics. Due to the ergodicity and randomness of chaotic sequence, the chaos search (CS) architecture can be solved the traditional optimization [4]. Much of engineering is concerned with the topic of optimization, and at the heart of much of the optimization are dynamical systems. Dynamical systems can be thought of either nonlinear continuous-time differential equation. Chaos occurs in dynamical systems, and frequently in engineering, which becomes the central fascination. Some of domestic scholars use chaos search for nonlinear optimization with constraints. The main advantage is to search the solution in certain field regardless of the continuous and difference of the function that can be solved complex optimization with constrains. Furthermore, it is easy to seek the optimal solutions (or near optimal solutions) without local optimization.

The main contribution of the paper is statement of a new hybrid algorithm combined clustering analysis and chaos search, together with a proof of their correctness [5]. Several clusters are divided by preference-fitting rules. Then the optimal route has been found using chaos in every cluster [6]. The simulation results show that it is an effective method. The paper is organized as follows.

Section 2 introduces the meaning and descriptions of LRP and describes the mathematical model of LRP. Section 3 emphasizes on the hybrid algorithm based on clustering analysis and chaos search for LRP. Section 4 shows the computational results and analysis. Finally, the last section gives conclusions and future work.

DESCRIPTION OF LRP

Location-routing problem (LRP) is one of the problems in integrated logistics optimizations. The LRP can be defined as follows. A feasible set of potential facility sites and location; and expected demands of each customer are given. Each customer is to be assigned to a facility, which will supply its demand. The shipments of customer demand are carried out by vehicles, which are dispatched from the facilities and operated on routes that include multiple customers. There is a fixed cost associated with opening a facility at each potential site, and a distribution cost associated with any routing of vehicles that includes the cost of acquiring the vehicles used in the routing, and the cost of delivery operations. The cost of delivery operations is linear in the total distance traveled by the vehicles [7]. The LRP is used to determine the location of the facilities and the vehicle routes from the facilities to customers, with the aim of minimizing the sum of the location and distribution costs, while ensuring that the vehicle capacities are not exceeded [8]. The expression of LRP is given in**Figure 1**.

A constraint-based model is presented in this paper for the location routing problem. The hypotheses are as follows:

The transportation is just in time.

The facility is both starting point and destination of circular vehicle routing; each facility serves more than two customers.

Nature of demand/supply: deterministic.1 There are multiple facilities.

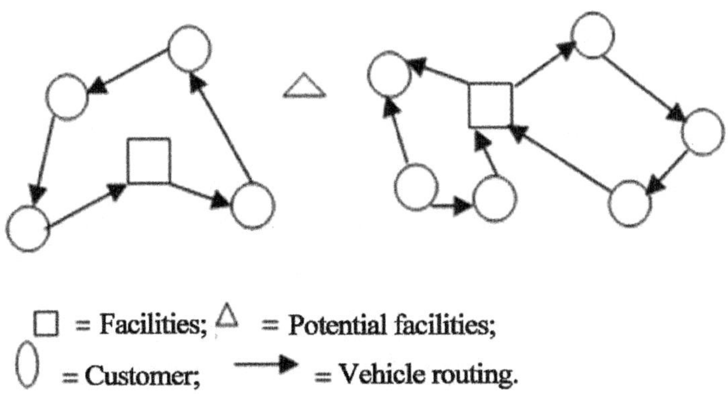

Figure 1: Expression of LRP.

Size of vehicle fleets: single vehicle (A facility is served by one vehicle; meanwhile, it is satisfied that the total demand of the route is no less than one vehicle's capacity).

Vehicle capacities are determined. The total quantity (amount) of goods is limited in every route by each vehicle's capability.

Facility capacities are undetermined; not all facilities have been chosen in every decision.

Each customer is served by one and only one vehicle. Considering the complexity of supply/demand markets, it is assumed that they are retail markets.

Each facility is considered a separate entity, not linked to the other facilities.

Objective: minimize total costs.

Decision Variables Considered Are as Follows

$$x_{ijk} = \begin{cases} 1, \text{if vehicle } k \text{ goes from customer } i \text{ to customer} \\ j, i \in S, j \in S, k \in V, i \neq j, \\ 0, \text{otherwise} \end{cases}$$

$$Z_r = \begin{cases} 1, \text{if a facility is establishe } d \text{ atsite } r, r \in G \\ 0, \text{otherwise} \end{cases}$$

Model Parameters Are Given as Follows

$$G\{r|r = 1,\cdots,m\}$$

is the set of m feasible sites of candidate facility,

$$H\{i|i = m+1,\cdots,m+N\}$$

is the set of N customers to be served,

$$S\{G\} \cup \{H\}$$

is the set of all feasible sites and customers (it is also referred to nodes),

$$V\{v_k|k = 1,\cdots,K\}$$

is the set of K vehicles available for routing from the facilities, C_{ij} average annual cost per distance traveling from

node i to node $j, i \in S, j \in S$,

F_r annual cost of establishing and operating a facility at site $r(r = 1,\cdots,m)$, q_j average number of units demands by customer $j(j \in H)$, Q_k capacity of vehicle

$k(k=1,\cdots,K)$, d_{ij} distance from node i to node j, X_{rik}, X_{rjk} traveling by vehicle k from node r to customer i or to customer j, respectively.

Model of a Special LRP Is Defined as Follows

The objective function of LRP is

$$f(x)=\min\sum_{i\in S}\sum_{j\in S}\sum_{k\in V}C_{ij}X_{ijk}d_{ij}+\sum_{r\in G}F_rZ_r \tag{1}$$

Subject to

$$\sum_{k\in V}\sum_{i\in S}X_{ijk}=1,\forall j\in H, \tag{2}$$

$$\sum_{i\in H}\sum_{j\in S}q_jX_{ijk}\le Q_k,\forall k\in V, \tag{3}$$

$$\sum_{i\in S}X_{ipk}-\sum_{j\in S}X_{pjk}=0,\forall k\in V,p\in S, \tag{4}$$

$$\sum_{r\in G}\sum_{j\in H}X_{rjk}\le 1,\forall k\in V, \tag{5}$$

$$\sum_{k\in V}X_{rjk}+Z_r+Z_j\le 2,\forall m=1,\cdots,R,r\in G, \tag{6}$$

$$\sum_{k\in V}\sum_{j\in H}X_{rik}-Z_r\ge 0,\forall r\in G, \tag{7}$$

$$\sum_{j\in H}X_{rik}-Z_r\le 0,\forall k\in V,r\in G, \tag{8}$$

$$X_{ijk}=0 \text{ or } 1,\forall i,j\in S,k\in V, \tag{9}$$

$$Z_r=0 \text{ or } 1\ \forall r\in G. \tag{10}$$

In this model, the objective function minimizes the total cost of routing, and establishing and operating the facilities. Constraint (2) ensures that each customer is served by one and only one vehicle. Constraint (3) ensures that the vehicle capacity constraints are not exceeded for any of the vehicles used in routing, while (4) is the route continuity constraint, which implies that the same vehicle should leave every point that is entered by a vehicle. Constraint (5) guarantees that each vehicle is routed from one depot. Constraint (6) guarantees that there are no links between any two depots. Constraints (7) and

(8) require that a vehicle can only from a depot if that depot is opened. The last two of constraints are the integer constraints.

A HYBRID ALGORITHM BASED ON CLUSTERING ANALYSIS AND CHAOS SEARCH FOR LRP

Clustering Analysis of LRP

Clustering analysis is one of the major techniques in pattern recognition. It is a technique for grouping data and finding where similar data are assigned to the same cluster whereas dissimilar data should belong to different clusters. The conventional clustering methods generate partitions, in a partition; each pattern belongs to one and only one cluster.

The customer aggregate C is divided to cluster by the euclidean measure. The steps of optimal facility location are performed in the following way, which have extensively analyzed and used in location theory to approximate distances between two spatial coordinates. For the sake of simplicity, it is assumed that each facility has its own set m of their coordinates of candidate sites for facility location, and no two facilities share the same location, which are denoted by

$$PF_1(X_1,Y_1),PF_2(X_2,Y_2),\cdots,$$
$$PF_i(X_i,Y_i),\cdots,PF_m(X_m,Y_m).$$

Every two coordinates of candidates sites for facility locations are denoted by $PF_i(X_i,Y_i)$ and $PF_j(X_j,Y_j)$, and which satisfied with

$$R_{ij}^{1/2}\left[\left(X_i-X_j\right)^2+\left(Y_i-Y_j\right)^2\right]^{1/2} \tag{11}$$

where R_{ij} is one half of the distance between $PF_i(X_i,Y_i)$ and $PF_j(X_j,Y_j)$.

Rules for classification are given as follows.

Rule1

$$r_{ip}<R_{ij} \tag{12}$$

C_p can only be divided into PF_i if formula (12) is true.

Rule 2 The coordinates of customers on the circumferences are satisfied that the distance to the nearest two customers is less than that to the sub nearest two points. sub-nearest meaning that they are not included a cluster by reason of distance).

Rule 3 If

$$r_{jq} > \max R_{ij} \tag{13}$$

where r_{jq} is the distance from any customer C_q to any facility PF_j, $\max R_{ij}$ is half of the most distance between any two facilities. Either these customers will not be served, or else the candidate's facility location will be abolished. Last two rules only relate to abnormal customers points.

where N is the numbers of customers,

$$C = [C_1, C_2, \cdots, C_N], C_1(x_1, y_1), \cdots,$$
$$C_p(x_p, y_p), \cdots, C_N(x_N, y_N)$$

are expressed the points and their coordinates, respectively.

r_{ip} is the distance between any candidate facility PF_i and any customer p. The tangent cycles $O_1, O_2, \ldots, O_i, O_j, \ldots, O_m$ are drawn with the radius R_{ij}, where $R_1, R_2, \ldots, R_i, R_j, \ldots, R_m$ are their radiuses, respectively.

An Algorithm of Chaos Search Optimization for LRP

The fundamental idea is that chaotic sequences belonging to the ergodic points in [0,1] are constructed by the chaos system [9]. They transform to chaotic ergodic variables in solution space of optimization and then search for a good solution. Chaotic sequences are produced by the chaotic equation, of which the Logistic difference equation is one of the most typical models [10]. The Logistic difference equation is defined as.

$$f : x_{k+1} = \lambda \cdot x_k \cdot (1 - x_k) \tag{14}$$

where λ is the so called driving parameter(a parameter capturing such factors as fertility rate) initial living area, etc. The initial value, or seed x is in $[0,1]$.$(\lambda \in (2,4])$, There $f[0,1] \to [0,1]$ if $0 \le \lambda \le 4$. From about x is restricted to $[0,1]$ and λ to $(0,4)$.

The characteristic of Logistic map is given as follow:

1) When λ to $(2,3)$, there are stable points in Logistic map.

2) When λ to[3,3.57], two sequences with seeds only slightly apart diverge very rapidly, that is to say, there exists in the period-doubling bifurcation of the solutions.
3) When λ to(3.57,4), chaos start to come out, and Logistic map isn't in stable cycling orbit. It is difficult to decide which points with Lebesgueis more than zero in chaotic sequences.
4) When λ to 4, there is a stable chaotic sequence.

In the stable chaotic sequence, any orbit f is dense, that is to say, the points of one orbit f is included in open ball $B\in(x,\varepsilon)$, when $\forall\varepsilon>0$, and $\forall x\in[0,1]$. On the contrary, all points that belong to $[0,1]$ are pressed on towards any orbit f as precision as possible. In this paper when $\lambda=4$, the population sequence (X_k) is chaotic and ergodic and gives search optimal solutions (or near optimal solutions) solutions. A chaotic system is defined as one where two different, but close, seeds result in widely different population sequences. An ergodic system is one where the sequences (X_k) will approach every possible point arbitrarily, little by little (more formally, for every pair of element of $[0,1]$, there is an xk between them). Thus, an ergodic system is one that is "trying to converge to everything". The Logistic model can be solved not only for the continuous variable optimization, but also discrete variable optimization.

A Hybrid Algorithm for LRP

The hybrid algorithm includes two steps. First, preference-fitting rules to solve location-allocation cluster the customers. Second, the optimal routes is obtained by chaos search to be minimized the function. The Logistic chaotic model is transformed for discrete optimization of LRP [11]. A legal route will be determined with the numbers customer of r in one cluster, that is to say, the order of the customer clusters will be arranged [12]. The variable d indicates the route distance in any cluster to be minimized.

$$\min\sum_{i=1}^{r} d\left(C_i,C_{i+1}\right) \qquad (15)$$

where

$$C_{i+1}=C_1,d\left(C_i,C_{i+1}\right) \qquad (16)$$

denoted the distance between customer C_i and customer C_{i+1}.

Steps with the hybrid algorithm based on clustering analysis and chaos search are described as follow.

Step 1 Customers are clustered by preference-fitting rules.

1) Initial calculations for R_{ij} and r_{ip} are performed.
2) If $r_{ip} < R_{ij}$, the customer p is divided into cluster O_i.
3) Otherwise, return step (1.1).

Step 2 $X(i)(i=1,2,\cdots,r)$ is given by initial random value. The best optimal value of the function $opt\_f=0$, given with the control parameter k gifted with a larger value.

Step 3 Put the results of order arranged $X(i)$ into $Y(i)$. It is obtainable to a current searching route, and to a current legal route by hybrid algorithm, compared with $X(i),Y(i)$.

Step 4 Calculate the function values f of the current legal route.

Step 5 Remember the minimum function values $opt\_f : f < opt\_f$, and $opt\_f = f$, $\text{step} = 0$, and also remembering the present legal route.

Step 6 Decide whether the cycle is stopped or not. If step = k, stop.

Step 7 A set of $X(1,m)$ is generated by Logistic map, step = step + 1.

Step 8 Return steps 3, and cycling.

ANALYSIS OF THE COMPUTATIONAL RESULTS

In the following example three potential facilities, and ten customs were analyzed. Each was served. With the same type of distribution vehicle that had a carrying capacity equal to 30 ton. The annual cost establishing and operating a facility F_r is equal to 160 yuan/each, The transportation cost C_{ij} is equal to 2 yuan per ton, per km. The hybrid algorithm has been discussed to search the optimal route based on the data in Table 1.The programming is edited by VC++. The results of clustering analysis and running by 50,000 times are given in Table 2.

The simulation results are shown in Table 2. As seen in this table, it is an effective method based on clustering analysis and chaos search proposed in this paper. This method can be sought optimal solution (or near optimal solutions) adaptively and quickly for many candidate facilities and customers. The computational cost is dramatically reduced by chaos search to solving the smallscale question. The proposed optimal algorithm provides a new path for large-scale optimization in practical integrated logistics

EMPIRICAL ANALYSIS

The empirical analysis is given in relationship between with regional logistics and regional economy. The model

Table 1. Coordinate position of three potential facilities and ten customs.

Items	No	Coordinate Positions		Points Position
Potential Facilities	PF1	0	0	PF1
	PF2	5.03	0.23	PF2
	PF3	5.86	6.20	PF3
Customs Numbers	1	0.20	1.86	A
	2	1.07	3.45	B
	3	1.52	2.87	C
	4	1.76	2.69	D
	5	4.18	2.23	E
	6	4.30	1.09	F
	7	5.30	1.56	G
	8	4.72	2.77	H
	9	2.68	6.09	I
	10	4.76	6.03	J

Table 2. Simulation results.

Problem scale	Results by clustering	Good solution/Near good solution	Function value
3/10	1\A B C D 2\F G H E 3\I J	PF1-C-B-D-A-PF1, PF2-H-F-E-G-PF2, PF3-I-J-PF3	1936.30

with analysis is proposed for the relationship with modern logistics and GDP. The factors are chosen include GDP, logistics production value (WLCZ), freight turnover (HZL), logistics mileage (WLC), and which the data is chosen from 1978 to 2010. The unchangeable price in1978 is given with each variable. The quantitative software is accepted with EVIEWS 6.0, and which the trend of each variable is give in **Figure 2** including GDP, WLCZ, HZL, WLC.

Unit Root Test

Test type (C, T, K), The letters of C, T, K means intercept, Trend, lagging numbers. The first four are contains intercept and trend; The after four are not contains trend.

From the **Table 3**, it is known that the original data in time trend, after an order difference or second order difference, the data is smooth.

Co-integration Test

The Johansen co-integration test is accepted for GDP, WLCZ, HZL, WLC. This test results is presented in Table 4.

From **Table 4**, refused to no co-integration variables of the null hypothesis in 5% level of trace statistics.

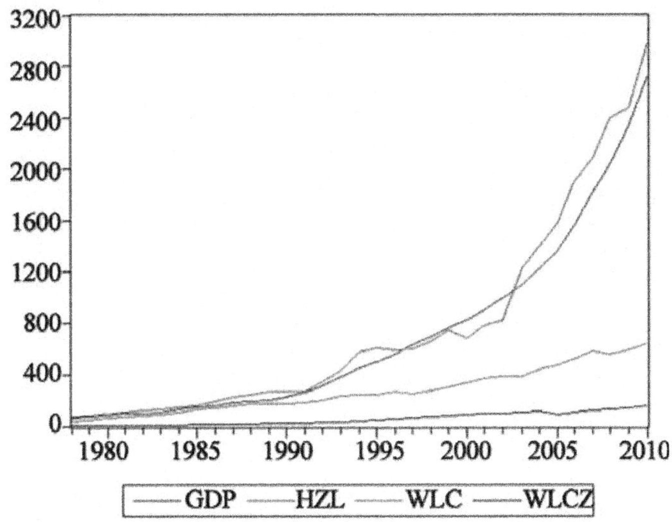

Figure 2: The trend of each variable.

Table 3: ADF unit root test results of each variable.

	Test type (C, T, K)	ADF Test statistic	5% level	10% level
GDP	(C, T, 1)	3.28	–3.56	–3.21
WLCZ	(C, T, 1)	–1.11	–3.55	–3.21
HZL	(C, T, 1)	2.14	–3.55	–3.21
WLC	(C, T, 1)	–0.28	–3.55	–3.21
D^2GDP	(C, T, 1)	–4.45	–3.57	–3.22
DWLCZ	(C, T, 8)	–5.58	–3.56	–3.21
DHZL	(C, T, 8)	–4.96	–3.56	–3.21
D^2WLC	(C, T, 8)	–6.65	–3.59	–3.23

Table 4: Johansen test results.

Null hypothesis	Trace statistic	5% level	Eigenvalue
$r = 0^*$	52.09	47.85	0.58
$r \leq 1$	2.85	29.79	0.41
$r \leq 2$	18.14	15.49	0.23
$r \leq 3$	0.04	3.84	0.001

Note: The letters of r means numbers of co-integration equation, *means reject null hypothesis.

This demonstrates that the variable is at least one cointegration equation between the relationships, that there is a long-term equilibrium relationship between variables.

$$\text{GDP} = -1.784\text{WLCZ} + 1.364\text{WLC} - 1.91\text{HZL} \qquad (2.22) \qquad (0.88)$$

(0.24) (17)

The above co-integration equation reflects the longrun equilibrium relationship between variables, but cannot fully reflect the short-term changes relationship between variables. Therefore, it is necessary to establish the error correction model (ECM), to estimate short-term changes between variables. Estimated ECM is give as following:

$$\Delta GDP = 4.141 \times \Delta WLCZ + 0.498 \times \Delta HZL + 0.846 \times \Delta WLC - 0.472 \times ECMt-1 \quad (3.01)\ (6.84)\ (1.90)\ (-2.58) \tag{18}$$

Adjusted R – squared = 0.747

From the equation, T statistics test of all variables are significant in level 10%, the error correction of ECMt-1 coefficient is negative, accord with reverse revision mechanism, thus further shows that the error correction model is good, with residual sequence is the white noise.

By comparing the co-integration equation and ECM, We can conclusion that t compared on short-term, freight turnover (HZL) have negative influence on the GDP in long term, but its elasticity coefficient is greater than the short-term elastic coefficient; Output value of logistics (WLCZ) have short-term elastic significantly greater than its long-term flexibility to GDP, and both have different symbols, changes of direction that does not accordance, it indicates that in recent years the logistics infrastructure investment lead to the logistics industry close to a saturation stage in Fujian province, although output value of logistics (WLCZ) promote the growth of GDP in short-term, but due to the expansion of the regular, the GDP growth plays a role of obstacles in the long term.

CONCLUSIONS

In this present paper, it has been proposed that the effective method for LRP using a hybrid algorithm based on clustering analysis and chaos search. Through computer simulations the method has been shown to solve largescale LRP. Using the clustering technique, the computational cost is dramatically reduced. Due to the feature of chaotic global ergodicity, the optimal routes are randomly seeked in one cluster, which ensures search efficiently and produces optimal solutions (or near optimal solutions) without local optimization. Thus, the emphasis is directed towards developing a methodology, which has been improved the quality of solution.

The application to other combinatorial optimization problems should be investigated in integrated logistics. The relationship is given with regional logistics and its development. There exists a long-term equilibrium relation between economics growth and logistics development in Fujian province.

ACKNOWLEDGMENTS

This paper was supported in part by National Natural Science Foundation of China under grant number 71040009, in part by Program for New Century Excellent Talents in University under grant number NCET-10- 0118, in part by Youth Foundation of Chinese Education by HuoYing-dong under grant number 104009.

REFERENCES

1. J. H. Bookbinder and K. E. Reece, "Vehicle Routing Considerations in Distribution System Design," European Journal of Operation Research, Vol. 37, No. 2, 1988, pp. 204-213.doi:10.1016/0377-2217(88)90330-X
2. D. B. Bruno, F. Vincent, S. Paul, K. Philip and P. Patrick, "Solving Vehicle Routing Problems Using Constraint Programming and Metaheuristics," Journal of Heuristics, Vol. 6, No. 4, 2000, pp. 501-523. doi:10.1023/A:1009621410177
3. V. Campos and E. Mota, "Heuristic Procedures for the Capacitated Vehicle Routing Problem," Computational Optimization and Applications, Vol. 16, No. 3, 2000, pp. 265-277. doi:10.1023/A:1008768313174
4. L. Cooper, "The Transportation-Location Problem," Operations Research, Vol. 20, No. 1, 1972, pp. 94-108. doi:10.1287/opre.20.1.94
5. G. Dijin, "Hybrid Evolutionary Method for Capacitated Location-Allocation Problem," Computer and Industrial Engineering, Vol. 33, No. 3-4, 1997, pp. 577-580.
6. V. Fernando and A. Joaquin, "Homoclinic Chaos in Inverted Pendula," Proceeding of the 39th IEEE Conference On Decision and Control, Sydney, 12-15 December 2000, pp. 4821-4822.
7. H. Heung-Suk, "Design of Supply-Chain Logistics System considering Service Level," Computers and Industrial Engineering, Vol. 43, No. 7, 2002, pp. 283-297.doi:10.1016/S0360-8352(02)00075-X
8. G. Laporte, "Hamiltonian Location Problems," European Journal of Operation Research, Vol. 12, No. 1, 1983, pp. 80-87. doi:10.1016/0377-2217(83)90182-0
9. G. Laporte and Y. Nobert, "An Exact Algorithm for Minimizing Routing and Operating Costs in Depot Location," European Journal of Operation Research, Vol. 6, No. 2, 1986, pp. 224-226. doi:10.1016/0377-2217(81)90212-5
10. J. Ra, "Dissipative Enterprises, Chaos, and the Principles of Lean Organizations," Omega, Vol. 26, No. 6, 1998, pp. 397-340.
11. C. Watson-Gandy and P. Dohrn, "Depot Location with van Salesmen—A Practical Approach," Omega, Vol. 1, No. 3, 1973, pp. 321-329. doi:10.1016/0305-0483(73)90108-4

12. P. N. William and J. B. Wesley, "Solving the Pickup and Delivery Problem with Time Windows Using Reactive Tabu Search," Transportation Research Part B, Vol. 34, No. 2, 2000, pp. 107-121. doi:10.1016/S0191-2615(99)00016-8

Chapter 2

SIMULATION AND OPTIMIZATION TECHNIQUES FOR SAWMILL YARD OPERATIONS—A LITERATURE REVIEW

Asif Rahman[1,2], Siril Yella[1], Mark Dougherty[1]

[1]Department of Computer Engineering, Dalarna University, Borlänge, Sweden;

[2]Department of Information Technology, Örebro University, Örebro, Sweden.

ABSTRACT

Increasing costs and competitive business strategies are pushing sawmill enterprises to make an effort for optimization of their process management. Organizational decisions mainly concentrate on performance and reduction of operational costs in order to maintain profit margins. Although many efforts have been made, effective utilization of resources, optimal planning and maximum productivity in sawmill are still challenging to sawmill industries. Many researchers proposed the simulation models in combination with optimization techniques to address problems of integrated logistics optimization. The combination of simulation and optimization technique identifies the optimal strategy by simulating all complex behaviours of the system under consideration including objectives and constraints. During the past decade, an enormous number of studies were conducted to simulate operational inefficiencies in order to find optimal solutions. This paper gives a review on recent developments and challenges associated with simulation and optimization techniques. It was believed that the review would provide a perfect ground to the authors in pursuing further work in optimizing sawmill yard operations.

INTRODUCTION

The optimization problems in any domain are specific concern for managers and decision makers. Even though there are technological developments in the field of optimization techniques, the need for more efficient techniques which can solve such problems will always exist. With increasing competitiveness in real world scenarios, mathematical models of such problems are getting more and more complex. In order to deal with such problems, sophisticated tools and techniques are needed which can efficiently solve such problems. In recent years, many researches have focussed on combinatorial optimization problems using simulations and optimization techniques to improve profits and performance. Accordingly, many areas of supply chain management have adapted optimization techniques to improve and optimize their resources efficiently.

The supply chain logistics is considered as an integrated process in which a group of several organizations, such as suppliers, producers, distributors and retailers, work together [1]. One of the most important problems in supply chain logistics is the optimizations of resources and it is a very complex problem that deals with the planning, handling and control, for the storage of raw material between arrival and production [2]. Further, supply chain management includes logistics transportation, resource allocations, storage bin arrangements, scheduling and routing. Several areas of logistics transportation, resource allocation and arrangements (storage area/containers etc.) within organization demand proper planning and optimization techniques in order to attain significant reduction in operational costs. This review is an attempt to gain a better understanding of the techniques or methodologies used in supply chain management and to provide a basis for research in sawmill yard operations. In most of the sawmill industries, efficient planning and optimization of resources are key factors for successful reduction in operational costs. Storage bins in sawmill yard have a significant impact on the resources and production supply, whereas, resources are essential equipment for transportation, loading and unloading of logs from one place to another within sawmill yard. Inefficient utilization of resources results in unnecessary operational costs. In order to attain significant reduction in operational costs, a comprehensive and systematic approach for optimization of operations in sawmill yard is required. There isn't enough of literature available on the problems encountered in the sawmill yard, i.e. between arrival and production, whereas problems encountered in sawmill production line, automatic debarking and grading have been investigated in the past. However, the problems in sawmill yard are rather similar to most of the problems encountered in supply chain logistics. Therefore, we try to map the literature, which are immediately relevant to current area of investigation and study.

Most of the researchers are focussing towards developing new methodologies and techniques which can not only provide reasonable good solutions but are also computationally fast. Existing researches in supply chain domain using various Artificial Intelligence, Simulations and Mathematical optimization techniques have been investigated. AI techniques are one of the most popular and fascinating tools to solve the real world optimization problems. Further, they are more feasible to solve complex optimization problems compared to traditional techniques.

The aim of this paper is, therefore, to make a review of the work carried out thus far on operational optimization using various optimization and simulation techniques. As an enormous amount of literature available in domain of optimisation, we will focus mainly on techniques used for decision making within the organisation that improves overall performance. The main issues to address include: what are the main problems within operational routines that have been investigated using combinatorial optimization (CO) and discrete event simulation (DES) techniques? What optimization techniques have been employed? What are the main findings and achievements up to date? And finally, discussion concerning the possible use of such techniques in current areas of investigation is presented for further research directions. Only the classified techniques that are immediately relevant to the area were covered in this review, due to the wide scope of the subject.

This paper is organised in five sections. Subsequent to the introduction in Section 1, the review methodology is mentioned in Section 2. AI and optimization techniques are briefed in Section 3. Section 4 describes the current area of investigation and examines the research of applying AI and DES techniques to various optimization problems in sawmill. Then key findings are concluded. Finally, a summary of existing research and a discussion on the future research directions are provided in Section 5.

REVIEW METHODOLOGY

The search process was carried out with scientific databases. The review papers were sorted out from more than 30 journals (Table 1) such as European Journal of Operation Research, International Journal of Production economics, Computers & Industrial Engineering, IEEE journals and Expert systems with Applications etc. Given the specific interest on how Optimization and Artificial Intelligence techniques were used in combination with simulation models to optimize the supply chain logistics operations. More than 100 papers including the journals, conference, book chapters and books were reviewed for this study.

Scope

The concept of simulation optimization has been analysed by many researchers from various perspectives. However, it is beyond the scope of this paper to address all problems in detail. In an attempt to provide a more intensive review of existing papers in this area, this paper mainly focuses on problems which are immediately relevant to problems in sawmill yard logistics. The problems related to automatic grading, production scheduling and efficient debarking techniques will not be discussed in this paper.

Journals	*References*
Advance Engineering & Informatics	1
Applied Mathematics and Computation	1
Applied Soft Computing	2
Automation in Construction	2
Computer and Chemical Engineering	6
Computer and Industrial Engineering	11
Decision Support System	1
Environmental Management	2
European Journal of operation Research	13
Expert System with Application	5
IEEE Journals	5
Int. Journal of Production Economics	6
Int. Journal of Production Research	2
Int. Journal of Science & Applied IT	1
Journal of Intelligent Leaning Systems & Applications	2
Operation Research Spectrum	1
Production Planning and Control	2
Renewable Energy	1
Robotics and Computer Integrated Manufacturing	2
Simulation Modelling Practice and Theory	3
Transport Research Part E	1
Others	11
Total	80

Table 1: List of journals reviewed.

OPTIMIZATION TECHNIQUES

According to [3] optimization models are used to quantitatively tackle many emerging logistics challenges. The challenges in supply chain design problem comprises the decisions regarding the number and location of production facilities, the amount of capacity at each facility, the assignment of each market region to one or more locations, and supplier selection for sub-assemblies, components and materials [4]. The supply chain design is related to the definition of the structure of the chain, i.e. the sequential links between different sourcing, production and distribution activities or processes [5].

Enormous amount of literature can be found on supply chain logistics with different objectives and using various methodologies. However, the goal of each methodology is to optimize the supply chain process. The different factors which have been considered, namely: costs, location, allocation, capacity, routing, agility and transportation.

An efficient algorithm to optimize complex real world problems has been a challenge to researches over the years in different domains. Simulation-optimization methods have proved to be powerful tool to reduce operational costs and facilitate effective design operations for optimization problems, and have become an area of extensive investigation over the past decade [6]. Optimization problems are normally expensive and time-consuming. It is then important to develop tools that can help identify the most cost-efficient ways of implementing and solving optimization problems. Simulation models, when used alone, cannot guarantee that the design strategy is optimal. They can handle what-if scenarios/questions, but are weak in seeking answers to which one is the best [7]. The optimization techniques can be categorised into three groups: 1) Linear and Integer Programming; 2) Artificial Intelligence based methods; and 3) Discrete event simulations incorporating optimization techniques (Table 2).

Linear and Integer Programming

Mathematical programing has been used in a number of supply chain optimization and management problems. Ryu et al. [8] presented a bi-level modelling approach comprising two linear programming models, one for production planning and one for distribution planning. Ozgur and Fusun [9] provided a probabilistic linear programming model to make strategic resource planning decisions in supply chain. Oh and Karimi [10] put forward a linear programming model that integrates production and distribution planning for a multinational firm in the chemical sector.

Table 2: Field of Study

Subject Area	*Number of Papers*
Ant Colony Optimization	9
Discrete Event Simulation	33
Genetic Algorithm	19
Linear & Integer Programming	11
Multi-Agent System	12
Neural Networks	4
Simulated Annealing	12
Swarm Optimization	6
Others	22
Total	113

Yue Wu [11] studied a time staged linear programming model for modelling production loading problem with import quota limits, where, decision regarding quantities of used resources, machines, labour, initial quotas, inventory and outsourcing levels were considered. Jung et al. [12] compared linear programming models for centralized and decentralized production and transport planning environments. Kanyalkar and Adil [13] used a linear programming model for aggregated and detailed production and dynamic distribution planning in a multiproduct and multi-plant supply chain. Kallrath [14] presented a mixed integer linear programming model for production, distribution and sales planning with different time scales for business and production aspects. Silke et al. [15] presents a linear mixedinteger modelling approach for basic components in a biomass supply chain. The main focus of this work was to represent the relationship between moisture and energy content of different kinds of biomass and to handle long-term processes in the optimization like passive drying effects. Yavuz et al. [16] introduced a generalized mixed integer programming (MIP) formulation that allows iterative interaction with a simulation model by taking into account the impact of uncertainty on the objective function value of previous solutions. Talluri and Baker [17] used a multi-phase mathematical programming approach for effective supply chain network design to meet the constantly changing needs of customers at low cost, high quality, short lead times and high verity. Panagiotis and Lazaros [18] proposed an integrated model based on a detailed mathematical programming formulation that addresses some of the complex issues related to the design and operations of global supply chain networks.

The above mentioned studies used Linear or Integer programming algorithms for dealing supply chain optimizations problems. Most of the authors focused on financial and tactical operational aspects within the organization taking into account production operations, distribution efficiency, resource planning, operational costs, customer satisfaction and product quality etc.

AI Based Methods

Artificial Intelligence is basically a branch of computer science concerned with making computers behave like humans or emulation of human behaviour (McCarthy, 1956). AI is nothing but the ability of a computer program to evaluate information and make a decision according to pre-established criteria. Recently, there has been a growing interest in using artificial intelligence techniques to solve optimization problems. This is mainly because the challenge existing in the solution process that usually requires incorporating complex simulation models into optimization framework. This leads to the difficulty in seeking the globally optimal solution with conventional non-linear optimization algorithms [6].

AI techniques are suitable in addressing the qualitative aspects of decision support system, which are not adequately captured when using analytical models [19]. AI techniques has been used in different fields, however, this paper focuses on the problems related to supply chain logistics and transportation. The artificial intelligence techniques, in particular Genetic Algorithm (GA), Ant Colony optimization (ACO), Particle Swarm Optimization (PSO), Simulated Annealing (SA), Neural Networks (NN) and Multi-agent systems has been analysed to solve optimization problems in supply chain logistics for reduction in operational costs and increase efficiency.

Genetic Algorithm

Genetic algorithm (GA) is a stochastic searching method that mimics natural evolution and is based on the concept of "survival of the fittest". GA is implemented as a computer simulation in which a population of abstract representation of candidate solution to an optimization problem evolves towards better solutions [20]. Evolutionary algorithms were invented to mimic some of the process observed in natural evolution. Evolution occurs on chromosomes organic devices for encoding the structure of living beings. Processes of natural selection then drive those chromosomes that encode successful structure to reproduce more frequent than those encode failed structures. By using simple encoding and reproduction mechanism, the algorithm can then display complicated behaviour and turn out to solve some extreme difficult problems [21].

Genetic Algorithms (GA) are robust adaptive optimization method based on biological principals, GA is an efficient algorithm to solve combinatorial optimization problems by exploring a large and complex search space in an adaptive way guided by the equivalent biological evolution mechanisms of reproduction, crossover and mutation [22]. The most important advantage of GA is their ability to use accumulative information about the initial unknown search space in order to move the next search into useful spaces [23]. In the past decade, a large number of combinatorial optimization problems and solution methods have been presented in the literature on integrated supply chain management [24].

Gjerdrum et al. [25] proposed a mathematical programming model to reduce operational costs, while maintaining customer order fulfilment at high level of for a supply chain network. Amodeo et al. [26] presents a simulation based multi-objective optimization method for the optimization of inventory policies of supply chains considering total inventory cost and service level.

Reza and Mahsa [27], introduced a new model for a distribution network in a three-echelon supply chain, which not only minimizes the total costs but also follows Just-intime distribution purposes in order to better represent the real-world situations. Fulya et al. [28] have used a steadystate genetic algorithm (ssGA) for the single-source multiproduct multi-stage supply chain network design problem, to satisfy the customers demand with minimum costs. D'Souza et al. [29] developed a Genetic Algorithm and Artificial Immune System to tackle the "Pickup and Delivery" problem effectively giving optimal solutions. By using these meta-heuristic methods the authors have optimize overall transport between different destinations leading to better utilization of time and resources. David et al. [30] considered the problem of finding an optimized schedule for the just-in-time production and delivery of ready-mixed concrete. The scheduling algorithm combines a GA and a set of constructive heuristics, which guarantee the determination of a feasible schedule for any given set of requests. Lau et al. [31] uses a stochastic search technique called fuzzy guided genetic algorithm (FLGA) to simultaneously optimize travelling time and travelling distance in a multiple depot vehicle routing problem. Daniels and Parsons [32] proposed the use of hybrid algorithm that combines an agent-based approach with GA global search operator to show that it is both faster and more effective than either a pure GA or a pure agent-based simulation to space arrangement problems. David et al. [33] proposed a hybrid algorithm based on GA and LP to minimize the distribution costs for a pulp producer. GA are the most studied and applied meta-heuristic in the optimization field, because GA are very easy to implement in all sorts of problems, and usually guarantee good solutions, whatever the type of solution space [2].

A variety of problems concerning the simulation based optimization have been widely handled using Genetic Algorithm. One of the promising methods of optimizing problems were performance can be evaluated by a simulation model is the use of GAs [34]. Shin et al. [35] proposed a simulation model incorporating genetic algorithm for optimal hoist planning in building constructions, uses simulation to test the optimal hoist plan whereas GA searches for a set of hoists that optimizes rental costs. Yeh and Lin [36] proposed a GA method to obtain a near optimal solution for hospital emergency department. Further, simulation has been used to evaluate how system is impacted by various nursing schedules. Shin and Iima [37] proposed a GA o find the best classroom seating arrangement based on the questionnaire results and analysis of the observation of behaviour between students. According to authors, the satisfaction of students for the proposed method is higher than the traditional class room seating arrangements.

Simulated Annealing

In traditional problem solving strategies, some of them guarantee finding a global solution, others don't, but they all share a common pattern. Either they guarantee discovering the global solution, but are too expensive (i.e., too time consuming) for solving typical real world problems, or else they have a tendency of getting stuck in local optima. Since there is no chance to speed up algorithms that guarantee finding the global solution, i.e., there is no chance finding polynomial time algorithm for most real problems as they tend to be NP hard, the other remaining option is to design algorithms that are capable of escaping local optima. After reaching a local optimum, a new starting point is generated and the search is commenced all over again. Two main approaches that we'll discuss here are based on an additional parameter called (temperature) that changes the probability of moving from one point of search space to another or a memory, which forces the algorithm to explore new area of search space. These two approaches are called Simulated Annealing and Tabu Search. We will focus on Simulated Annealing. It's a method that simulates the physical process of annealing. Annealing is where a material is heated and then cooled usually for softening and making the material less brittle. Simulate annealing; therefore, exposes a solution to heat and cools producing a more optimal solution [33, 38].

Simulated annealing is the deterministic search approach based on the annealing process of metal. This procedure can be considered a hybrid method of local search and usually employs a downhill algorithm unless probability is decreases to 0. In this situation, uphill moves would be applied for increasing probability [39]. Simulated annealing (SA) is inspired in the physical annealing process. The randomized nature of the procedure permits asymptotic convergence to optimal solution under mild conditions [40]. Vaidyanathan and Anthony [41] studied the applicability of simulated annealing as a very effective and useful solution approach to complex problems involved in supply chain management to reduce costs and improve efficiency. Chibeles-Martins et al. [42] presents a meta-heuristic approach, based on the simulated annealing to optimize the computational burden of supply chain network. Kathryn et al. [43] proposed a hyper heuristic driven simulated annealing approach to minimise the packaging waste by optimizing storage and transportation shipper sizes. Hamid and Keivan [44] uses a simulated annealing approach for optimizing the fleet size and freight car allocation problem wherein car demands and travel times were assumed to be deterministic and unmet demands were backordered. Kathryn et al. [43] developed a simulated annealing based hyper heuristic model for determining shipper sizes for storage and transportation. Allaoui and Artiba [45] considered a multiobjective optimization problem taking into consideration the system constrains. Suman [46] used a Pareto dominated simulated annealing for multi-objective optimization problems where he has made extensive comparisons of multi-objective simulated annealing algorithms. Mahmoud et al. [47] presented a simulated annealing algorithm for solving multi-objective stochastic optimization problem. The algorithm uses hill-climbing feature to escape the local minimalistic trap. A new Pareto set for multi-objective stochastic optimization is also introduced to assist the simulated annealing algorithm attain optimality. Further, simulation is used to estimate the objective function values.

Ant Colony Optimization

Ant colony optimization (ACO) is a technique of problem solving inspired by the behaviour of ants in finding paths from the nest to food and a new search metaphor for solving combinatorial optimization problems, and has been unexpectedly successful in recent years [48,49]. Ant colony optimization is a relatively new optimization technique that has been gaining followers in the past decade, since it is especially suited for problems with dynamic behaviour [48]. The ants cooperate using an indirect form of communication mediated by pheromone trails of scent and find an optimal solution to their tasks guided by both information (exploitation) which has been acquired and search (exploration) of the new route. Therefore the premature convergence probability of the system is lower [50].

Silva et al. [51] proposed an ant colony optimization technique to optimize the different operational activities of generic supply chain. Further, simulation results suggested that new methodology is more efficient than a simple decentralized methodology for different instances of a supply chain. Luis et al. [52] uses a multi-objective ant colony optimization to supply chain design to optimize total supply chain cost while keeping the total lead times within required delivery due dates. Wei and Arun [53] presents a meta-heuristic ant colony optimizations (ACO) approach for solving the logistics problem arising in disaster relief activities, such as planning, dispatching and evacuating. Chaharsooghi and Kermani [54] studied an effective ant colony optimization for multi-objective resource allocation problems in order to minimize the costs of resources and maximize the overall efficiency. Felix and Niraj [55] uses a multiple ant colony optimization (MACO) approach to design a balanced and efficient supply chain network that maintains the best balance of transit time and customers service. Rami et al. [56] proposed an Ant colony optimization algorithm to solve the transportation problem of cross-docking network to minimizing the transportation cost in a network by loading trucks in the supplier locations and then route them either directly to the customers or indirectly to crossdocking facilities so the loads can be consolidated.

Aaron [57] proposed a parametric study of ACO applied to Water Distribution System Optimization. Using the derived parameter guidelines, iteration-best ant system, has been seen to perform extremely competitively when compared with other optimization algorithms.

Particle Swarm Optimization

Particle Swarm Optimization (PSO), an evolutionary computation technique firstly introduced by Kennedy and Eberhart in 1995 [58,59], is based on the collective motion of a flock of particle; the particles swarm. In the simplest version of PSO, each member of the particle swarm is moved through a problem space by two elastic forces. One attracts it with random magnitude towards the best location so far encountered by the particle. The other attracts it with random magnitude towards the best location encountered by any member of the swarm. The position and velocity of each particle are updated at each time step until the swarm as a whole converges to an optimum [60]. The swarm optimization algorithms have been used in different domains, however, only topics immediately relevant to the area of investigation are presented below.

Swarm Optimization (PSO) are increasingly been used in optimization problems related to TSP, VRP, Shortest Path, Resource Allocation, Scheduling and logistics distribution etc. Further, simulations and hybrid algorithms were also used to enhance the results. Wei et al. [61], presents the model of logistics distribution centres location problem considering setup cost, turnover cost and demands of the customers as a combinatorial optimization problems. For the convenience of solving the model, some mathematical properties of the model were also obtained. Further, discrete particle swarm optimization algorithm, tabu search algorithm and sequential quadratic algorithm were integrated to seek the minimum of total cost and assignment plan of the distribution centres. Lie and Fanhua, [62] presented a 4-stage production distribution problem and used the improved PSO algorithm to solve the problem. Further, simulation has also been used to show that the algorithm is effective and efficient in both computation and convergence. Yue-Jiao et al. [63] presents a set-based PSO to solve vehicle routing problem with time windows, the proposed algorithm provides better results than the existing best known results. Yanfang and Hengqing [64] proposed a new method for solving the shortest path problem in stochastic traffic networks. The approach based on combining fluid neural network model with PSO. This algorithm is simple and easy and can find quickly the shortest path without falling into local minimums, which will occur if energy function is used to solve the same problem. Chilukuri et al. [65]

have proposed a PSO-based fair Resource allocation and scheduling algorithm for the IEEE 802.16 System. And have compared the results with the static FDMA algorithm and have found it better delay characteristics with increasing number of users while still maintaining the fairness and throughput utilization. This makes the proposed scheme absolutely useful for real-time applications.

Neural Networks

An artificial neural network consists of an interconnected group of artificial neurons. Such a network performs computation and manipulates information based on the connectionist approach in a similar but simpler fashion than the brain would perform [66]. In terms of modelling, remarkable progress has been made in the last few decades to improve artificial neural networks (ANN). Artificial neural networks are strongly interconnected system of so called neurons which have simple behaviour, but when connected they can solve complex problems. Changes may be made further to enhance its performance [67].

Combinatorial optimization problems are NP-complete and have received lot of interest from (Artificial Neural Networks) ANN researchers, many researchers have tried to solve it, but no method can solve it completely. As a result, a number of neural networks have been developed to solve wide range of optimization problems. The majority of the existing studies are based on Hopfield networks or its extensions. This may be partly due to successful applications of Hopfield network to the Travelling Salesman Problem (TSP) and other optimization problems [68]. It can be shown that usually among existing ANNs techniques, Hopfield neural networks and SOM have applied for optimization problems. Some of the many ANN applications have been reported for optimization problems include Assignment Problems, clustering Problems, Scheduling Problems and Travelling Salesman Problem (TSP) [69].

Multi-Agent System

Intelligent agents are able to perceive their environment and respond in a timely fashion, exhibit goal directed behaviour by taking the initiative, and capable of interacting with other agents, in order to satisfy their design objectives. Intelligent agents are autonomous, smart and active. They decide themselves whether or not to reform an action on request from another agent. Agents are different form objects, you order objects to do things, but you request agents to cooperate. The objects don't have any of the behaviours either. A multi agent system is inherently multi-threaded and each agent is assumed to have at least one thread of active control [70].

Multi Agent systems (MAS) are increasingly been used in complex problems related to logistic planning, air traffic control, military demining, financial portfolio management, among others [71]. Multi agent systems offer such a useful feature as parallelism, robustness and scalability. They are highly applicable in domains and problems, where centralized approaches meet their limits. Multi agent based approaches are well suited for domains, which require the integration and interaction of multiple source of knowledge, the resolution of interest and goal conflicts or time bounded processing of data [72].

Multi-agent technologies start to penetrate into transportation and logistics domain only recently. Intelligent agent represents the organizations within the logistics domain; model their logistics functions, processes, expertise and interaction with other organizations. Due to similarity in characteristics between intelligent agents and organizations, agent technology is an appropriate choice of modelling organization in the logistics domain [73]. Multi-agent or agent based systems that mainly are used to solve typical problems in the transportations and logistics domain such as decision support system, logistics planning system, and simulation and modelling system, which support both decision-making and planning, where experts can make their decisions [74].

The problems in which agent technologies have been applied within transport logistics are planning, scheduling, management etc. Many researchers have focused on the field of multi agent systems. A number of significance advances have been made in both the design and implementation of autonomous agents. Neagu et al. [75] have reported Agent-Based system design to solve the Pickup and Delivery Problem with Time Windows (PDPTW) using constraint satisfaction techniques (CST). Kazami et al. [76] solves a production and distribution planning problem and compare the results through the use of genetic algorithm to the ones coming from employing a multi agent decision support. Vita and Janis [74] proposed an overview of the problem solving methods in transportation and logistics domain. Two different multi-agent architectures has been worked out and used in the real life problem within the transportation and logistics domain. This architecture support both decision making and planning. Persson et al. [77] analyses the strengths and weakness of agent-based approaches and mathematical optimizations techniques. According to the author's qualitative evaluation, agent-based approaches tend to be preferable when, the problem domain is large, probability of node or link failures is high, time scale of the domain is short, sensitive information that should be kept locally, and mathematical optimization techniques when, the cost of communication is high, the domain is monolithic in nature, quality of solution can be guaranteed, it is important that a system optimal or near optimal solution is found. Combining the approaches into a hybrid approach can be a promising way.

The potential of agent technology is also suitable for modelling practical supply chain management problems, where analytic or optimization results cannot be easily applied or global information sharing and central coordination are difficult to achieve [78]. Liang and Huang [79] developed a multi-agent system to simulate a supply chain, where agents represent the different actors within a supply chain. Agents are coordinated to control inventory and minimize the total cost by sharing information and forecasting knowledge. Ahn & Lee [78] present a practical collaboration framework for supply chain management wherein multi-agent systems form dynamic information network and co-ordinate their production and order planning according to synchronised estimation of market demand. Caridi and Cavalieri [80] provide a critical analysis of agent based approaches applied to supply chain management domain, pointing out the lack of real world applications and the low maturity level of agent based manufacturing technology. Akanle and Zhang [19] proposed an Agent based model for optimizing supply chain configurations to meet customers demand with minimum costs.

Simulation Techniques

Supply chain simulation is often carried out in general purpose simulation software as reported in [81] or using special purpose analysing tools as reported in [82]. Simulation, as a generic technique, has been identified as a suitable tool for the analysis of complex and dynamics systems such as supply networks [83]. Computer simulation and simulation models can be used to model complex supply networks close to real systems, execute those models, and observe system behaviour. Simulation also provides the ability to simulate the effect of particular events on supply network performance and experimenting with different scenarios without huge investments (time and money) and process disruption [84]. Simulation can be defined as the process of designing an abstract model of a real system and conducting experiments with this model for the purpose of either understanding system behaviour or evaluating various strategies within the limits imposed by a set of criteria for the operation of the system [85]. Simulation enables managers to study and evaluate alternative solutions to analyse the effect of process or operation modification [86].

The simulation based approaches to assess the operational routines within the domain and the impact of logistics is not a new concept. Many simulations models are increasingly being used for optimization and decision making purposes in most of the organizations. The commercially available simulation packages have also being increased. Many newer simulation packages now offer object-oriented modelling architecture; therefore the simulation applications are more portable and scalable than before [87]. Simulation models have been successfully used in different domains, such as, berth planning, resources optimization and resource allocation at container terminal [88,89], manufacturing activities in an automobile supply chain, automotive industry [90,91], fright transportation analysis [92], optimization of supply chain network, management and demand uncertainty [93-100]. Xiatao et al. [101] and Amalia & Marianthi [102] emergency department in health care [36,103] water resource management [104], multiple performance measures [105], scheduling system, production line in industry and flexible transport services [106-108].

Discrete Event Simulations

Discrete event simulation can be particularly useful in non-stable environments such as supply networks that are characterized by a high degree of interdependences and where logistic processes play important role, and also there is a need for process optimization [109]. However, as a descriptive method, discrete event simulations can only be used to perform optimization through "what if" case studies involving the comparison of several cases or scenarios [101]. Simulations are often developed to check the validity of the model or to evaluate its efficiency. It permits examination of the behaviour of real world system by developing simplified models of real system. It can be used to explore the behaviour of artificial systems in order to predict what might happen should such a system come into existence in the real world [110]. Simulation allows researchers to study processes in ways nature prohibits, given that it can be run many times with the values of the model parameters modified in each run and changes observed in outputs. According to, computer simulation can be applied for different purposes, namely prediction, performance, entertainment, training, education, confirmation and discovery [111].

Recently, other techniques have been combined with discrete event simulations techniques in supply chain domain. Jaroslav and Pavel [96] proposed an integer based simulation optimization, where, local search algorithm minimizes the number of evaluations and simulation optimizes the objective function. Carvalho et al. [97] presents a supply chain simulation study for an automotive supply chain. The study was to evaluate alternative supply chain scenarios for improving supply chain resilience to a disturbance and understanding how mitigation strategies affect each supply chain entity performance. Taejong et al. [98] proposed a hybrid algorithm for discrete event simulation to reduce the computational load and to improve the efficiency of supply chain. Tu Hoang et al. [99] combines Integer programming and GA to produce supply chain configuration design. Moreover, a reference model of supply chain has been established based on distributed simulation and business process modelling techniques [100]. Xiatao et al. [101] presents an extension to the simulation-based optimization framework for optimizing supply chains under uncertainty. Amalia and Marianthi [102] proposed a hybrid simulation optimization approach, where, combination of mathematical programming and agent based simulation model has been investigated to minimize the summation of production cost, transportation cost, inventory holding and shortage costs, subject to capacity and inventory balance constraints in a supply chain management. Further, Georgiana [112] presents a use of simulation and multiple linear regressions to improve operational inefficiencies of a small scale process industry. Ye et al. [113] proposed a simulation-optimization approach to improve clinical supply chain management including patient demand and demand scenario forecast. Young et al. [114] proposed a combined discrete and continuous simulation model for evaluating and optimizing supply chains as supply chain systems are neither completely discrete nor continuous. Fengli et al. [115] describes the development of a simulation model for biofuel supply chain, where harvesting or processing, transportation and storage were the key activities in the model in order to minimize feedstock cost, energy consumption and GHG emissions.

SAWMILL YARD OPERATIONS

Bergkvist Insjön AB—A Case Study

Bergkvist Insjön AB[1] is one of the large privately owned sawmills in Sweden with an annual production volume of around 400,000 cubic meters. A brief description of the work routine has been provided to the sake of clarity. Typical operations at the Bergkvist Insjön AB begin with the arrival of logs. Trucks loaded with timber arrive at sawmill where they are unloaded by a log stacker. Unloaded timber is allowed to pass through a measuring station in which a human inspector checks for the quality of the timber. Further the timber passes through a 3-D scanner that will determine the type and dimension of the wood. Depending upon the type and dimensions, timber gets placed into appropriate storage bins which are ranged alongside a long transport band. Depending on the production requirements, timber from specific storage area or storage bins are emptied and fed to the saw mill with the aid of a second log stacker. Timber is then passed through a second 3-D scanner. The role of the second scanner is to double check that the right type and dimension of timber is being selected for production. Lack of proper control at this stage seriously compromises quality of the end product. Timber that is rejected by the second scanner is sent back to the measurement station. Such timber are reassigned a new (timber) class and appropriate storage bins as per the new grade and the process continues. Upon acceptance, the logs are processed and are directed to further stages of production.

To be able to meet with the production and supply challenges, Bergkvist Insjön AB has been constantly updating its resources (for handling operations) over time. In spite of the adaptive nature, Bergkvist Insjön AB has been facing severe challenges that are threatening the overall efficiency and thereby limiting profit margins of sawmill. A particular challenge concerns the unexpected arrival of trucks loaded with timber. The trucks arriving at sawmill have to be unloaded before they can leave. Further Bergkvist-Insjön AB is bound to pay certain amount of waiting fee to the truck drivers for keeping them waiting beyond certain time limit. Unexpected arrival of several trucks within intervals is imposing enormous amount of pressure on the internal resources (log stackers). Unexpected arrival of trucks is not an internal problem that can be regulated by Bergkvist Insjön AB due to the presence of large number of actor's such as small enterprises making delivery of timber, truck drivers, timber procurers etc. Informing the truck drivers (prior to their arrival) about possible serving times i.e. the best time that one could come in to avoid delays, could be seen as a way to cope up with the problem. However the ability to provide the drivers with arrival information is largely dependent on the efficiency and predictability of internal operations such as loading and unloading procedures, storage capacity, arrangement of storage bins, number of resources, measuring and processing speed etc.

Bergkvist Insjön is facing difficulties in managing their resources, especially when many trucks (loaded with timber) arrive unexpectedly at the same time. Achieving better routine to carry out internal operations is essential. Hence, the main objective of this review is to investigate and propose a means of optimizing all internal operations at Bergkvist, thereby helping them to improve overall performance and promote efficiency. Finally, it is expected that such an improvement will help the truck drivers in planning better timber delivery routines.

Future Work

The efficient optimization of sawmill yard operations is a complicated task, involving a variety of factors with storage bin arrangements, unexpected arrivals and rejections. Many traditional solutions based on experiences and empirical analysis, which may not represent optimal strategy at sawmill yard.

Simulation and optimization models based on sawmill operational routines is an immediate requirement to examine and evaluate sawmill yard operations, bin arrangements and behaviour of log stackers. The ultimate goal here is to minimize usage of resources and maximize supply of logs to sawmill for production. However, the possibility for accomplishing this goal could be affected by many constraints, and limitations that exit in sawmill yard and its operational routines.

The operational inefficiencies of sawmill yard seriously threatens the overall efficiency and profit margins of sawmill. Several types of optimization techniques and simulation models can be used to investigate problems causing operational inefficiency in sawmill yard. Further, new strategies and behaviors can be adopted to improve the overall performance in sawmill yard. Problems concerning the sawmill yard operations were studied recently using discrete event simulations [116]. Further, discrete event simulation can also be combined with optimization algorithms to improve and optimize the operational routines. As there are several optimization algorithms and each technique have their own benefits, so it is left to uses to make a choice of best suitable technique.

Optimal arrangements of storage a bin is one of most important operational inefficiency among others in sawmill yard. It takes enormous amount of time to simulate all possible arrangements, as n becomes large exhaustive search is computationally infeasible. Therefore, an effective simulation model incorporating Genetic Algorithm (GA) to optimize the arrangement could be a promising alternative, as GA has been used very efficiently by many researchers. Further, agent based simulation model or discrete event simulation can be used in combination with GA.

CONCLUSIONS

The papers reviewed were from more than 30 different Journals (Table 1); it is due to the fact that optimization problems are application driven. Hence, most of the papers reporting the use of Simulation and AI techniques to solve optimization problems were found to be published in Operation Research Journals and other Journals immediately relevant to the area compared to pure Computing or AI based Journals. The current literature review has been done with the intention of focussing on the latest advancements and trends provided by employing AI techniques to solve operational optimization problems. AI based approaches are the most popular techniques adopted to solve optimization problems, particularly in transportation and logistics domain. Further, simulations are also used in combination with the techniques to solve uncertainty of a system to predict performance under various scenarios. Techniques such as neural networks, multi-agent simulation, genetic algorithm, ant colony algorithm, simulated annealing, discrete event simulation and mathematical programming etc. have been used to handle a variety of problems in supply chain domain.

Identifying the best optimization technique that can be combined with simulation models to solve operational inefficiencies is impossible since all optimization algorithms, which have been used so far exhibits benefits of their own. Therefore it is left to the users to make a choice for the most suitable algorithm that can be incorporated into simulation model. Moreover, selection of appropriate technique depends on the usage, application and complexity of the task. It is hoped that this review provides a good platform for researchers interested in pursuing further work in optimizing sawmill yard operations.

An interesting aspect that has been recognised during the review was that the combinations of AI techniques with simulation models greatly enrich the flexibility of problem solving. Therefore, there may be a great potential for further research either to improve the efficiency and effectiveness of existing practice or to create a new paradigm by integrating more practical algorithms. The operational inefficiencies in different domains have been addressed by Simulation and AI based optimization techniques, there are some specific areas which have not yet been well explored. The in-house operations of sawmill yard logistics, therefore, it is expected that this paper can stimulate more research in the field of sawmill yard operations.

ACKNOWLEDGEMENTS

The authors would like to thank the anonymous reviewers for their valuable comments that greatly helped us to improve the contents of this paper.

REFERENCES

1. M. Josefa, P. David, D. M. Manuel and V. Eduardo, "Mathematical Programming Models for Supply Chain Production and Transport Planning," European Journal of Operation Research, Vol. 204, No. 3, 2010, pp. 377-390.http://dx.doi.org/10.1016/j.ejor.2009.09.008
2. C. A. Silva, J. M. Sousa and T. A. Runkler, "Rescheduling and Optimization of Logistics Process Using GA and ACO," Journal of Engineering Application of Artificial Intelligence, Vol. 21, No. 3, 2008, pp. 343-352. http://dx.doi.org/10.1016/j.engappai.2007.08.006
3. M. C. Aakil, N. Xiaofeng and P. Shaligram, "Optimization Models in Emergency Logistics: A Literature Review," Journal of Socio-Economic Planning Sciences, Vol. 46, No. 1, 2012, pp. 4-13. http://dx.doi.org/10.1016/j.seps.2011.04.004

4. S. Chopra and P. Meindl, "Supply Chain Management: Strategy, Planning, and Operations," 2nd Edition, Prentice Hall, New York, 2003.
5. F. Persson and J Olhager, "Performance Simulation of Supply Chain Designs," International Journal of Production Economics, Vol. 77, No. 3, 2002, pp. 231-245.http://dx.doi.org/10.1016/S0925-5273(00)00088-8
6. X. S. Qin, G. H. Huang and L. He, "Simulation and Optimization Technologies for Petroleum Waste Management and Remediation Process Control," Journal of Environmental Management, Vol. 90, No. 1, 2009, pp. 54- 76.http://dx.doi.org/10.1016/j.jenvman.2008.07.002
7. G. H. Huang and J. Xia, "Barriers to Sustainable WaterQuality Management," Journal of Environmental Management, Vol. 61, No. 1, 2001, pp. 1-23.http://dx.doi.org/10.1006/jema.2000.0394
8. J. H. Ryu, V. Dua and E. N. Pistikopoulos, "A Bi-Level Programming Framework for Enterprise-Wide Process Networks under Uncertainty," Journal of Computers and Chemical Engineering, Vol. 28, No. 6-7, 2004, pp. 1121- 1129.http://dx.doi.org/10.1016/j.compchemeng.2003.09.021
9. K. Ozgur and U. Fusun, "Probabilistic Linear Programming Approach for Supply Chain Networking Decisions," European Journal of Operation Research, Vol. 209, No. 3, 2011, pp. 253-264. http://dx.doi.org/10.1016/j.ejor.2010.09.025
10. H. C. Oh and I. A. Karimi, "Global Multiproduct Production—Distribution Planning with Duty Drawbacks," AICHE Journal, Vol. 52, No. 2, 2006, pp. 595-610.http://dx.doi.org/10.1002/aic.10669
11. W. Yue, "A Time Staged Linear Programming Model for Production Loading Problem with Import Quota Limit in a Global Supply Chain," Journal of Computers & Industrial Engineering, Vol. 59, No. 4, 2010, pp. 520-529.http://dx.doi.org/10.1016/j.cie.2010.06.007
12. H. Jung, B. Jeong and C. G. Lee, "An Order Quantity Negotiation Model for Distributor-Driven Supply Chains," International Journal of Production Economics, Vol. 111, No. 1, 2008, pp. 147-158. http://dx.doi.org/10.1016/j.ijpe.2006.12.054
13. A. P. Kanyalkar and G. K. Adil, "An Integrated Aggregate and Detailed Planning in a Multi-Site Production Environment Using Linear Programming," International Journal of Production Research, Vol. 43, No. 20, 2005, pp. 4431-4454.http://dx.doi.org/10.1080/00207540500142332
14. J. Kallrath, "Combined Strategic and Operational Planning—An MILP Success Story in Chemical Industry," Journal of Operation Research Spectrum, Vol. 24, No. 3, 2002, pp. 315-341. http://dx.doi.org/10.1007/s00291-002-0102-6

15. V. D. Silke, H. B. Bjorn and I. S. Hans, "Linear MixedInteger Models for Biomass Supply Chains with Transport, Storage and Processing," Journal of Energy, Vol. 35, No. 3, 2010, pp. 1338-1350. http://dx.doi.org/10.1016/j.energy.2009.11.017

16. A. Yavuz, N. K. Sukran and M. D. Jamison, "Incorporating Uncertainty in Optimal Decision Making: Integrating Mixed Integer Programming and Simulation to Solve Combinatorial Problems," Journal of Computers & Industrial Engineering, Vol. 56, No. 1, 2009, pp. 106-112. http://dx.doi.org/10.1016/j.cie.2008.04.003

17. S. Talluri and R. C. Baker, "A Multi-Phase Mathematical Programming Approach for Effective Supply Chain Design," European Journal of Operational Research, Vol. 141, No. 3, 2002, pp.544-558. http://dx.doi.org/10.1016/S0377-2217(01)00277-6

18. T. Panagiotis and G. P. Lazaros, "Optimal Production Allocation and Distribution Supply Chain Network," International Journal of Production Economics, Vol. 111, No. 2, 2008, pp. 468-483. http://dx.doi.org/10.1016/j.ijpe.2007.02.035

19. O. M. Akanle and D. Z. Zhang, "Agent Based Model for Optimizing Supplu-Chain Configurations," International Journal of Production Economics, Vol. 115, No. 2, 2008, pp. 444-460. http://dx.doi.org/10.1016/j.ijpe.2008.02.019

20. D. Goldberg, "Genetic Algorithm in Search, Optimization and Machine Learning," Addison-Wesley, Reading, 1989.

21. L. Davis, "Handbook of Genetic Algorithm," Van Nostrand Reinhold, New York, 1991.

22. A. A. Javadi, R. Farmani and T. P. Tan, "A Hybrid Intelligent Genetic Algorithm," Journal of Advance Engineering Informatics, Vol. 19, No. 4, 2005, pp. 255-262.http://dx.doi.org/10.1016/j.aei.2005.07.003

23. H. Ceylan and H. K. Ozturk, "Estimating Energy Demand of Turkey Based on Economic Indicators Using Genetic Algorithm Approach," Journal of Energy Conversion and Management, Vol. 45, No. 15-16, 2004, pp. 2525-2537.

24. L. Songsong and G. P. Lazaros, "Multi-Objective Optimization of Production, Distribution and Capacity Planning of Global Supply Chain in the Process Industry," Journal of Omega, Vol. 41, No. 2, 2013, pp. 369-382. http://dx.doi.org/10.1016/j.omega.2012.03.007

25. J. Gjerdrum, S. Nilay and G. P. Lazaros, "A Combined Optimization and Agent Based Approach to Supply Chain Modelling and Performance

Assessment," Journal of Production Planning and Control, Vol. 12, No. 1, 2001, pp. 81-88.

26. L. Amodeo, H. Chen and A. El Hadji, "Multi-Objective Supply Chain Optimization: An Industrial Case Study," Applications of Evolutionary Computing, Vol. 4448, 2007, pp. 732-741.

27. Z. F. Reza and E. Mahsa, "A Genetic Algorithm to Optimize the Total Cost and Service Level for Just-in-Time Distribution in a Supply Chain," International Journal of Production Economics, Vol. 111, No. 2, 2008, pp. 229- 243.http://dx.doi.org/10.1016/j.ijpe.2006.11.028

28. A. Fulya, G. Mitsuo, L. Lin and K. Ismail, "A SteadyState Genetic Algorithm for Multi-Product Supply Chain Network Design," Journal of Computers and Industrial Engineering, Vol. 56, No. 2, 2009, pp. 521-537 http://dx.doi.org/10.1016/j.cie.2007.05.012

29. C. D'souza, S. N. Omkar and J. Senthilnatj, "Pickup and Delivery Problem Using Meta-Heuristic Techniques," Journal of Expert System with Applications, Vol. 39, No. 1, 2012, pp. 328-334. http://dx.doi.org/10.1016/j.eswa.2011.07.022

30. N. David, S. Michele, T. Biagio and K. Uzay, "Genetic Algorithms for Supply-Chain Scheduling: A Case Study in the Distribution of Ready-Mixed Concrete," European Journal of operation Research, Vol. 177, No. 3, 2007, pp. 2069-2099.

31. H. C. W. Lau, T. Chan, W. T. Tsui and W. K. Pang, "Application of Genetic Algorithm to Solve the Multi Depot Vehicle Routing Problem," IEEE Transactions on Automation Science and Engineering, Vol. 7, No. 2, 2010, pp. 383-392.http://dx.doi.org/10.1109/TASE.2009.2019265

32. A. S. Daniels and M. G. Parson, "Development of Hybrid Agent Genetic Algorithm Approach to General Arrangements," Proceedings of Computer Applications and Information Technology in the Maritime Industries, Cortona, 2007, pp. 197-209.

33. B. David, C. Dick and R. Mikael, "A Hybrid Algorithm for Distribution Problems," IEEE Intelligent Systems, Vol. 20, No. 4, 2005, pp. 19-25. http://dx.doi.org/10.1109/MIS.2005.59

34. F. Azadivar and J. Wang, "Facility Layout Optimization Using Simulation and Genetic Algorithm," International Journal of Production Research, Vol. 38, No. 17, 2000, pp. 4369-4383. http://dx.doi.org/10.1080/00207540050205154

35. Y. Shin, H. Cho and K. Kang, "Simulation Model Incorporating Genetic Algorithm for Optimal Temporary Hoist Planning in High-Rise Building

Construction," Automation in Construction, Vol. 20, No. 5, 2011, pp. 550-558.http://dx.doi.org/10.1016/j.autcon.2010.11.021

36. J. Y. Yeh and W. S. Lin, "Using Simulation Techniques and Genetic Algorithm to Improve the Quality Care of a Hospital Emergency Department," Expert System with Applications, Vol. 32, No. 4, 2007, pp. 1073-1083. http://dx.doi.org/10.1016/j.eswa.2006.02.017
37. K. Shin-ike and H. Iima, "A Method for Determining Classroom Seating Arrangements by Using Genetic Algorithm," 2011 Proceedings of SICE Annual Conference, Tokyo, 13-18 September 2011, pp. 161-166.
38. L. Davis and M. Steenstrup, "Genetic Algorithm and Simulated Annealing an Overview," Pitman, 1987.
39. Z. Xin, Y. Hongnian and A. Anthony, "An Overview of Simulation in Supply Chains," Advanced Design and Manufacture to Gain a Competitive Edge, Book Chapter 3, Springer, London, 2008, pp. 407-416.
40. C. Reeves, "Modern Heuristic Techniques for Combinatorial Problems," John Wiley & Sons, Chichester, 1990.
41. J. Vaidyanathan and R. Anthony, "A Simulated Annealing Methodology to Distribution Network Design and Management," European Journal of Operation Research, Vol. 144, No. 3, 2003, pp. 629-645. http://dx.doi.org/10.1016/S0377-2217(02)00153-4
42. N. Chibeles-Martins, T. Pinto-Varela, A. P. Barbosa-Povoa and A. Q. Novais, "A Simulated Annealing Algorithm for the Design and Planning of Supply Chains with Economic and Environmental Objectives," Computer Aided Chemical Engineering, Vol. 30, 2012, pp. 21-25. http://dx.doi.org/10.1016/B978-0-444-59519-5.50005-8
43. A. D. Kathryn, S. Eric and B. Edmund, "A Simulated Annealing Based Heuristic for Determining Shipper Sizes for Storage and Transportation," European Journal of Operation Research, Vol. 179, No. 3, 2007, pp. 759-774.http://dx.doi.org/10.1016/j.ejor.2005.03.058
44. R. S. Hamid and G. Keivan, "A Simulated Annealing Approach for Multi-Periodic Rail-Car Fleet Sizing Problem," Journal of Computers & Operation Research, Vol. 36, No. 6, 2009, pp. 1789-1799. http://dx.doi.org/10.1016/j.cor.2008.05.004
45. H. Allaoui and A. Artiba, "Integrating Simulation and Optimization to Schedule a Hybrid Flow Shop with Maintenance Constraints," Computers & Industrial Engineering, Vol. 47, No. 4, 2004, pp. 431-450. http://dx.doi.org/10.1016/j.cie.2004.09.002
46. B. Suman, "Study of Simulated Annealing Based Algorithms for Multiobjective Optimization of a Constrained Problem," Computers

and Chemical Engineering, Vol. 28, No. 9, 2004, pp. 1849-1871. http://dx.doi.org/10.1016/j.compchemeng.2004.02.037

47. A. Mahmoud, D. Ali, A. Ameen, A. Raid and F. M. Nishat, "Simulated Annealing for Multi Objective Stochastic Optimization," International Journal of Science and Applied Information Technology, Vol. 2, No. 2, 2013, pp. 18- 21.
48. M. Dorigo and T. Stutzle, "Ant Colony Optimization," MIT Press, Cambridge, 2004.
49. S. C. Zhan, J. Xu and J. Wu, "The Optimization Selection on the Parameters of the Ant Colony Algorithm," Bulletin of Science and Technology, Vol. 19, No. 5, 2013, pp. 381- 386. http://dx.doi.org/10.1007/b99492
50. J. Yang and Y. Zhuang, "An Improved Ant Colony Optimization for Solving Complex Combinatorial Optimization Problem," Applied Soft Computing, Vol. 10, No. 2, 2010, pp. 653-660. http://dx.doi.org/10.1016/j.asoc.2009.08.040
51. C. A. Silva, J. M. C. Sousa, T. A. Runkler and J. M. G. Sa da Costa, "Distributed Supply Chain Management Using Ant Colony Optimization," European Journal of Operation Research, Vol. 199, No. 2, 2009, pp. 349-358.http://dx.doi.org/10.1016/j.ejor.2008.11.021
52. A. Luis, M. Moncayo and Z. Z. David, "Multi-Objective Ant Colony Optimization: A Meta-Heuristic Approach to Supply Chain Design," International Journal of Production Economics, Vol. 131, No. 1, 2011, pp. 407-420.http://dx.doi.org/10.1016/j.ijpe.2010.11.026
53. Y. Wei and K. Arun, "Ant Colony Optimization for Disaster Relief Operations," Transportation Research Part E, Vol. 43, No. 6, 2007, pp. 660-672.http://dx.doi.org/10.1016/j.tre.2006.05.004
54. S. K. Chaharsooghi and A. H. M. Kermani, "An Effective Ant Colony Optimization Algorithm for Multi-Objective Resource Allocation Problem," Journal of Applied Mathematics and Computation, Vol. 200, No. 1, 2008, pp. 167-177.http://dx.doi.org/10.1016/j.amc.2007.09.070
55. T. S. C. Felix and K. Niraj, "Effective Allocation of Customers to Distribution Centres: A Multiple Ant Colony Optimization Approach," Robotics and Computer-Integrated Manufacturing, Vol. 25, No. 1, 2009, pp. 1-12.http://dx.doi.org/10.1016/j.rcim.2007.05.002
56. M. Remi, A. Jean-Paul and J. Hosang, "Ant Colony Optimization Algorithm to Solve for the Transportation Problem of Cross-Docking Network," Computers & Industrial Engineering, Vol. 59, No. 1, 2010, pp. 85-92. http://dx.doi.org/10.1016/j.cie.2010.03.002

57. A. C. Zecchin, A. R. Simpson, H. R. Maier and J. B. Nixon, "Parametric Study of an Ant Algorithm Applied to Water Distribution System Optimization," IEEE Transactions on Evolutionary Computation, Vol. 9, No. 2, 2005, pp. 175- 191.http://dx.doi.org/10.1109/TEVC.2005.844168
58. J. Kennedy and R. C. Eberhart, "Particle Swarm Optimization," Proceedings of IEEE International Conference on Neural Networks, Perth, 27 November-1 December 1995, pp. 1942-1948. http://dx.doi.org/10.1109/ICNN.1995.488968
59. K. Andreas and D. Andreas, "Facility Location Models for Distribution System Design," European Journal of Operational Research, Vol. 162, No. 1, 2005, pp. 4-29.http://dx.doi.org/10.1016/j.ejor.2003.10.031
60. W. B. Langdon and R. Poli, "Evolving Problems to Learn about Particle Swarm Optimizers and Other Search Algorithms," IEEE Transactions on Evolutionary Computation, Vol. 11, No. 5, 2007, pp. 561-578. http://dx.doi.org/10.1109/TEVC.2006.886448
61. G. Wei, K. S. Huang and Y. C. Guan, "Automation and Logistics," IEEE International Conference on Digital Object Identifier, 2007, pp. 2830-2835.
62. W. Lei and M. Fanhua, "Natural Computation," ICNC 4th International Conference on Natural Computation, Vol. 7, Jinan, 18-20 October 2008, pp. 659-663.
63. Y. J. Gong, J. Zhang, O. Liu, R. Z. Huang, H. S. H. Chung and Y. H. Shi, "Optimizing the Vehicle Routing Problem with Time Windows: A Discrete Particle Swarm Optimization Approach," IEEE Transactions on, Systems, Man and Cybernetics-Part C: Applications and Reviews, Vol. 42, No. 2, 2012, PP. 254-267.
64. Y. f. Deng and H. Q. Tong, "Dynamic Shortest Path Algorithm in Stochastic Traffic Networks Using PSO Based on Fluid Neural Network," Journal of Intelligent Learning Systems and Applications, Vol. 3, No. 1, 2011, pp. 11-16.http://dx.doi.org/10.4236/jilsa.2011.31002
65. K. C. Chilukuri and R. Prasad, "Particle Swarm Optimization Based Approach for Resource Allocation and Scheduling in OFDMA Systems," International Journal of Communications, Network and System Sciences, Vol. 3, No. 5, 2010, pp. 466-471.http://dx.doi.org/10.4236/ijcns.2010.35062
66. N. S. Fadi, I. H. Hafsa, M. N. Raja, D. Salima and S. Shaikha, "Iris Recognition Using Artificial Neural Network," Journal of Expert System with Applications, Vol. 38, No. 5, 2011, pp. 5940-5946. http://dx.doi.org/10.1016/j.eswa.2010.11.029

67. K. Mark, T. Ashutosh and M. Jörn, "A Review of Soft Computing Applications in Supply Chain Management," Journal of Applied Soft Computing, Vol. 10, No. 3, 2010, pp. 661- 674. http://dx.doi.org/10.1016/j.asoc.2009.09.004

68. S. Ihsan and G. Burckaan, "A Neural Network Model for Scheduling Problems," European Journal of Operation Research, Vol. 93, No. 2, 1996, pp. 288-299.http://dx.doi.org/10.1016/0377-2217(96)00041-0

69. A. R. Soroush, I. N. Kamal-Abadi and A. Bahreininejad, "Review on Applications of Artificial Neural Networks in Supply Chain Management and Its Future," Journal of World Applied Sciences, Vol. 6, No. 1, 2009, pp. 12-18.

70. N. Nikotan, J. Han and M. Behesthi, "Software Project Scheduling Using a Multi-Agent System," Proceedings of 2011 8th International Conference on Information Technology: New Generations, Las Vegas, 11-13 April 2011, pp. 212-213.

71. R. C. Cavalcante, I. I. Bittencourt, A. P. Da Silva, M. Silva, E. Costa and R. Santos, "A Survey of Security in Multi Agent Systems," Journal of Expert System with Applications, Vol. 39, No. 5, 2012, pp. 4835-4846.

72. K. Kyoungmin and K. Kyong Ju, "Multi-Agent-Based Simulation System for Construction Operations with Congested Flow," Journal of Automation in Construction, Vol. 19, No. 7, 2010, pp. 867-874. http://dx.doi.org/10.1016/j.autcon.2010.05.005

73. D. Perugini, S. Wark, A. Zschorn, D. Lambert, L. Sterling and A. Pearce, "Agent in Logistics Planning-Experiences with the Coalition Agent Experiment Project," 2nd International Joint Conference on Autonomous Agents and Multi-agent Systems (AAMAS 2003), Melbourne, 14-18 July 2003.

74. G. Vita and G. Janis, "Technologies and Multi-Agent System Architecture for Transportation and Logistics Support: An Overview," International Conference on Computer Systems and Technologies-CompSysTech' 2005, Varna, 16-17 June 2005, pp. IIIA.6-1-IIIA.6.6.

75. N. Neagu, K. Dorer and M. Calisti, "Solving Distributed Delivery Problems with Agent-Based Technologies and Constraint Satisfaction Problems," 2006.

76. A. Kazemi, M. H. Fazel and S. M. Husseini, "A MultiAgent System to solve the Production-Distribution Planning Problem for Supply Chain: A Genetic Algorithm Approach," International Journal of Advance Manufacturing Technology, Vol. 44, No. 1-2, 2009, pp. 180-193.

77. J. A. Persson, P. Davidsson, S. J. Jhansson and F. Werenstedt, "Agent-Based Approaches and Classical Optimization Techniques for Dynamic Distributed Resource Allocation: A Preliminary Study," 2003.

78. H. Ahn and H. Lee, "Agent Based Dynamic Network for Supply Chain Management," BT Technology Journal, Vol. 22, No. 2, 2004, pp. 18-27. http://dx.doi.org/10.1023/B:BTTJ.0000033467.83300.c0

79. W. Y. Liang and C. C. Huang, "Agent-Based Demand Forecasting in Multi-Echelon Supply Chain," Journal of Decision Support Systems, Vol. 42, No. 1, 2006, pp. 390- 407.

80. M. Caridi and S. Cavalieri, "Multi-Agent System in Production Planning and Control: An Overview," Journal of Production Planning and Control, Vol. 15, No. 2, 2004, pp. 106-118.http://dx.doi.org/10.1080/0953728041 0001662556

81. F. Person, "Supply Chain Simulation: Experiences from Two Case Studies," In: A. Verbraeck and V. Hlupic, Eds., Proceedings from the 15th European Simulation Symposium, Delft, 26-29 October 2003, pp. 399- 404.

82. J. Dong, H. Ding, C. Ren and W. Wang, "IBM Mart SCORA SCOR Based Supply Chain Transformation Platform through Simulation and Optimisation Techniques," Proceedings of the 2006 Winter Simulation Conference, California, 3-6 December 2006, pp. 650-659.

83. D. R. Towill, "Supply Chain Dynamic," International Journal of Computer Integrated Manufacturing, Vol. 4, No. 4, 1991, pp. 197-208.http://dx.doi. org/10.1080/09511929108944496

84. S. Dusan, S. Nenad and B. Radenkovic, "Supply Network Modelling and Simulation Methodology," Journal of Simulation Modelling and Practice and Theory, Vol. 17, No. 4, 2009, pp. 743-766. http://dx.doi. org/10.1016/j.simpat.2009.01.001

85. E. R. Shannon, "Systems Simulation: The Art and Science," Prentice-Hall, Upper Saddle River, 1975.

86. S. G. Bell, "Simulation: A Data-Driven Tool to Lower Costs ASCET," Vol. 4, Montgomery Research, San Francisco, 2002.

87. E. J. Chen, Y. M. Lee and P. L. Selikson, "A Simulation Study on Logistic Activities in a Chemical Plant," Journal of Simulation Modelling Practice and Theory, Vol. 10, No. 3-4, 2002, pp. 235-245.

88. L. Pasquale and M. M. Rina, "Berth Planning and Resource Optimization at a Container Terminal via Discrete Event Simulation," European Journal of Operational Research, Vol. 133, No. 3, 2001, pp. 537-547. http:// dx.doi.org/10.1016/S0377-2217(00)00200-9

89. H. Zhang and H. Li, "Simulation-Based Optimization for Dynamic Resource Allocation," Automation in Constructions, Vol. 13, No. 3, 2004, pp. 409-420.

90. P. Henri, B. Romain and C. Christophe, "A Continuous Simulation Approach for Supply Chains in the Automotive Industry," Simulation Modelling Practice and Theory, Vol. 15, No. 2, 2007, pp. 185-198. http://dx.doi.org/10.1016/j.simpat.2006.09.019

91. G. M. Maziar, K. Behrooz and D. Mohammad, "A Simulation Study of Logistics and Manufacturing Activities in an Automobile Supply Chain," The 41th International Conference on Computer and Industrial Engineering, Los Angeles, 23-25 October 2011, pp. 74-80.

92. F. Hanno, "Simulation of Logistics in Food Retailing for Freight Transportation Analysis," 12th World Conference on Transport Research, Lisbon, 11-15 July 2010.

93. K. S. Sanjay, M. K. Tiwari, H. D. Wan and S. Ravi, "Optimization of the Supply Chain Network: Simulation, Taguchi and Psychoclonal Algorithm Embedded Approach," Computers and Industrial Engineering, Vol. 58, No. 1, 2010, pp. 29-39.http://dx.doi.org/10.1016/j.cie.2009.07.016

94. Y. J. June, B. Gary, F. P. Jospeh, V. R. Gintaras and E. David, "A Simulation Based Optimization Approach to Supply Chain Management under Demand Uncertainty," Computer and Chemical Engineering, Vol. 28, No. 10, 2004, pp. 2087-2106.http://dx.doi.org/10.1016/j.compchemeng.2004.06.006

95. P. Dobrila, R. Rajat and P. Radivoj, "Modelling and Simulation of a Supply Chain in an Uncertain Environment," European Journal of Operational Research, Vol. 109, No. 2, 1998, pp. 299-309. http://dx.doi.org/10.1016/S0377-2217(98)00058-7

96. S. Jaroslav and P. Pavel, "Integer Simulation Based Optimization by Local Search," Journal of Procedia Computer Science, Vol. 1, No. 1, 2010, pp. 1341-1348.

97. H. Carvalho, A. P. Barroso, V. H. Machado, S. Azevedo and V. Cruz-Machado, "Supply Chain Redesign for Resilience Using Simulation," Journal of Computers and Industrial Engineering, Vol. 62, No. 1, 2012, pp. 329-341.http://dx.doi.org/10.1016/j.cie.2011.10.003

98. Y. Taejong, C. Hyunbo and Y. Enver, "Hybrid Algorithm for Discrete Event Simulation Based Supply Chain Optimization," Journal of Expert System with Applications, Vol. 37, No. 3, 2010, pp. 2354-2361. http://dx.doi.org/10.1016/j.eswa.2009.07.039

99. T. H. Truong and F. Azadivar, "Simulation Optimization in Manufacturing Analysis: Simulation Based Optimization for Supply Chain Configuration Design," The 35th Conference on Winter Simulation: Driving Innovation, New Orleans, 7-10 December 2003, pp. 1268-1275.

100. M. Rabe, F. W. Jaekel and H. Weinaug, "Reference Models for Supply Chain Design and Configuration," Proceedings of the 38th Conference on Winter Simulation, Winter Simulation Conference, Monterey, 3-6 December 2006, pp. 1143- 1150.

101. W. Xiatao, F. P. Joseph and V. R. Gintaras, "SimulationBased Optimization with Surrogate Models-Application to Supply Chain Management," Journal of Computers & Chemical Engineering, Vol. 29, No. 6, 2005, pp. 1317- 1328.http://dx.doi.org/10.1016/j.compchemeng.2005.02.018

102. N. Amalia and G. L. Marianthi, "Hybrid Simulation Based Optimization Approach for Supply Chain Management," Journal of Computers & Chemical Engineering, Vol. 47, 2012, pp. 183-193. http://dx.doi.org/10.1016/j.compchemeng.2012.06.045

103. A. A. Mohamed and M. A. Talal, "Simulation Optimization for an Emergency Department Healthcare Unit in Kuwait," European Journal of Operation Research, Vol. 198, No. 3, 2009, pp. 936-942. http://dx.doi.org/10.1016/j.ejor.2008.10.025

104. Y. Huang, Y. P. Li., X. Chen, A. M. Bao and M. Zhou, "Simulation-Based Optimization Methods for Water Resources Management in Tarim River Basin, China," Procedia Environmental Sciences, Vol. 2, 2010, pp. 1451-1460.http://dx.doi.org/10.1016/j.proenv.2010.10.158

105. L. R. Scott, M. H. Catherine and T. T. Mark, "Optimization of Systems with Multiple Performance Measures via Simulation: Survey and Recommendations," Computers & Industrial Engineering, Vol. 54, No. 2, 2008, pp. 327- 339.http://dx.doi.org/10.1016/j.cie.2007.07.004

106. F. Marcus, H. C. N. Amos and M. Philip, "A SimulationBased Scheduling System for Real-Time Optimization and Decision Making Support," Robotics and ComputerIntegrated Manufacturing, Vol. 27, No. 4, 2011, pp. 696- 705.http://dx.doi.org/10.1016/j.rcim.2010.12.006

107. D. Berna, B. Tolga and U. A. Eren, "Simulation Optimization Based DSS Application: A Diamond Tool Production Line in Industry," Simualtion Modelling Practice and Theory, Vol. 14, No. 3, 2006, pp. 296-312. http://dx.doi.org/10.1016/j.simpat.2005.07.001

108. C. Pasquale, M. Daniele, R. Giuseppe and T. Marco, "A Dynamic Simulation Model of a Flexible Transport Services for People in

Congested Area," Procedia-Social and Behavioral Sciences, Vol. 54, No. 4, 2012, pp. 357-364.

109. J. Banks, S. Buckley, S. Jain, P. Lendermann and M. Manivannan, "Panel Session: Opportunities for Simulation in Supply Chain Management," Proceedings of the 2002 Winter Simulation Conference, IEEE, San Diego, 8-11 December 2002, pp. 1652-1658.

110. C. Goldspink, "Methodological Implications of Complex Systems Approach to Sociality: Simulation as a Foundation for Knowledge," Journal of Artificial Societies and Social Simulation, Vol. 5, No. 1, 2002, pp. 1-19.

111. V. Albino, N. Carbonara and I. Gainnoccaro, "Supply Chain Cooperation in Industrial Districts: A Simulation Analysis," European Journal of Operation Research, Vol. 117, No. 1, 2007, pp. 261-280. http://dx.doi.org/10.1016/j.ejor.2005.12.007

112. B. Georgiana, "Simulation and Optimization in Supply Chain," Journal of Procedia Economics and Finance, Vol. 3, 2012, pp. 635-641. http://dx.doi.org/10.1016/S2212-5671(12)00207-9

113. C. Ye, M. Linas, O. Seza and V. R. Gintaras, "SimulationOptimization Approach to Clinical Trial Supply Chain Management with Demand Scenario Forecast," Journal of Computers & Chemical Engineering, Vol. 40, No. 11, 2012, pp. 82-96.http://dx.doi.org/10.1016/j.compchemeng.2012.01.007

114. H. L. Young, K. C. Min, J. K. Seo and B. K. Yun, "Supply Chain Simulation with Discrete-Continuous Combined Modelling," Journal of Computers & Industrial Engineering, Vol. 43, No. 1-2, 2002, pp. 375-392.

115. F. L. Zhang, D. M. Johnson and M. A. Johnson, "Development of a Simulation Model of Biomass Supply Chain for Biofuel Production," Journal of Renewable Energy, Vol. 44, 2012, pp. 380-391. http://dx.doi.org/10.1016/j.renene.2012.02.006

116. B. Daniel, L. B. Luc and A. S. Mohamed, "Discrete Event Simulation to Improve Log Yard Operations," INFOR, Vol. 50, No. 4, 2012, pp. 175-185.

Chapter 3

USING NON-ADDITIVE MEASURE FOR OPTIMIZATION-BASED NONLINEAR CLASSIFICATION

Nian Yan[1,2], Zhengxin Chen[1], Yong Shi[1,3], Zhenyuan Wang[4], Guimin Huang[5]

[1]College of Information Science and Technology, University of Nebraska at Omaha, Omaha, USA

[2]Nebraska Furniture Mart, Berkshire Hathaway Company, Omaha, USA

[3]Research Center on Fictitious Economy and Data Sciences, Chinese Academy of Sciences, Beijing, China

[4]Department of Mathematics, University of Nebraska at Omaha, Omaha, USA

[5]Guilin University of Electronic Technology, Guilin, China

ABSTRACT

Over the past few decades, numerous optimization-based methods have been proposed for solving the classification problem in data mining. Classic optimization-based methods do not consider attribute interactions toward classification. Thus, a novel learning machine is needed to provide a better understanding on the nature of classification when the interaction among contributions from various attributes cannot be ignored. The interactions can be described by a non-additive measure while the Choquet integral can serve as the mathematical tool to aggregate the values of attributes and the corresponding values of a non-additive measure. As a main part of this research, a new nonlinear classification method with non-additive measures is proposed. Experimental results show that applying non-additive measures on the classic optimization-based models improves the classification robustness and accuracy compared with some popular classification methods. In addition, motivated by well-known Support Vector Machine approach, we transform the primal optimization-based nonlinear classification model with the signed non-additive measure into its dual form by applying Lagrangian optimization theory and Wolfes dual programming theory. As a result, $2^n - 1$ parameters of the signed non-additive measure can now be approximated with m (number of records) Lagrangian multipliers by applying necessary conditions of the primal classification problem to be optimal. This method of parameter approximation is a breakthrough for solving a non-additive measure practically when there are a relatively small number of training cases available ($m \ll 2^n - 1$). Furthermore, the kernel-based learning method engages the nonlinear classifiers to achieve better classification accuracy. The research produces practically deliverable nonlinear models with the non-additive measure for classification problem in data mining when interactions among attributes are considered.

INTRODUCTION

Classic optimization-based methods formulate classification problems by modeling data with standard optimization techniques using objectives and constraints. Mathematical programming provides general solution to the zoptimization problem. For example, references [1,2] proposed two classification models based on reducing the misclassification through minimizing overlaps or maximizing the distance of two data points in a linear system. A method named Multiple Criteria Linear Programming (MCLP) [3,4] has been initialized to compromise the objectives of models in [1] and [2] simultaneously and achieved a better data separation in a linear system. Alternatively, a quadratic model can be used to deal with linearly inseparable situation [5]. The key idea of those approaches is to separate data when they are in different classes as well as pull data together when they are in the same class. Initiated by [6], another well-known optimization-based classification method is Support Vector Machine (SVM), which mathematically constructs hyperplanes by support vectors. Furthermore, SVM separates data nonlinearly by introducing so-called nonlinear kernel functions.

Although these optimization-based methods separate data linearly or nonlinearly, they do not consider contributions from the interaction among attributes. In this paper, we use a nonadditive measure to model data with interactions and propose new nonlinear classification models. Nonlinear integrals can be used as tools to aggregate unknown parameters in the non-additive measure and values of attributes. As one of nonlinear integrals, the Choquet integral [7] is chosen as the aggregation tool for data modeling for classification problem. In addition, we investigate the direction of constructing nonlinear objectives by developing kernel functions in nonlinear classification models, a technique taken by SVM.

The rest of this paper is organized as follows: In Section 2, an overview of classic optimization-based classification methods is provided. Section 3 reviews definitions of non-additive measures and the Choquet integral. In

Section 4, a new optimization-based classification model with a non-additive measure is proposed. Section 5 describes the Lagrangian optimization approach to solve the issue of dealing limited training samples with the proposed nonlinear classification model. Section 6 shows performance of the proposed models in experimental results. Finally, Section 7 provides conclusions from this research.

PRELIMINARY

In this section, we provide an overview of classic optimization-based classification methods.

Consider that a dataset consists of n attributes and m records. Let X = {x_1, x_2, ···, x_n} denote the set of feature attributes and y be the class label, where $y_j \in \{-1,1\}$ for a two classes dataset. The dataset has a form as follows:

x_1	x_2	···	x_n	y
f_{11}	f_{12}	···	f_{1n}	y_1
f_{21}	f_{22}	···	f_{2n}	y_2
		···		
f_{m1}	f_{m2}	···	f_{mm}	y_m

where following elements

f_{j1}	f_{j2}	···	f_{jn}

are the values of attributes x_1, x_2, ···, x_n for the j-th record in the dataset, denoted by f_j, j = 1, ···, 2, m. Note that f_j can be regarded as a vector. In addition, y_j is the corresponding class label in the j-th record.

The mathematical programming or optimization-based approach have been widely used for many applications. Particularly, numerous mathematical programming methods based on optimization techniques have been proposed for solving classification problem [1-3,6]. In classification, the concept of classes is generally expressed as $wf_j - b$, where w, f_j, and b represent attribute weights, values, and classification critical value respectively. Therefore, wf_j is the weighted sum of all the attributes. For a dataset with two classes, the decision function for the classes are defined as:

$$y_j\left(\mathbf{w}f_j - b\right) \le 0$$

where $y_j = 1$ if the j-th record belongs to class 1 and $y_j = -1$ if the j-th record belongs to class 2.

The two linear classification methods [1,2] based on the idea of reducing misclassification by minimizing the overlaps or maximizing the sum of distances in a linear system. One approach is to maximize the sum of minimum distances (MMD) of data from the critical value. Another approach separates the data by minimizing the sum of deviations (the overlapped distances between classes) (MSD) of data from the critical value. These two classic linear classification models can easily be described with a standard form of optimization, i.e.

$$\begin{aligned} &\text{Maximize} && \sum_{j=1}^{m} \beta_j \\ &\text{Subject to} && y_j\left(\mathbf{w}f_j - b\right) \ge \beta_j \\ & && \beta_j \ge 0, \mathbf{w} \text{ is unrestricted} \end{aligned} \qquad \text{(MMD)}$$

$$\begin{aligned} &\text{Minimize} && \sum_{j=1}^{m} \alpha_j \\ &\text{Subject to} && y_j\left(\mathbf{w}f_j - b\right) \le \alpha_j \\ & && \alpha_j \ge 0, \mathbf{w} \text{ is unrestricted} \end{aligned} \qquad \text{(MSD)}$$

where α_j denotes the degree of the overlapping of the two classes and β_j denotes the distances from the observation to the critical classification value b. The weights w are optimized by linear programming, a typical optimization technique. The critical value b is given as a constant non-zero real number.

The above two linear classification models provide the basic idea of data separation, which pulls the data apart from the boundary (maximize the sum of β_j in MMD) or to make the smallest data overlapping area (minimize the sum of α_j in MSD). However there are some optimization difficulties in those approaches. For example, the MMD model cannot be optimized because the value of β_j can reach as large as possible since the goal is to maximize the sum of β_j. Thus, in the implementation of MMD model, β_j is bounded as $\beta_j \le \beta^*$, where β^* is a given positive constant. The MMD classification model is only able to classify linearly separable dataset. Similarly, the α_j in MSD model has to be bounded to a very small positive value α^* as $\alpha_j \ge \alpha^*$.

Efforts have been made to improve optimization-based linear classification for better dealing with linearly inseparable. For example, MCLP approach was initiated by compromising two objectives of MMD and MSD simultaneously and achieved a better classification within a linear system [3]. MCLP model compromises objectives as [3]:

$$\begin{aligned}&\text{Minimize} && \sum_{j=1}^{m}(\alpha_j - \beta_j)\\ &\text{Subject to} && y_j(\mathbf{w}f_j - b) = \alpha_j - \beta_j\\ & && \beta_j, \alpha_j \geq 0; \beta_j \leq \beta^*; \beta^* \neq 0; \mathbf{w} \text{ is unrestricted}\end{aligned} \tag{MCLP}$$

where β^* is a positive constant to restrict the upper bound of β_i.

Another direction of improving optimization-based classification is to develop nonlinear models by constructing nonlinear objectives, such as the Multiple Criteria Quadratic Programming (MCQP), a nonlinear optimization classification [5].

NON-ADDITIVE MEASURES

A common characteristic of the methods described above is that the modeling is based on the assumption that contributions from all attributes toward classification are the sum of contributions of each attribute. None of those methods considers the interactions among attributes toward classification, which may provide a better understanding of the nature of classification and achieve more satisfactory results. In addition, the model should be able to represent the underlying phenomenon of applications such as classification in a more adequate manner because attributes are not completely isolated from each other. Such a model should have the potential of increased robustness, defined as the ability of maintaining effective performance on both training and testing results on a diversity of datasets. Particularly, a classification model is said to be robust when the performance of its testing results is not significantly distant from training.

The theory of non-additive measure can achieve increased robustness and better performance in classification. The bases of non-additive measures and the nonlinear integrals are briefly reviewed in the rest of this section.

Definition of Non-Additive Measures

The attribute interactions can be represented by a non-additive measure. The concept of non-additive measure (also referred to as fuzzy measure theory) was initiated in the 1950s and has been well developed since 1970s [7-9].

Let finite set $X = \{x_1, \cdots, x_n\}$ denote the attributes in a multidimensional dataset. Several important types of non-additive measure are defined as the followings [8]:

Definition 1. A non-additive measure μ defined on X is a set function $\mu:\mathcal{P}(X)\rightarrow[0,\infty)$ satisfying $\mu(\varnothing)=0$, where $\mathcal{P}(X)$ denotes the power set of X, μ is monotone if it satisfies $\mu(\varnothing)=0$ and $E\subseteq F$ if $E\subseteq F$, where E, F are any sets in $\mathcal{P}(X)$.

Definition 2. A signed non-additive measure μ is defined on X is a set function $\mu:\mathcal{P}(X)\rightarrow(-\infty,\infty)$.

The values of non-additive measure μ are unknown parameters. The signed non-additive measure is adopted to develop optimization-based nonlinear classification models.

Choquet Integral

Nonlinear integrals are used as data aggregation tools to integrate the values of attributes with respect to a nonadditive measure. As one of nonlinear integrals, the Choquet integral is more appropriate for applications such as classification because it provides very important information in interactions among attributes [10].

Now let the values of $f=\{f(x_1),f(x_2),\cdots,f(x_n)\}$ denote the values of each attribute in the dataset; let μ be the non-additive measure. The general definition of the Choquet integral, with function $f:X\rightarrow(-\infty,\infty)$, based on signed non-additive measure μ, is defined in formula 1 as

$$(c)\int f d\mu=\int_{-\infty}^{0}\left[\mu(F_\alpha)-\mu(X)\right]d\alpha+\int_{0}^{\infty}\mu(F_\alpha)d\alpha \tag{1}$$

where $F_\alpha=\{x\,|\,f(x)\geq\alpha\}$ is called α-cut set of f, for $\alpha\in(-\infty,\infty)$, n is the number of attributes in the dataset.

Choquet integral may be calculated as [11]:

$$(c)\int f d\mu=\sum_{j=1}^{2^n-1}z_j\mu_j \tag{2}$$

where

$$z_j=\begin{cases}\min\limits_{i:frc\left(\frac{j}{2^i}\right)\in[0.5,1)}\left(f(x_i)\right)-\max\limits_{i:frc\left(\frac{j}{2^i}\right)\in[0,0.5)}\left(f(x_i)\right) & \text{if} >0 \text{ or } j=2^n-1\\ 0 & \text{otherwise}\end{cases} \tag{3}$$

$$frc\left(\frac{j}{2^i}\right)$$

is the fractional part of $\frac{j}{2^i}$

and the maximum operation on the empty set is zero. Let $j_n, j_{n-1}, \cdots, j_1$ represent the binary form of j, the i in formula 3 is determined as following:

$$\left\{i \middle| frc\left(\frac{j}{2^i}\right) \in [0.5,1)\right\} = \{i \mid j_i = 1\}$$

and

$$\left\{i \middle| frc\left(\frac{j}{2^i}\right) \in [0,0.5)\right\} = \{i \mid j_i = 0\}$$

It is important to emphasize that the concept of Choquet integral is not equal to a nonadditive measure μ. Rather, it is a mathematical tool aggregating the values of attributes with respect to the signed non-additive measure μ; as such, it is similar to the linear weighted sum to aggregate the corresponding attribute with respect to the weights in a linear model such as MSD.

OPTIMIZATION-BASED NONLINEAR CLASSIFIERS WITH NON-ADDITIVE MEASURES

The idea of using non-additive measure in classification problem is not new. In the fuzzy measure community, non-additive measures have been utilized for modeling attribute interactions for data separation purpose. For example, reference [12] used the Choquet integral with respect to the non-additive measure on statistical pattern classification based on possibility theory, an optimization-based classification model was later proposed with a non-additive measure [13]. Reference [14] proposed the k-Interactive (k = 2) classification with feature selections based on a pattern matching algorithm similar to [12]. Classification can also be achieved by directly separating the date using the weighted Choquet integral projection [15] or using a penalized signed fuzzy measure [16]. A detailed discussion of geometric meaning of the contributions from feature attributes in nonlinear classification can be found in [17].

There are limitations on above methods, notably: (a) Impractical: Due to the complexity of the non-additive measure, the methods were only applicable for datasets with small number of attributes, generally less than 5. (b) Limited performance: the classification accuracy was not promising compared to other popular methods [13] due to lack of better learning algorithms for determining unknown parameters of a non-additive measure. For instance, classification model in [15] with the Choquet integral has infinite number of solutions and the proposed method can only determine one of them. To address these limitations, this current research intends to provide a more practical and powerful solution toward nonlinear classification with a non-additive measure.

In addition, early studies of non-additive measure for classification also show limitations on classification accuracy and scalability. For example, although classification model in [12] is well developed in theory (similar to Bayesian classifier), the classification did not show any benefits of using non-additive measure and it is even more difficult to obtain good results on small Iris dataset, a benchmark dataset from UCI Machine Learning Repository (http://archive.ics.uci.edu/ml/). An optimizationbased nonlinear classification model [13] with a nonadditive measure was later proposed and studied. The results show it even performs worse than linear classifier on iris dataset and only competitive to fuzzy k-NN classifier on other datasets. The research [13] suggests a better non-additive identification algorithm is needed.

An improvement for nonlinear optimization-based classifiers with non-additive measure might be the optimization process of the critical value for classification. Inn MCLP model, the classification critical value b is not optimized but arbitrarily chosen. A better method to determine b could be updating b with the average of the lowest and largest predicted scores [15] during learning iterations. Alternatively, the critical value b in MCLP also can be replaced with soft-margin b$\pm$1 similar to SVM which constructs a separation belt instead of a single cutting line. With this technique, it is guaranteed to produce a unique solution to the model because the goal of the optimization is to find the cutting line which is most close to the misclassified data points on both sides. The MCLP model can be extended to a linear programming solvable problem with optimized b and the signed non-additive measure, as shown below:

$$\text{Minimize} \quad \sum_{j=1}^{m} \alpha_j$$

$$\text{Subject to} \quad y_j\left(\left(c\right)\int f d\mu - b\right) \leq 1 + \alpha_j$$

$$\alpha_j \geq 0; \mu, b \text{ are unrestricted} \qquad \text{(M1)}$$

where $y_i \in \{-1,1\}$, μ, b and α are critical values to be determined. Model M1 can be solved by standard linear programming techniques such as the simplex method. However, the number of μ, which is $2^n - 1$, is exponentially related to the number of attributes (n) because of power set operation in the Choquet integral. When the number of records (m) in the training set is relatively small (e.g. $m \ll 2^n - 1$), the model is difficult to be solved. In the next section, we propose a compromised solution to deal with this situation.

NONLINEAR CLASSIFICATION WITH THE SIGNED NON-ADDITIVE MEASURE BY LAGRANGIAN OPTIMIZATION

As mentioned, it is hard to optimize the non-additive optimization-based classification models when there are not enough observations ($m \ll 2^n - 1$). The existing approaches such as hierarchical Choquet integral [18] and the k-Interactive measure [14] ignored some values of non-additive measure μ to some extent. As a solution, the Lagrangian optimization theory can be incorporated to transform the model into practically solvable form with the best approximation of parameters of non-additive measure μ. The Karush-Kuhn-Tucker (KKT) conditions [19] applied in the Lagrangian optimization process are the necessary conditions to guarantee an optimizationbased classification model to reach optimum. To develop a nonlinear classifier which can deal with this situation, a quadratic non-additive optimization-based model is constructed and transformed.

Lagrangian Theory for Optimization

The Lagrangian theory is intended to provide the necessary conditions for a given nonlinear optimization problem to reach an optimal solution. The KKT conditions in the Lagrangian optimization provide the necessary conditions for the proposed classification model to have an optimal solution. Generally, an optimization problem can be presented as [20]:

Definition 3. Given functions f, g_i, i = 1, ⋯, k, and h_i, i = 1, ⋯, m, defined on vector $w \in \Omega \subseteq \Re^n$, the primal optimization problem is defined as:

Minimize $f(w)$

Subject to $g_i(w) \le 0, i = 1, \cdots, k$

$h_i(w) = 0, i = 1, \cdots, m$

The generalized Lagrangian function corresponding to the definition 3 is

$$L(\mathbf{w}, \boldsymbol{\lambda}, \boldsymbol{\delta}) = f(w) + \sum_{i=1}^{k} \lambda_i g_i(w) + \sum_{i=1}^{m} \delta_i h_i(w)$$

where $\boldsymbol{\lambda}=\langle \lambda_1,\lambda_2,\cdots,\lambda_i \rangle$, $\boldsymbol{\delta}=\langle \delta_1,\delta_2,\cdots,\delta_i \rangle$ are the Lagrangian multipliers and $\lambda_i,\delta_i \in \Re^n$.

Instead of solving the primal optimization problem, an alternative is to optimize the Lagrangian dual problem, which is defined as

Maximize $\theta(\lambda,\delta)$,

Subject to $\lambda \geq 0$,

where $\boldsymbol{\theta}(\boldsymbol{\lambda},\boldsymbol{\delta}) = \mathbf{inf}_{\mathbf{w}\in\Omega}\, \boldsymbol{L}(\mathbf{w},\boldsymbol{\lambda},\boldsymbol{\delta})$, the greatest lower bound on Lagrangian function L over Ω. The optimal solution for the objective function is called the value of the problem.

The following Karush-Kuhn-Tucker theorem Tucker 1951 gives the necessary conditions for an optimal solution to general optimization problem.

Theorem 1. Given an optimization problem as defined in definition 3, the necessary and sufficient conditions for a point w^* to be an optimum are, $\exists \lambda^*,\delta^*$, such that

$$\frac{\partial L(w^*,\lambda^*,\delta^*)}{\partial w}=0$$

$$\frac{\partial L(w^*,\lambda^*,\delta^*)}{\partial \delta}=0$$

$$\lambda_i^* g_i(w^*)=0, i=1,\cdots,k.$$

$$g_i(w^*)\leq 0, i=1,\cdots,k.$$

$$\lambda_i^* \geq 0, i=1,\cdots,k.$$

The first two conditions are also the necessary conditions for the optimization problem to reach optimal. The third condition is called KKT complementary condition. The first two conditions are also the necessary conditions for the optimization problem to reach optimal. The sufficient condition is true only if function (L) of w is convex. In this research, since the convexity of the primal problem is yet to be proved, only the necessary conditions can be considered. Since the constraints of the primal optimization problem does not have the condition of equality, only the first two conditions and $\lambda_i^* \geq 0$ are to be applied. Thus, the lagrangian function for this particular optimization problem is described as:

$$L(\mathbf{w},\boldsymbol{\lambda}) = f(w) + \sum_{i=1}^{k} \lambda_i g_i(w)$$

A necessary condition for a normal point w^* to be a minimum of f(w) subject to $g_i(w) = 0$, i = 1, ⋯, k, is Cristianini2000:

$$\frac{\partial L(w^*, \lambda^*, \delta^*)}{\partial w} = 0$$

for some values of λ^*.

Quadratic Non-Additive Optimization-Based Classification

We extend model M1 to a quadratic programming form and rewrite to model M2, as follows:

$$\text{Minimize} \quad \frac{1}{2}\|\boldsymbol{\mu}\|^2 + C\sum_{i=1}^{m} \alpha_i$$

$$\text{Subject to} \quad y_j\left((c)\int f d\mu + b\right) \geq 1 - \alpha_j$$

$$\alpha_j \geq 0; \mu, b \text{ are unrestricted} \qquad \text{(M2)}$$

where $y_i \in \{-1, 1\}$

It is important to note that $\frac{1}{2}\|\boldsymbol{\mu}\|^2$ is a constructed objective for modeling purpose. The constant C is normally set to be very large to minimize the impact from the constructed objective.

Nonlinear Classifier with the Non-Additive Measure

The optimization problem M2 can be transformed into its corresponding dual problem. Similar to the optimization process of Support Vector Machine Cristianini2000, firstly, the primal Lagrangian is:

$$L(\boldsymbol{\alpha}, \boldsymbol{\mu}, b) = \frac{1}{2}\|\boldsymbol{\mu}\|^2 + C\sum_{i=1}^{m} \alpha_i$$

$$-\sum_{i=1}^{m} \lambda_i \left[y_i\left((c)\int f d\mu + b\right) - 1 + \alpha \right]_i$$

where λ_i are the Lagrangian multipliers and C is a given relatively large positive constant. According to the general definition of the Choquet integral:

$$(c)\int f d\mu = \sum_{k=1}^{2^n-1} z_k \mu_k$$

Define vectors z_i, α, λ, μ as

$$\mathbf{z_i} = \left\langle z_{i1}, z_{i2}, \cdots, z_{i2^n-1} \right\rangle$$

$$\boldsymbol{\alpha} = \left\langle \alpha_1, \alpha_2, \cdots, \alpha_m \right\rangle$$

$$\boldsymbol{\lambda} = \left\langle \lambda_1, \lambda_2, \cdots, \lambda_m \right\rangle$$

$$\boldsymbol{\mu} = \left\langle \mu_1, \mu_2, \cdots, \mu_{2^n-1} \right\rangle$$

Thus,

$$(c)\int f d\mu = \left\langle \mathbf{z_k} \cdot \boldsymbol{\mu_k} \right\rangle$$

Now the necessary KKT conditions are applied on the primal Lagrangian function L by taking the partial derivative of L with respect to μ, α and b to be zero as follows:

$$\frac{\partial L(\boldsymbol{\alpha}, \boldsymbol{\mu}, b)}{\partial b} = -\sum_{i=1}^{m} \lambda_i y_i = 0$$

$$\frac{\partial L(\boldsymbol{\alpha}, \boldsymbol{\mu}, b)}{\partial \boldsymbol{\alpha}} = \mathbf{C} - \boldsymbol{\lambda} = \mathbf{0}$$

where C is a constant vector and is defined as:

$$\mathbf{C} = \left\langle \underbrace{C, C, \cdots, C}_{\mathrm{m}} \right\rangle$$

Therefore, the following conditions hold:

$$\sum_{i=1}^{m}\lambda_i y_i = 0$$

$$\boldsymbol{\lambda} = \mathbf{C}$$

$$\boldsymbol{\mu} = \sum_{i=1}^{m}\lambda_i y_i \mathbf{z_i}$$

Thus, after applying the necessary conditions on L, the primal becomes

$$L(\boldsymbol{\alpha},\boldsymbol{\mu},b)$$

$$= \frac{1}{2}\|\boldsymbol{\mu}\|^2 + C\sum_{i=1}^{m}\alpha_i - \sum_{i=1}^{m}\lambda_i\left[y_i\left(\sum_{k=1}^{2^n-1} z_{ik}\mu_{ik} + b\right) - 1 + \alpha_i\right]$$

$$= \frac{1}{2}\sum_{i,j=1}^{m}\left(y_i y_j \lambda_i \lambda_j (\mathbf{z_i}\cdot\mathbf{z_j})\right) + C\sum_{i=1}^{m}\alpha_i$$

$$- \sum_{i,j=1}^{m}\left(y_i y_j \lambda_i \lambda_j (\mathbf{z_i}\cdot\mathbf{z_j})\right) - \sum_{i=1}^{m}\lambda_i y_i b + \sum_{i=1}^{m}\lambda_i - \boldsymbol{\lambda}\sum_{i=1}^{m}\alpha_i$$

$$= \sum_{i=1}^{m}\lambda_i - \frac{1}{2}\sum_{i,j=1}^{m}\left(y_i y_j \lambda_i \lambda_j (\mathbf{z_i}\cdot\mathbf{z_j})\right)$$

where $(\mathbf{z_i}\cdot\mathbf{z_j})$ denotes the inner product of z_i and z_j.

The primal problem can be transformed into its dual problem according to Wolfe's dual programming theory, as the following shows:

Maximize $\quad \sum_{i=1}^{m}\lambda_i - \frac{1}{2}\sum_{i,j=1}^{m}\left(y_i y_j \lambda_i \lambda_j (\mathbf{z_i}\cdot\mathbf{z_j})\right)$

Subject to $\quad \sum_{i=1}^{m} y_i \lambda_i = 0$

$\quad \lambda_i \geq 0$

where $y_i \in \{-1,1\}$.

Since λ = C and elements in C are constants, and the model can be further simplified as:

$$\text{Maximize} \quad -\frac{1}{2}\sum_{i,j=1}^{m}\left(y_i y_j \lambda_i \lambda_j \left(\mathbf{z_i} \cdot \mathbf{z_j}\right)\right)$$

$$\text{Subject to} \quad \sum_{i=1}^{m} y_i \lambda_i = 0$$

$$\lambda_i \geq 0$$

where $y_i \in \{-1,1\}$.

Model M3 can be regarded as a general optimizationbased nonlinear classifier with the signed non-additive measure. In addition, the inner product can be further replaced with kernel functions to deliver more accurate classification.

It is observed that for constructing optimal separation in a feature space, one does not need to consider the feature space in explicit form, but only has to calculate the inner products between support vectors and the vectors of the future space [21]. Thus, the inner product operation can be replaced with kernel functions K, the function that corresponds to an inner product in the expanded feature space. Nonlinear kernel functions are able to map the data into hyper space to achieve better classification.

The three well-known kernel functions have been adopted by Model M3. They are linear, Polynomial and RBF kernel functions.

$K_{linear}(z_i, z_j) = (z_i, z_j)$ (innerproduct/linear Kernel)

$K_{Polynomial}(z_i, z_j) = (1 + (z_i, z_j))^k$ (Polynomial Kernel)

$K_{RBF}(z_i, z_j) = e^{-\|z_i - z_j\|^2 / 2\sigma^2}$ (RBF Kernel)

By solving M3 with standard optimization technique, the number of parameters to be optimized is reduced from level of 2^n to m.

Model M3 with kernel functions can be solved by Sequential Minimal Optimization (SMO) [22]. SMO algorithm decomposes the original QP (Quadratic Programming) problem into smaller QP problem by heuristically choosing two Language multipliers, which makes the smallest possible optimization problem. For each small QP problem, an analytic method is used for solving the two Language multipliers. Through altering two Language multipliers at each step, the objective function will be decreased and the convergence is guaranteed according to Osuna's theorem Osuna [23].

In conclusion, we point out that the applied KKT conditions are the necessary conditions for the classification model to reach optimum. Model M2 is transformed into its dual form M3 during the Lagrangian optimization to deal with the case of learning with small training dataset ($m \ll 2^n - 1$). Through this compromised solution, the $2^n - 1$ parameters of the signed non-additive measure can now be approximated by m Lagrangian multipliers.

APPLICATIONS

The proposed Model M3 has been applied on both artificial and UCI machine learning datasets for classification purpose and compared with performance of other methods.

There are two artificial datasets with two classes randomly generated according to the definition of the Chouqet integral. One dataset has two dimensions (2D) and the other has three (3D). The 5-fold cross-validation is used for classification evaluation. Model M3 is also compared to other popular classification methods, such as SVMs, Decision Tree (J48), Logistic Regression and Naive Bayes. The average classification accuracy in percentage is summarized in **Table 1** for evaluating testing sets in all 5 folders. As a result, M3 performs best on both artificial datasets when nonlinear kernel methods are used although performance of different kernel methods varies. The results confirm the theoretical assumption that models with non-additive measure can deal with attribute interactions because the datasets were created based on features of the Choquet integral.

To better understand this nonlinear classification, we visually represent how Model M2 (the primal problem) to perfectly classify the two dimensional artificial dataset in **Figure 1**. The example was taken from fold-1 training set of the two dimensional artificial dataset. This training set contains 160 data points, including 85 in class 1 and 75 in class 2. Model M2 creates a three dimensional decision space (x_1, x_2, y), where x_1, x_2 are the attributes of the two dimensional dataset and y is the decision score of M2. The model classifies data as class 1 when y > b, otherwise class 2. **Figure 1**presents one solution from the cross validation. In **Figure 1**(a) data points belonging to two classes are represented by asterisks (.) and dots (.), respectively.

Methods	3D	2D
M2	97.2	100
M3 (Linear)	98.8	96.0
M3 (Poly)	99.5	98.0
M3 (RBF)	99.7	99.0
M3 (Sigmod)	62.9	95.0
LibSVM (Linear)	97.7	82.0
LibSVM (Poly)	62.9	95.0
LibSVM (RBF)	98.8	98.0
LibSVM (Sigmod)	97.4	67.0
Decision Tree (J48)	96.3	98.5
Logistic Regression	97.7	84.0
Naïve Bayes	94.1	92

Table 1: Classification accuracy on two artificial datasets.

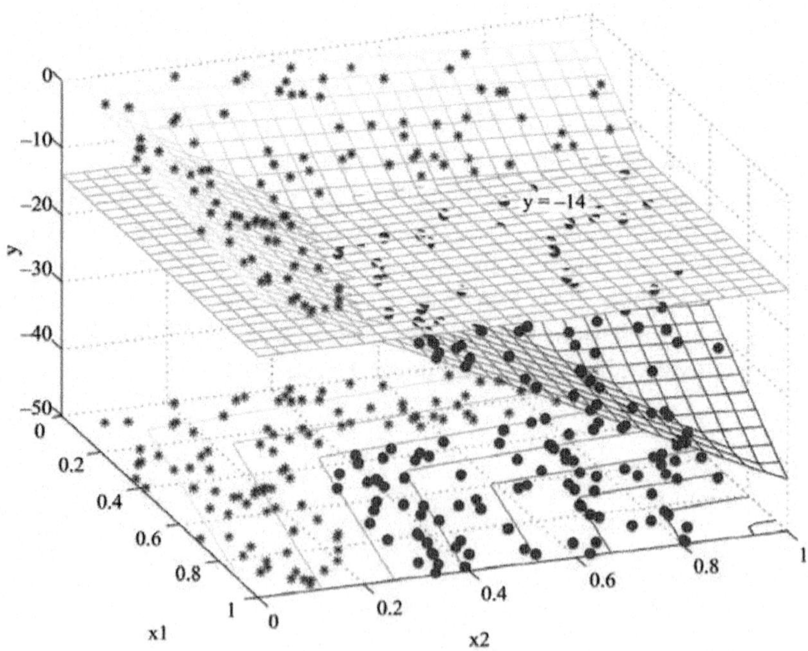

(a)

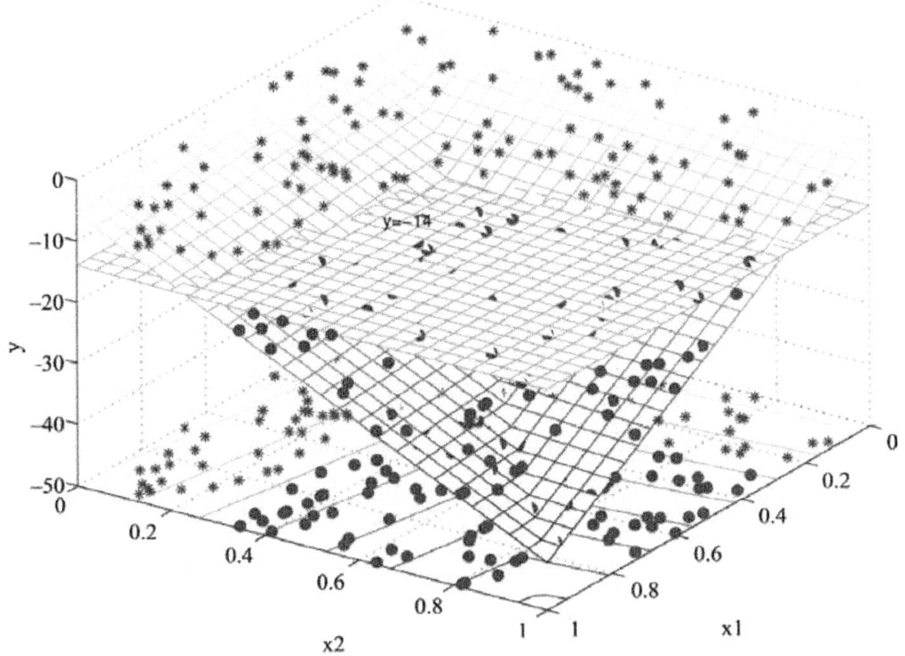

(b)

Figure 1: Nonlinear classification by Model M2. (a) View 1; (b) View 2.

The data points shown on the bottom of the figure depict the original 2D data, which are apparently not linearly separable. After applying Choquet integral to create a third dimension y, the corresponding 3D data points are now located in two different 2D planes, and are now linearly separable. **Figure 1**(b) represents the same data set but provides a different perspective to view the data. The linearly inseparable two dimensional dataset x_1, x_2 is lifted into a hyper space (x_1, x_2, y) by M2 and then can be easily classified by the decision boundary $y = -14$ (value of the critical value b).

In addition, the data cannot be perfectly classified by the linear model MSD as **Figure 2** shows. After applying MSD model, the corresponding 3D data points are still located in one flat surface in the three dimensional space and the two classes cannot be linearly separated. In MSD model, the critical value b is set to 1 and the MSD classification model separates data by decision function $y = w_1f_1 + w_2f_2$ ($y > 1$ indicates class 1 and $y < 1$ for class 2), with the solution $w_1 = 0.92$, $w_2 = 0.70$.

Classification of UCI Datasets

The UCI's Pima Indian Diabetes and the Australian Credit Approval datasets were classified with model M3. The Australian dataset contains two classes (approval or not) and it has 14 attributes and 690 instances. Both datasets were transformed into [–1, 1] with z-score normalization and the 5-folder cross-evaluation was conducted for the application. The constant variable c was set to 100000 for all the experiments. **Table 2** is the summarization of the results compared with the SVM classifier with RBF kernel.

The above results show that M3 outperformed SVM with RBF kernel on the Australian credit dataset which indicates the model is more robust when the dataset has more feature attributes, in the sense of that the performance of the testing is not significantly worse then the training. Our experiences also show that the use of Lagrangian optimization makes it feasible to solve nonadditive measure when the number of attributes is up to 14. The use of kernel functions also ensured the classification accuracy of the nonlinear model with the signed non-additive measure.

CONCLUSION

We have proposed a new classification approach based on optimization-based models while the attributes interactions are considered. The theory of non-additive measures were utilized to model the data with interactions.

Table 2. Classification results on two UCI datasets

Methods	Diabetes	Australian Credit
SVM (RBF)	76.17	78.70
M3 (RBF)	75.00	84.06
M3 (Linear)	73.18	84.35
M3 (Poly)	75.00	84.20
M3 (Sigmod)	73.96	82.90

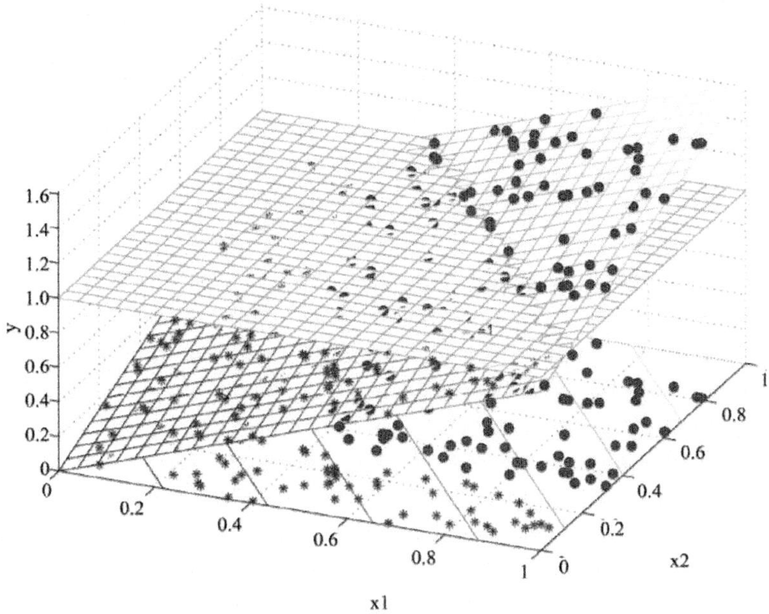

(a)

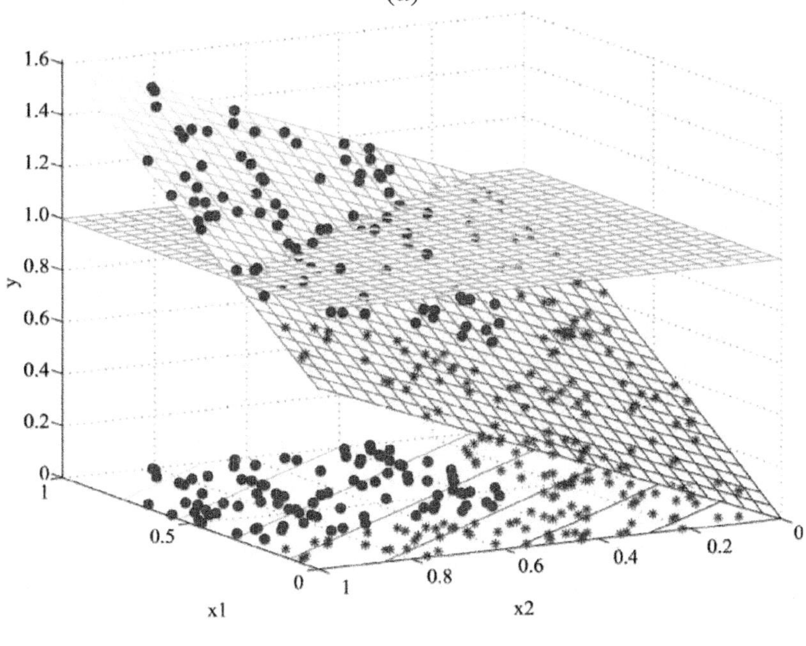

(b)

Figure 2: Linear classification by Model MSD. (a) View 1; (b) View 2.

Traditionally, nonlinear integrals are the aggregation tools for non-additive measures. The Choquet integral is good for data modeling purpose. We have demonstrated the value of using non-additive measure on optimizationbased classification and proposed a more efficient nonlinear model M3, which can classify data by solving less number of parameters. The $2^n - 1$ parameters of the signed non-additive measure can now be approximated by m Lagrangian multipliers. The optimization of the dual model M3 is guaranteed by KKT conditions, which are the necessary conditions for the nonlinear programming to be optimal. This method of parameter approximation is useful when the training set has limited number of samples. The proposed approach is thus suitable for classification applications where training sample is small comparing with the number of attributes. The experiment on the artificial dataset demonstrated the geometric meaning and profound theory of the nonlinear classification models. Applications on UCI datasets have shown that this nonlinear approach increases model robustness as the classification accuracy is stable and the accuracy of testing results is close to that of the training results. We are now in the process of applying our approach to various data mining applications.

ACKNOWLEDGEMENTS

This work has been partially supported by grants from Nebraska EPSCoR Program, USA, Nebraska Furniture Market Co., USA, BHP Billiton Co, Australia, Innovative Group from National Natural Science Foundation of China (#70921061, #70621001, #70901011), China and the CAS/SAFEA International Partnership Program for Creative Research Teams, China.

REFERENCES

1. N. Freed and F. Glover, "Simple but Powerful Goal Programming Models for Discriminate Problems," European Journal of Operational Research, Vol. 7, No. 1, 1981, pp. 44-60.doi:10.1016/0377-2217(81)90048-5
2. N. Freed and F. Glover, "Evaluating Alternative Linear, Programming Models to Solve the Two-Group Discriminate Problem," Decision Science, Vol. 17, No. 2, 1986, pp. 151-162.doi:10.1111/j.1540-5915.1986.tb00218.x
3. Y. Shi, "Multiple Criteria and Multiple Constraint Levels Linear Programming: Concepts, Techniques and Applications," World Scientific Pub Co Inc., New Jersey, 2001.
4. G. Kou, Y. Peng, Z. Chen and Y. Shi, "Multiple Criteria Mathematical Programming for Multi-Class Classification and Application in Network

Intrusion Detection," Information Sciences, Vol. 179, No. 4, 2009, pp. 371-381. doi:10.1016/j.ins.2008.10.025

5. Y. Peng, G. Kou, Y. Shi and Z. Chen, "A Multi-Criteria Convex Quadratic Programming Model for Credit Data Analysis," Decision Support Systems, Vol. 44, No. 4, 2008, pp. 1016-1030. doi:10.1016/j.dss.2007.12.001
6. V. Vapnik, "The Nature of Statistical Learning Theory," Springer-Verlag, New York, 1995.
7. G. Choquet, "Theory of Capacities," Annales de l'Institut Fourier, Vol. 5, 1954, pp. 131-295. doi:10.5802/aif.53
8. Z. Wang and G. J. Klir, "Fuzzy Measure Theory," Plenum, New York, 1992.
9. Z. Wang and G. J. Klir, "Generalized Measure Theory," Springer, New York, 2008.
10. Z. Wang, K.-S. Leung and G. J. Klir, "Applying Fuzzy Measures and Nonlinear Integrals in Data Mining," Fuzzy Sets and Systems, Vol. 156, No. 3, 2005, pp. 371-380.doi:10.1016/j.fss.2005.05.034
11. Z. Wang and H. Guo, "A New Genetic Algorithm for Nonlinear Multiregressions Based on Generalized Choquet Integrals," The 12th IEEE International Conference on Fuzzy Systems (FUZZ'03), Vol. 2, 25-28 May 2003, pp. 819-821.
12. M. Grabisch and M. Sugeno, "Multi-Attribute Classification Using Fuzzy Integral," IEEE International Conference on Fuzzy System, San Diego, 8-12 March 1992, pp. 47-54.
13. M. Grabisch and J.-M. Nicolas, "Classification by Fuzzy Integral: Performance and Tests," Fuzzy Sets System, Vol. 65, No. 2-3, 1994, pp. 255-271. doi:10.1016/0165-0114(94)90023-X
14. L. Mikenina and H. J. Zimmermann, "Improved Feature Selection and Classification by the 2-Additive Fuzzy Measure," Fuzzy Sets and Systems, Vol. 107, No. 2, 1999, pp. 197-218.doi:10.1016/S0165-0114(98)00429-1
15. K. Xu, W. Z., P. Heng and K. Leung, "Classification by Nonlinear Integral Projections," IEEE Transactions on Fuzzy Systems, Vol. 11, No. 2, 2003, pp. 187-201.doi:10.1109/TFUZZ.2003.809891
16. H. Fang, M. Rizzo, H. Wang, K. Espy and Z. Wang, "A New Nonlinear Classifier with a Penalized Signed Fuzzy Measure Using Effective Genetic Algorithm," Pattern Recognition, Vol. 43, No. 4, 2010, pp. 1393-1401. doi:10.1016/j.patcog.2009.10.006

17. J. Chu, Z. Wang and Y. Shi, "Analysis to the Contributions from Feature Attributes in Nonlinear Classification Based on the Choquet Integral," 2010 IEEE International Conference on Granular Computing (GrC), San Jose, 14-16 August 2010, pp. 677-682.

18. T. Murofushi, M. Sugeno and K. Fujimoto, "Separated Hierarchical Decomposition of the Choquet Integral," International Journal of Uncertainty, Fuzziness and Knowledge-Based Systems, Vol. 5, No. 5, 1997, pp. 563-585. doi:10.1142/S0218488597000439

19. H. Kuhn and A. Tucker, "Nonlinear Programming," Proceedings of 2nd Berkeley Symposium on Mathematical Statistics and Probabilistics, 1951, pp. 481-491.

20. N. Cristianini and J. Shawe-Taylor, "An Introduction to Support Vector Machines," Cambridge University Press, Cambridge, 2000.

21. B. Boser, I. Guyon and V. N. Vapnik, "A Training Algorithm for Optimal Margin Classifiers," 5th Annual Workshop on Computational Learning Theory, 1992, pp. 144- 152.doi:10.1145/130385.130401

22. J. Platt, "Fast Training of Support Vector Machines Using Sequential Minimal Optimization," Technical Report, Microsoft Research, 1998.

23. E. Osuna, R. Freund and F. Girosi, "An Improved Training Algorithm for Support Vector Machines," Neural Networks for Signal Processing [1997] VII. Proceedings of the 1997 IEEE Workshop, Amelia Island, 24-26 September 1997, pp. 276-285.

Chapter 4

MAPPING WHAT THEY KNOW: CONCEPT MAPS AS AN EFFECTIVE TOOL FOR ASSESSING STUDENTS' SYSTEMS THINKING

Jaklin Tripto, Orit Ben-Zvi Assaraf, Miriam Amit

Department of Science and Technology Education, Ben Gurion University of the Negev, Beer Sheva, Israel

ABSTRACT

In 2006 a new biology curriculum called "Human Biology: Emphasizing the Role of Homeostasis" was introduced into the Israeli high school system. Complex systems like those that make up the human body have become increasingly important as a focus of high school education. This study examines the effectiveness of the concept map as a assessment tool for students' systems thinking, a tool that provides researchers with a detailed picture of the systems thinking development taking place within individual students. The content of the students' concept maps was translated into information about students' system thinking using the Systems Thinking Hierarchy (STH) model, a model in which system thinking is categorized according to eight hierarchical characteristics or abilities. The goal was to use the maps to characterize Israeli high school students' understanding of the body's systemic nature. To do this, we identified the extent to which the students understand three central elements of systems, namely hierarchy, homeostasis, and dynamism, and then analyzed this understanding according to its place within the hierarchical stages of the STH model. The extensive qualitative data analysis of 48 concept maps made by 11th grade biology majors suggest that the strength of the concept map is in its ability to describe the first two levels of system thinking (analysis and synthesis). However, it proved less successful in eliciting evidence of the third and highest level, particularly of students' understanding of patterns, of homeostasis and their capacity for temporal thinking.

INTRODUCTION

Systems in Science Education Today

In 2006 a new biology curriculum called "Human Biology: Emphasizing the Role of Homeostasis" was introduced into the Israeli high school system. One of the arguments against the previous curriculum was that it taught human biology as separate components, which led to compartmentalized learning. As [1] has noted, to understand (biological) systems we must "shift our notion of 'what to look for' in biology from a mere examination of the system's components, to an understanding of its structure and dynamics. This is due to the fact that a system is not just an assembly of genes and proteins; its properties cannot be fully understood merely by drawing diagrams of their interconnections" (p. 1662). It was thought that unifying human biology around homeostasis would provide students with a more complete picture of the human body, allowing them to integrate its multiple components. Homeostasis also enables a deeper understanding of the complexity of the human body, as it explains both the interactions between the body and its environment, and the processes that occur on its different organizational levels.

The science education curriculum today has become increasingly focused on the study of complex systems, which is unsurprising considering the nature of the world our students live in—one increasingly governed by complex systems that are dynamic, self-organizing, and continually adapting [2,3]. Accordingly, the curricula of such fields as biology [4], and earth sciences [5,6] all require students to understand complex dynamic processes like eco-processes, fractals, laser beams, heart rhythms, and weather patterns, so that processes of this sort have become a unifying cross-disciplinary construct [7].

As a result of this focus on students' comprehension of systems, science education has seen a substantial growth in research into complex systems and complexity theories, as well as into students' abilities to deal with such complex systems of natural, social, and technological systems. The complex systems research community, though diverse, nevertheless shares the following common assumptions: a) Many natural systems operate at multiple distinct levels of organization; b) such systems involve nonlinear interactions among the system's elements including positive and negative feedback loops; c) even when the only interactions that exist in a system are among its individual elements, important macroscopic descriptions can still be applied to the system as a whole and are critical for understanding its patterns; d) systemlevel patterns can emerge without any force explicitly striving for the pattern, through the self-organized activity of many interacting elements; and e) the same system pattern can often be found in diverse domains, and it is useful to describe systems in sufficiently general terms such that these commonalities can be revealed [8].

The innovation of our study was in its use of concept maps as a tool with which to characterize, in the best and fullest possible way, students' system thinking [9]. It also stands out in the relatively large size of its population (48) and in the advanced state of their biology studies. Through an extensive analysis of their concept maps, this study assessed the system thinking of a large group of high school biology majors-students who have completed two years of extended biology studies that include indepth study and discussion of the human body. Their curriculum was based on a systems approach; it included concepts on both the macro and the micro levels, and at this point in their studies the students are expected to make connections between the two, and to understand the importance of the human body as a system. In this article we summarize our results and use them to assess the overall benefits and limitations of the concept map as a tool for teaching and assessing students' system thinking.

THEORETICAL BACKGROUND

The Body as a Biological System

Biological knowledge of the human body consists of a wide variety of facts and principles. Nevertheless, in the context of systems this multitude is customarily centered upon the following three system characteristics: a) hierarchy, b) homeostasis, and c) dynamism. Thus, it is references to these specific elements (defined below) that we look for in our analysis of students' system comprehension.

Hierarchy—To understand biological systems, students must comprehend their levels of organization, since a system is characterized by hierarchies and it is impossible to understand one organization level without understanding the level beneath it [10]. Such hierarchies have two general relationships: between the macro and micro [11,12] and between different levels in the hierarchy of the human body [13]. Kresh (2006) describes the relationships between systems and their components in terms of a dual status, pointing out that a living system's components also function at the same time as "subwholes", i.e. smaller, complete systems in themselves. Thus, at any given time, entities within the hierarchical structure of life (from microscopic entities such as cells and molecules, to larger ones such as organs, families and tribes) exist both as "dependent parts" of a larger system, and as "independent wholes" with subordinated parts of their own (p. 6) [14]. This means that to understand biosystems one must refer to interactions occurring both between the system's parts and between various systems.

Homeostasis—refers both to the maintenance of a stable internal environment and to the regulatory processes (operating via feedback) leading to that stability, and these meanings can be difficult to assimilate for many students. Understanding homeostasis is difficult because some processes are hidden to the eye and/or involve dynamic perception [15,16]. For example, preserving stability is (partly) based on temperature regulation. Thus, students know that when it is hot we perspire (i.e. they are aware of the proximate reason leading to the change), but the physiological processes responsible for this result are more difficult to visualize and comprehend [17].

Dynamism—Hmelo-Silver, Holton, & Kolodner (2000) and Whitner (1985) define a dynamic system as a coherent whole comprised of components interacting with each other both within single systems and between systems. The mechanism responsible for this interaction is based upon matter transportation between all the levels of a body's hierarchy from the single cells to the entire body. Moreover, [18] also raises the importance of synergic properties of a system, which emerge from the system's dynamic nature. Wilson (2006) suggests that a major obstacle to dynamic thinking is connected to the ability to follow matter as it is transported through a system [19]. Even college students find it difficult to understand this process in plants, which prevents a basic comprehension of photosynthesis. Additionally, [20] have reported a lack of understanding about genetic systems amongst high school students stemming from their inability to relate to causal/mechanistic explanations.

Understanding Complex Systems—Problems and Solutions

As the complexities and difficulties described above show understanding the structural and behavioral aspects of complex systems is a challenging intellectual endeavor for both scientists and science students [2]. Teachers too have become increasingly aware of both the importance of studying complex systems and the difficulty students have in understanding them, and several methods have consequently been developed for improving the accomplishments of students in this area. Studies suggest that the main difficulty of systems learning is the significant effect of individuals' prior knowledge on learning outcomes, both quantitatively and qualitatively [20-23]. Computer based simulations of complex phenomena are becoming an increasingly common method of generating within students the necessary prior knowledge of what a system is and how it works [24]. Computer simulations have many distinct advantages when it comes to acquainting students with the basic components, functions and behaviors that make up a system. In a simulation, the learner is presented with and explores a sophisticated model (built by an expert) of a subject domain, and can adjust various parameters of the model and explore the consequences of these changes to the system as a whole in an immediate and visible manner generally unavailable in "real" systems outside the classroom. While the ability to run simulations (or pre-built models) interactively is a vast improvement over static textbook-based learning with its emphasis on formulae and the manipulation of mathematical tokens, [25] has pointed out that they do have some limitations too. For instance, systems designed for computer models by experts and the causes and effects the systems present are "channeled" according to the phenomenon these experts wish to stress. Without access to the underlying structures controlling the system as a whole (but only to specific causes and effects) learners cannot explore the implications of changing these structures on the behavior of the system.

A central conjecture of Shore's research is that for learners to make powerful use of models, they must first build their own models and design their own investigations [25]. A number of powerful computer modeling environments have been designed to address this idea and support users in building their own useful models (e.g., STELLA—[26,27], Starlogo—[28,29], Agentsheets—[30], KidSim—[31] STELLA—a Structural Thinking, Experiential Learning Laboratory with Animation [26] is a modeling environment in which learners can both build and manipulate system models. Another program that enables modeling for dealing with understanding complex systems is NetLogo. NetLogo [32] is a modeling language and environment that supports creating, exploring and analyzing agent-based models, exploring models of chemical systems, which integrate multiple representations (visual representations of both the microand macro-levels and symbolic representation of its properties) [33-37].

Working to design various approaches to computerized learning environments has brought forth a need for a more general consensus as to the basic conceptual framework through which we wish to present systems to students. One idea for such a framework is the Systems Thinking Hierarchy (STH) model developed by [9]. This is the model we used to assess students' understanding of systems. Ben-Zvi Assaraf & Orion (2005) suggested that thinking about and understanding a system can be categorized according to eight hierarchical characteristics or abilities, which are evinced by students in an ascending order [9]. These eight characteristics compose the STH model, which was developed following a study of 8th grade students. The model's characteristics are arranged in ascending order of advancement into three sequential levels: 1) analyzing the system components (characteristic 1); 2) synthesizing system components (2, 3, 4, 5); and 3) implementation (6, 7, 8). Each lower level is the basis for developing the next level's thinking skills.

The characteristics are as follows (for examples specific to human body systems, see Appendix 1):

1) Identifying the components and processes of a system (level A);
2) Identifying simple relationships among a system's components (level B);
3) Identifying dynamic relationships within the system (level B);
4) Organizing the systems' components, processes, and their interactions, within a framework of relationships (level B);
5) Identifying matter and energy cycles within the system (level B);
6) Recognizing hidden dimensions of the system (i.e. understanding phenomena through patterns and interrelationships not readily seen) (level C);
7) Making generalizations about a system and identifying patterns (i.e. a) Hierarchy b) Homeostasis c) Dynamism—matter transitions) (level C);
8) Thinking temporally (i.e. employing retrospection and prediction) (level C).

Concept Maps as a Research Tool in Science Education

Concept maps are graphical tools for organizing and representing knowledge. They are one of several techniques recommended by National Science Education Standards to promote student thinking, because "Students need opportunities to present their abilities and understanding and to use the knowledge and language of science to communicate scientific explanations and ideas" [38]. Concept maps were developed in 1972 in the course of Novak's research program at Cornell, where he sought to follow and understand changes in children's knowledge of science, but found it difficult to identify specific changes in the children's understanding of science concepts using only interview transcripts [39]. The necessity of finding a better way to represent children's conceptual understanding led to the idea of representing children's knowledge in the form of a concept map. Thus was born a new tool not only for use in research, but also for many other uses.

Novak & Gowin (1984) described hierarchical concept maps as tools for getting students to examine their prior knowledge before studying new materials [40]. They consist of spatially grouped nodes with key-words representing concepts, connection lines representing the semantic connection of concepts, labels on the connection lines to specify the kind of the semantic connection, and arrows on the connection lines to specify the directions of the relationships or to guide the reader through the map [41]. In this manner, concept maps function as an external network-like representation of students' knowledge structures.

The use of concept mapping is often linked to the "constructivist" view of learning, since a concept map makes a good starting-point for constructivist teaching. Hmelo-Silver & Azevedo (2006), for example, claim that in order to think systematically, students must be "scaffolded" for systems thinking [42]. Ben-Zvi Assaraf & Orion (2010) emphasize that to do this we must first gain insight into how students may best be "scaffolded", and into how their domain knowledge relates to their models of system thinking [43]. In this they echo [5] and [12], who also raise the question of what prior knowledge should be in place before we begin to engage students in the exploration of emergent systems. In this context, concept maps can be a powerful tool for researchers, providing them with this sought-after prerequisite information about what students' know and how they think. For example, by identifying misconceptions and recognizing different learning styles, concept maps allows researchers to examine the way learners restructure their knowledge [40,44-46]. Because concept maps focus on the structure and the links that the student perceives, mapping is a means of eliciting the relationships that each student perceives among concepts [47,48].

In addition to being a means of assessment, concepts maps can also actively promote students' learning and knowledge intergration. Novak & Gowin (1984) argued that writing concept maps leads students to increase the precision of their understanding of the concepts [40]. Since concept maps represent frameworks or models that people use to organize their knowledge [49], they can highlight commonalities across different domains, or in this case, different systems [50]. Therefore, they may support learning by guiding students to see the common frameworks in different problem domains. In complex learning systems, conceptual representations can improve understanding of the interrelationships in the system by highlighting key aspects.

Concept Maps as a Tool for Identifying and Assessing Systems Thinking

Several science education studies have used pre and post instructional CMs to assess students' conceptual understanding [44,51,52]. Cognitive mapping and systems thinking share the properties of structure, dynamism and hierarchy, and research indicates that increases in the number of concepts, connections and diversity in CMs are a reliable parameter for gauging students' systemic thinking [47,52]. In Biology science education [53], who investigated how students using concept maps understood the topic of blood sugar levels, reported that students comprehended the macro level better than they did the micro level. In another study, Concept Maps were used to characterize the system thinking skills among students in 10th grade in the context of human body, with similar results [54]. In the Earth science education literature, several previous studies used pre and post instructtional concept maps to assess students' conceptual understanding of natural systems. Related examples include [18], who studied marine life; [52], who studied the Water cycle; [55], on the interactions between polar Earth systems and [49], who studied global atmospheric change.

In addition to reflecting students' system thinking, concept maps have also been shown to facilitate that learning complex systems. Odom & Kelly (2000), for instance, explored the effectiveness of concept mapping, expository instruction, and a combination concept mapping/learning cycle in promoting conceptual understanding of diffusion and osmosis [56]. Similarly, [57] tested the effect of concept map construction on conceptual understanding in college chemistry laboratories. The authors reported that students had a strong positive attitude towards using concept maps for a better understanding of chemistry laboratory concepts.

RESEARCH GOALS

The goal of this paper is to characterize the concept map as a tool to test system thinking in the context of the human body. To do this, I will identify (by means of the concept map) the extent to which the students understand the three central elements described above (hierarchy, homeostasis, dynamism), analyzing the data produced by the maps using the hierarchical stages of the STH model.

The primary research question below is broken down into three more specific research questions, corresponding to the three levels of the STH model:

a) How—according to their concept maps—are the different components of the STH model manifested amongst high school biology students?

b) How and to what extent do the students identify the components and processes that exist in the human body system?

c) How and to what extent do the students identify dynamic relations within the system?

d) How and to what extent do the students generalize, identify patterns in the system, and identify the hidden and time dimensions of the system?

EXPERIMENTAL DESIGN & METHODS

Research Population

The research population consisted of 48 11th grade boilogy students (boys and girls). It was homogenous to the extent that all of the students were studying for a matriculation diploma, but its cultural and social-economic background varied from a medium to a high social-economic class. All of these students studied the same curriculum, since Israel has a centralized education system. The biology syllabus for this age group centers around a curriculum entitled "Biology of the Human Body" (emphasizing homeostasis) through a system approach. It includes three subtopics:

- Homeostasis;
- The cell;
- Ecological systems;
- The students were given the concept map assignment when they had completed all three subjects.

Methods and Tools

Making the concept maps involved the following three stages: First, the students were asked to write 15 concepts and associate each with words related to the human body. Second, the students were asked to connect, within any single sentence, two concepts. They could use the same concept more than once. Third, students were asked to make a new concept map concerning the human body.

We used the analysis of the students' concept maps to look for the following system thinking components: the ability to identify the system's components and processes (reflected by the number of concepts in the map and the presence of the different human body systems among these concepts); The ability to identify dynamic relationships within the system (reflected in the number of linkages); the ability to organize components and place them within a framework of relationships (reflected in the number of represented generalization concepts, which reflect a more holistic perception of the system using a concept map); the number of macro level elements represented; the number of micro level elements represented; the representation of retrospection and prediction. Our analysis was validated by the judgment of two additional experts in the field of science education.

Appendix 1 shows a step-by-step description of how concept maps can be read as indicators of system thinking, based on the correlation of their contents to the STH model. The description is divided according to the model's three basic levels, and further subdivided into the model's eight individual characteristics. (Note: The fifth characteristic "identifying matter and energy cycles" is not featured here, as it is not relevant to human body systems.)

RESULTS

The results are organized around the three research questions (which also reflect the three levels of the STH thinking model). They relate how each component of the system thinking model was expressed in the students' concept maps, using concrete examples from two representative concept maps belonging to students Tami and Eli.

Level A

The ability to identify the components of a system and the processes within the system: The data from the concept maps showed that the students focused overwhelmingly on concepts relating to the body's structure, with 741 of a total 922 concepts referring to structure and only 181 referring to process. Furthermore, the distribution of the concepts within the structure category showed that a great many of the total 741 concepts related to structures on the Structures level, such as "lungs" or "heart" (numbering 257 in all). Structures on the molecular level, like "oxygen" and "glucose" were also well represented (at 233). Far less prominent was the Cell level (i.e. concepts like "cell" and "nucleus") and the Systems level ("digestive system," "respiratory system"), which numbered only 118 and 108 concepts respectively, and the Tissue level with only 50 concepts. Most of the components mentioned on Tami's map are at the structure level, and references to micro elements are less prominent. In Eli's map, however, half of concepts refer to the structure level while the cell level is not represented at all.

The connections between the concepts were also divided into those that referred to processes (e.g. "insulin is secreted by the pancreas") and those that were essentially descriptive (for example "the circulatory system includes the heart"). Out of a total 1021 connections, 448 fell into the former category, and 573 belonged to the latter.

Level B

The ability to identify simple relationships between or among the system's components, and (3) the ability to identify dynamic relationships within the system. The relationships/interactions level (which refers to the effect of one factor on another) was very dominant in the connections made by the students in the concept maps (108 connections, such as "platelets are responsible for blood clotting", out of the total 448 connections that referred to processes). The students' ability to identify dynamic interactions in the system, which manifested in the concept maps as descriptions of the transition of matter in the system (like "oxygen enters the body through the lungs"), appeared in only 29 connections. Tami's concept map, for instance, shows a variety of descriptive connections, but—unlike Eli's map—it includes no references to matter transmission (Appendix 1).

On the whole, our analysis revealed 279 junctions, with an average of 6 per map. A comparison between Eli and Tami's maps shows that though the number of concepts in Tami's map is a high 51 to Eli's 21, the number of junctions is much closer (11 in Tami's vs 10 in Eli's). Eli's map presents a well-branched system of connections (Figures 1 and 2).

Level C

The ability to recognize hidden dimensions of the system—to understand natural phenomena; (7) The ability to make generalizations (and identify patterns)—to solve problems based on understanding systems' mechanisms. (8) The ability to think temporally (no evidence of this found in these maps). Of the 448 connections referring to processes, 57 referred to the human body systems' hidden dimensions (for example, "carbon dioxide is a product of cellular respiration") (Appendix 1). References to the three system patterns we identified (hierarchy, homeostasis, dynamism) were distributed as follows: Homeostasis was the most prominent of the three (46 connections), constituting general descriptions of the body's inner stability (such as "some homeostasis regulates water") (Appendix 1). Hierarchy was next with 43 connections, which included statements like "the human body is composed of body systems" (Appendix 1). Dynamism was rarest of the three, with only 24 connections (Appendix 1).

DISCUSSION AND CONCLUSIONS

The goal of this study was to assess the effectiveness of the concept map as a tool to test system thinking in the context of the human body. Systems thinking is an essential skill in learning biology. Our results indicate the many difficulties students have in forming a holistic understanding of the human body. The nature of these difficulties is reflected by the concept maps, through the conclusions that may be drawn from how the students characterized the human body as a system—their choices and their limitations—especially regarding the properties that involve higher levels of thinking.

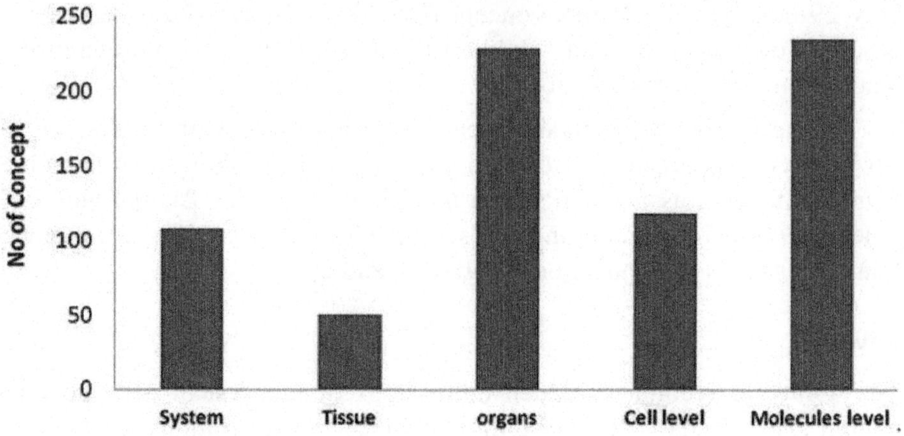

Figure 1: The distribution students of abilities in identifying the components in the human body system (N = 741).

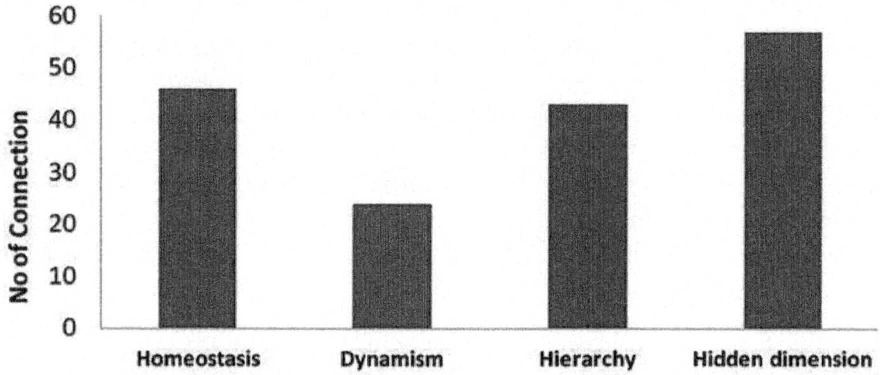

Figure 2: The distribution of student's abilities in generalizing and identifying patterns in the human body system (N = 170).

In addition to revealing students' difficulties, the maps reflect the role that students' domain knowledge plays in their understanding of dynamic systems. Though students may possess considerable domain knowledge, they do not necessarily possess the ways of thinking that can help them use it to analyze phenomena appropriately. Concept map analysis can provide insight into which elements of the domain knowledge students acquire throughout their biology education is successfully "put to use" in their understanding of complex systems, and how this application comes about. Based on our results, the strength of the concept map is in its ability to describe the first two levels of system thinking (analysis and synthesis). However, it proved less successful in eliciting evidence of the third and highest level, particularly of students' understanding of patterns, of homeostasis and their capacity for temporal thinking.

The students' concept maps emphasized the systems' structural components (most at the organ level) over the processes taking place within that system, likely because these were more difficult to grasp. This result is similar to both [42] and [21] who also noted students' tendency to concentrate on system parts, with little understanding of the way such parts interact within the system. In human biology, phenomena are interconnected at the anatomical, physiological and biochemical levels. This means that systemic thinking must account for these interconnections and the underlying mechanisms that drive a system [58]. Our subjects, however, had great difficulty in describing the mechanisms underlying the interactions between body components. Understanding mechanisms requires that a subject connect at least three components in a web of interaction, unlike processes, which can connect as few as two components. Possibly, the emphasis placed on learning structure and processes without explicit connections to mechanisms may create difficulties for students in integrating mechanisms into their conceptual structure [4,59].

The maps also showed that students had difficulty representing the dynamic nature of the human body. This difficulty has been recorded by other researchers as well, and may stem from the fact that their studies rarely expose them to the dynamic nature of a system, but mainly to its structures [60]. Others cite the high level of abstraction required to understand the dynamism of systems, due to the fact that their microscopic-and therefore not easily visible-level of their interactions [13,61].

Our subjects' concept maps barely mentioned "homeostasis" in their explanations—a result that also agree with those of other studies. These show that, due to its complexity, homeostasis is difficult to understand for students from high school to university age [15,62,63]. Simply put, homeostasis requires one to comprehend several processes taking place simultaneously, while relating each to the other. Studies attempting to discover the reason homeostasis poses such a challenge have concluded that understanding the mechanisms that maintain a stable inner environment is beyond an individual's life experience, and requires abstract thinking. The ability to think abstractly is in turn based on the capacity to achieve a certain level of higher order thinking. Indeed, [15] found a correlation between students' cognitive level and the extent to which they were able to comprehend homeostasis. Understanding homeostasis requires several cognitive abilities, such as discerning that multiple phenomena occur simultaneously and comprehending that every process is comprised of several stages. It is worth noting, however, that concept maps are not necessarily well suited to reveal this pattern, because it cannot be expressed through statements derived from the combination of only two concepts. It therefore only appears if students explicitly mention the term “homeostasis”, or processes clearly related to it, like “sugar regulation” or “water volume regulation”.

Both Eli and Tami’s maps lack an expression of thinking within the time dimension. Thinking temporally is based on the ability to make predictions, and to think retrospectively [64]. For example, one of the problems that the subjects addressed in this program was eating disorders. Coping with eating disorders requires an ability to think temporally, since it connects processes that occurred in the past to their continuing effects on the body in future. This level of thinking requires an awareness of interactions upon a timeline, and an understanding of how processes develop over time. Concept maps may not be suited to representing this particular ability, and accurately assessing and representing it will therefore require the addition and integration of narrative qualitative research tools such as interviews.

REFERENCES

1. H. Kitano, “Systems Biology: A Brief Overview,” Science, Vol. 295, No. 5560, 2002, pp. 1662-1664. doi:10.1126/science.1069492
2. M, J. Jacobson and U. Wilensky, “Complex Systems in Education: Scientific and Educational Importance and Implications for the Learning Sciences,” Learning Sciences, Vol. 15, No. 1, 2006, pp. 11-34. doi:10.1207/s15327809jls1501_4
3. R. Lesh, “Modeling Students Modeling Abilities: The Teaching and Learning of Complex Systems in Education,” Learning Sciences, Vol. 15, No. 1, 2006, pp. 45-52.doi:10.1207/s15327809jls1501_6
4. R. P. Verhoeff, A. J. Waarlo and K. T. Boersma, “Systems Modelling and the Development of Coherent Understanding of Cell Biology,” Science Education, Vol. 30, No. 4, 2008, pp. 543-568. doi:10.1080/09500690701237780
5. Y. Kali, N. Orion and B. Eylon, “The Effect of Knowledge Integration Activities on Students’ Perception of the Earth’s Crust as a Cyclic System,” Science Teaching, Vol. 40, No. 6, 2003, pp. 545-565. doi:10.1002/tea.10096
6. L. Tran, “Children and Adults’ Understanding of Ocean and Climate Sciences,” 2009. http://www7.nationalacademies.org/bose/NOAA_Education_Review_Homepage.html

7. M. Chi, "Commonsense Conceptions of Emergent Processes: Why Some Misconceptions Are Robust," Learning Sciences, Vol. 14, No. 2, 2005, pp. 161-199.doi:10.1207/s15327809jls1402_1

8. R. L. Goldstone and U. Wilensky, "Promoting Transfer by Grounding Complex Systems Principles," Learning Sciences, Vol. 17, No. 4, 2008, pp. 465-516.doi:10.1080/10508400802394898

9. O. Ben-Zvi Assaraf and N. Orion, "Development of System Thinking Skills in the Context of Earth System Education," Science Teaching, Vol. 42, No. 5, 2005, pp. 518- 560.doi:10.1002/tea.20061

10. M. C. P. J. Knipples, "Coping with the Abstract and Complex Nature of Genetics in Biology Education: The Yo-Yo Teaching and Learning Strategy," CD-[beta] Press, Utrecht, 2002.

11. R. G. Duncan and B. J. Reiser, "Reasoning across Ontologically Distinct Levels: Students' Understandings of Molecular Genetics," Science Teaching, Vol. 44, No. 7, 2007, pp. 938-959. doi:10.1002/tea.20186

12. D. A. Penner, "Explaining Systems Investigating Middle School Students' Understanding of Emergent Phenomena," Science Teaching, Vol. 37, No. 8, 2000, pp. 784- 806.doi:10.1002/1098-2736(200010)37:8<784::AID-TEA3>3.0.CO;2-E

13. C. E. Hmelo-Silver, D. L. Holton and J. L. Kolodner, "Designing Learning about Complex Systems," Learning Science, Vol. 9, No. 1, 2000, pp. 247-298.

14. J. Y. Kresh, "Integrative Systems View of Life: Perspectives from General Systems Thinking," In: S. Thomas Deisboeck and J. Yasha Kresh, Eds., Topics in Biomedical Engineering International Book Series, 2006, pp. 3- 29.

15. S. Westbrook and E. A. Marek, "A Cross Age Study of Student Understanding of the Concept Homeostasis," Journal of Research in Science Teaching, Vol. 29, No. 1, 1992, pp. 51-61.

16. E. Jungwirth and A. Dreyfus, "After This, Therefore Because of This: One Way of Jumping to Conclusions," Biological Education, Vol. 26, No. 2, 1992, pp. 139-142.doi:10.1080/00219266.1992.9655260

17. J. Budding, "Working with Personal Knowledge in Biology Classrooms on the Theme of Regulation and Homeostasis in Living Systems," Nato ASI Series of Computer and Systems Sciences, Vol. 148, 1996, pp.126-134.

18. P. A. Whitner, "Gestalt Therapy and General System Theory," University of Toledo, Ohio, 1985.

19. C. D. Wilson, C. W. Anderson, M. Heidemann, J. E. Merrill, B. W. Merritt, G. Richmond, D. F. Sibley and J. M. Parker, "Assessing Students' Ability to Trace Matter in Dynamic Systems in Cell Biology," Cell Biology Education, Vol. 5, No. 4, 2006, pp. 323-331.doi:10.1187/cbe.06-02-0142
20. M. H. Brown and R. S. Schwartz, "Connecting Photosynthesis and Cellular Respiration Preservice Teachers' Conceptions," Science Teaching, Vol. 46, No. 7, 2009, pp. 791-812.doi:10.1002/tea.20287
21. C. E. Hmelo-Silver and M. G. Pfeffer, "Comparing Expert and Novice Understanding of a Complex System from the Perspective of Structures, Behaviors, and Functions," Cognitive Science, Vol. 28, No. 1, 2004, pp. 127- 138. doi:10.1207/s15516709cog2801_7
22. K. Thompson and P. Reiman, "Patterns of Use of an Agent-Based Model and a System Dynamics Model: The Application of Patterns of Use and the Impacts on Learning Outcomes," Computers and Education, Vol. 54, No. 2, 2010, pp. 392-403.doi:10.1016/j.compedu.2009.08.020
23. U. Wilensky and M. Resnick, "Thinking in Levels: A Dynamic Systems Approach to Making Sense of the World," Science Education and Technology, Vol. 8, No. 1, 1999, pp. 3-19.doi:10.1023/A:1009421303064
24. M. Evagorou, K. Korfiatis, C. Nicolaou and C. Constantinou, "An Investigation of the Potential of Interactive Simulations for Developing System Thinking Skills in Elementary School: A Case Study with Fifth-Graders and Sixth-Graders," Science Education, Vol. 31, No. 5, 2009, pp. 655-674. doi:10.1080/09500690701749313
25. T. H. Shore, L. M. F. Shore and I. G. C. Thornton, "Construct Validity of Selfand Peer Evaluations of Performance Dimensions in an Assessment Center," Applied Psychology, Vol. 77, No. 1, 1992, pp. 42-54. doi:10.1037/0021-9010.77.1.42
26. B. Richmond and S. P. Stella, "High Performance Systems," Hanover, New Hampshire, 1990.
27. N. Roberts, "Teaching Dynamic Feedback Systems Thinking: An Elementary View," Management Science, Vol. 24, No. 8, 1978, pp. 836-843. doi:10.1287/mnsc.24.8.836
28. L. B. Resnick, "From Protoquantities to Operators: Building Mathematical Competence on a Foundation of Everyday Knowledge," In: G. Leinhardt, R. Putnam and R. A. Hattrup, Eds., Analysis of Arithmetic for Mathematics Teaching, Lawrence Erlbaum, Hillsdale, 1992, pp. 373-429.
29. U. J. Wilensky, "Connected Mathematics: Building Concrete Relationships with Mathematical Knowledge," 1993.

30. A. Repenning, "Agentsheets: A Tool for Building Domain-Oriented Dynamic Visual Environments," Conference on Human Factors in Computing Systems, Amsterdam, 24-29 April 1993, pp. 142-143.

31. D. C. Smith, A. Cypher and J. Spohrer, "Kidsim: Programming Agents without a Programming Language," Communications of the ACM, Vol. 37, No. 7, 1994, pp. 54-67.doi:10.1145/176789.176795

32. U. Wilensky, "GasLab: An Extensible Modeling Toolkit for Exploring Micro-and Macro-Views of Gases," In: N. Roberts, W. Feurzeig and B. Hunter, Eds., Computer Modeling and Simulation in Science Education, SpringerVerlag, Berlin, 1999. doi:10.1007/978-1-4612-1414-4_7

33. D. Ardac and S. Akaygun, "Effectiveness of Multimedia Based Instruction That Emphasizes Molecular Representations on Students' Understanding of Chemical Change," Science Teaching, Vol. 41, No. 4, 2004, pp. 317-337. doi:10.1002/tea.20005

34. R. Kozma, "Students Collaborating with Computer Models and Physical Experiments," In: C. Hoadley, Ed., Computer Support For Collaborative Learning Mahwah, Erlbaum, Mahwah, 2000, pp. 314-322.

35. J. W. Russell, R. B. Kozma, T. Jones, J. Wykoff, N. Marx and J. Davis, "Use of Simultaneous-Synchronized Macroscopic, Microscopic, and Symbolic Representations to Enhance the Teaching and Learning of Chemical Concepts," Chemical Education, Vol. 74, No. 3, 1997, pp. 330-334. doi:10.1021/ed074p330

36. J. Snir, C. L.Smith and G. Raz, "Linking Phenomena with Competing Underlying Models: A Software Tool for Introducing Students to the Particulate Model of Matter," Science Education, Vol. 87, No. 6, 2003, pp. 794-830. doi:10.1002/sce.10069

37. M. J. Van and J. T. De, "Supporting Students' Learning with Multiple Representations in a Dynamic Simulation-Based Learning Environment," Learning and Instruction, Vol. 16, No. 3, 2006, pp. 199-212. doi:10.1016/j.learninstruc.2006.03.007

38. National Research Council, NRC, "National Science Education Standards," National Academy, Washington, 1996.

39. J. D. Novak and D. Musonda, "A Twelve-Year Longitudinal Study of Science Concept Learning," American Educational, Vol. 28, No. 1, 1991, pp. 117-153.

40. J. D. Novak and D. B. Gowin, "Learning How to Learn," Cambridge University Press, Cambridge, 1984. doi:10.1017/CBO9781139173469

41. M. A. Ruiz-Primo, R. J., Li, M. Shavelson and S. E. Schultz, "On the Validity of Cognitive-Interpretations of Scores from Alternative Concept-

Mapping Techniques," Educational Assessment, Vol. 7, No. 2, 2001, pp. 99-141. doi:10.1207/S15326977EA0702_2

42. C. E. Hmelo-Silver and R. Azevedo, "Understanding Complex Systems: Some Core Challenges," Learning Sciences, Vol. 15, No. 1, 2006, pp. 53-61.doi:10.1207/s15327809jls1501_7
43. O. Ben-Zvi Assaraf and N. Orion, "Four Case Studies, Six Years Later: Developing System Thinking Skills in Junior High School and Sustaining Them over Time," Science Teaching, Vol. 47, No. 10, 2010, pp. 1253-1280. doi:10.1002/tea.20383
44. B. L. Martin, J. J. Mintzes and I. E. Clavijo, "Restructuring Knowledge in Biology: Cognitive Processes and Metacognitive Reflections," Science Education, Vol. 22, No. 3, 2000, pp. 303-323. doi:10.1080/095006900289895
45. C. L. Mason, "Concept Mapping: A Tool to Develop Reflective Science Instruction," Science Education, Vol. 76, No. 1, 1992, pp. 51-63. doi:10.1002/sce.3730760105
46. W. M. Roth, "Students Views of Collaborative Concept Mapping," An Emancipator Research Project, Science Education, Vol. 78, No. 1, 1994, pp. 1-34. doi:10.1002/sce.3730780102
47. R. White and R. Gunstone, "Probing Understanding," Falmer, London, 1992.
48. E. C. Zele, "Improving the Usefulness of Concept Maps as a Research Tool for Science Education," Science Education, Vol. 26, No. 9, 2004, pp. 1043-1064.doi:10.1080/1468181032000158336
49. R. Davis, H. Shrobe and P. Szolovits, "What is a Knowledge Representation?" Al Magazine, Vol. 14, No. 1, 1993, pp. 17-33.
50. L. R. Novick and C. E. Hmelo, "Transferring Symbolic Representations across Nonisomorphic Problems," Experimental Psychology: Learning Memory, and Cognition, Vol. 20, No. 6, 1994, pp. 1296-1321. doi:10.1037/0278-7393.20.6.1296
51. J. A. Rye and P. A. Rubba, "An Exploration of the Concept Map as an Interview Tool to Facilitate the externalization of Students' Understandings about Global Atmospheric Change," Science Teaching, Vo. 35, No. 5, 1998, pp. 521-546. doi:10.1002/(SICI)1098-2736(199805)35:5<521::AID-TEA4>3.0.CO;2-R
52. C. J. Songer and J. J. Mintzes, "Understanding Cellular Respiration: An Analysis of Conceptual Change in College Biology," Science Teaching, Vol. 31, No. 6, 1994, pp. 621-637. doi:10.1002/tea.3660310605

53. S. N. Chang and M. H. Chiu, "Probing Students' Conceptions Concerning Homeostasis of Blood Sugar via Concept Mapping," Proceedings of the Annual Meeting of the National Association for Research in Science Teaching, Vancouver/Canada, 1-4 April 2004.

54. O. Ben-Zvi Assaraf, J. Dodick and J. Tripto, "High School Students' Understanding of the Human Body System," Research Science Education, Vol. 43, 2013, pp. 33-56.

55. O. Ben-Zvi Assaraf and I. Orpaz, "The Life at the Poles; Study Unit Developing Junior High School Students' Ability to Recognize the Relations between Earth Systems," Science Education, Vol. 40, No. 4, 2010, pp. 525- 549.

56. A. L. Odom and P. V. Kelly, "Integrating Concept Mapping and the Learning Cycle to Teach Diffusion and Osmosis Concepts to High School Biology Students," Science Education, Vol. 85, No. 6, 2000, pp. 615-635.

57. P. G. Markow and R. A. Lonning, "Usefulness of Concept Maps in College Chemistry Laboratories: Students' Perceptions and Effects on Achievement," Science Teaching, Vol. 35, No. 9, 1998, pp. 1015-1029. doi:10.1002/(SICI)1098-2736(199811)35:9<1015::AID-TEA4>3.0.CO;2-G

58. L. Liu and C. E. Hmelo-Silver, "Promoting Complex Systems Learning through the Use of Conceptual Representations in Hypermedia," Science Teaching, Vol. 46, No. 9, 2009, pp. 1023-1040. doi:10.1002/tea.20297

59. R. P. Verhoeff, "Towards Systems Thinking in Cell Biology Education," Ph.D. Science and Mathematics Education, Universiteit Utrecht, Utrecht, 2003.

60. [61] L. Stern and J. E. Roseman, "Can Middle-School Science Textbooks Help Students Learn Important Ideas?" Research in Science Teaching, Vol. 41, No. 6, 2004, pp. 538-568. doi:10.1002/tea.20019

61. [62] F. Safayeni, N. Derbentseva and A. J. Canas, "A Theo retical Note on Concepts and the Need for Cyclic Concept Maps," Science Teaching, Vol. 42, No. 7, 2005, pp. 741-766.doi:10.1002/tea.20074

62. [63] R. Davis, H. Shrobe and P. Szolovits, "What is a Knowledge Representation?" Al Magazine, Vol. 14, No. 1, 1993, pp. 17-33.

63. [64] W. D. Simpson and E. A. Marek, "Understandings and Misconceptions of Biology Concepts Held by Students Attending Small High Schools and Students Attending Large High Schools," Science Teaching, Vol. 25, No. 5, 1988, pp. 361-374.doi:10.1002/tea.3660250504

64. [65] U. Wilensky and K. Reisman, "Thinking like a Wolf, a Sheep or a Firefly: Learning Biology through Constructing and Testing Computational Theories—An Embodied Modeling Approach," Cognition and Instruction, Vol. 24, No. 2, 2006, pp. 171-209.doi:10.1207/s1532690xci2402_1

APPENDIX 1

Correlation between Concept Maps and the STH Model

Below is a step-by-step description of how concept maps can be read as indicators of system thinking, based on the correlation of their contents to the STH model. The description is divided according to the model's three basic levels, and further subdivided into the model's eight individual characteristics. (Note: The fifth characteristic "identifying matter and energy cycles" is not featured here, as it is not relevant to human body systems.)

Level A: Analysis of System Components

Characteristic # 1: Identifying components and processes in the human body system. Characterizing system thinking at the components and processes level requires the following steps:

1) Selecting a suitable characteristic into which all the concepts written by the population may be pooled. In this study we chose "hierarchy in nature".
2) Dividing this "master-characteristic" into the categories—"Structure" and "Process"
3) Further dividing each of these into the sub-categories of "Microscopic" and "Macroscopic" levels.
4) Sorting the concepts written by the students into each of the categories now present under the mastercharacteristic "hierarchy in nature".
5) Counting all of the concepts provided by the population to arrive at an overall amount of concepts.
6) Counting the number of concepts in each category.
7) Calculating distributions for the estimation of the students' relative ability to represent system components vs system processes.

For a more thorough insight into the students' treatment of components vs processes, the maps should also be analyzed according to the connections students made between the concepts. This necessitates the following:

a) Counting all the connections made by the student. A connection is a word describing a connection between two concepts. For instance: (The veins) transfer (blood) from the (heart) to the (body). The underlined words represent the connections drawn between the concepts.

b) Analyzing the contents of the connections to derive statements. "Veins transfer blood from the heart to the body".

c) Sorting the resulting statements and removing those that are irrelevant to the study topic.

d) Sorting the statements into process/non-process related. A process-related statement refers to a string of actions or changes that are assigned a certain order within a gradual development. On the other hand, a merely descriptive statement would refer statically to an object's state or appearance.

e) Calculating distributions to compare process/nonprocess-oriented statements.

Level B: Synthesis of System Components

Characteristic # 2: Identifying simple relationships between system components. Evidence in concepts maps of relationships between system components can be gathered by identifying both the concepts in the students' body of knowledge, and the manner of their organization into meaningful connections. To do this one must:

a) Analyze the connections and translate them into statements.

b) Identify statements that address relationships between components, i.e. statements that address the effect of element "x" upon element "y".

Characteristic # 3: Identifying dynamic relationships in systems. This ability can be measured by the examination of the connection a student has formed between two concepts.

a) Analyze connections and translate them into statements.

b) Identify statements that express dynamism—i.e. statements in which the student refers to the transmission of a certain substance within the human body system.

Characteristic # 4: Organizing components and processes within a framework of relationships. Students' ability to connect a single component to a large number of other components can be assessed by examining the number of junctions on their concept map. A "junction" is a concept that has connections to at least three other concepts on the map. The number of junctions students mark between their concepts provides insight into the level of knowledge integration they have undergone. For this reason, the junctions in each map are to be counted.

Level C: Implementation

Characteristic # 6: Generalization and identification of patterns. Concept maps allow us to identify students' understanding of patterns in human body systems by analyzing the contents of their connections. To do this, the statements derived from these connections must be sorted, and those statements that relate to patterns identified. The three patterns to be looked for are: Homeostasis, Hierarchy and Dynamism. Homeostasis includes statements that generally describe the body's internal stability ("the concentration of urea and water in the body is regulated by homeostasis"). Hierarchy includes statements referring to scale in nature, while emphasizing one scale in relation to another ("the circulatory system includes capillaries"). Dynamism includes statements that address dynamic processes as system characteristics that occur in the human body ("oxygen enters the body through the lungs").

Characteristic # 7: Identifying hidden dimensions. To assess this characteristic, the statements derived from the map must be sorted, and those that refer to internal patterns and connections that are invisible on the body's surface must be identified.

Characteristic # 8: Temporal thinking. This includes both retrospective thinking (backwards) and projection (forwards). To identify a students' understanding that interactions taking place in the present can bring about and influence future events, those statements from the map in which there are temporal references must be identified.

Examples of the STH Model from Tami and Eli's concept map

STH model	Examples from Tami's concept map:
a) Junctions	Hormones, Homeostasis, Glucose regulation, food, digestive system, blood, blood vessels, body systems, cell respiration, circulatory system, blood cells.
b) Interaction between components	Thrombocytes are responsible for clotting; White blood cells have a role in the immune system; White blood cells produce antibodies; Hormones are responsible for homeostasis
c) Pattern Homeostasis	Hormones are responsible for homeostasis; Glucose regulation is part of homeostasis; Water regulation is part of homeostasis; Temperature regulation is part of homeostasis
d) Pattern Hierarchy e) Pattern Dynamism None	The human body is composed of body systems; the digestive system is one of the body's systems; the circulatory system is one of the body's systems; the respiratory system is one of the body's systems; The nervous system is one of the body's systems; Duodenum is an organ the digestive system; Stomach is an organ the digestive system; Mouth is an organ the digestive system; Cells build organs; Organs build body systems; Kidneys are an organ in the Urinary system; Lungs are an organ in the respiratory system; Blood is composed of plasma
f) Hidden dimensions	Hormones are secreted from glands; White blood cells produce antibodies; Carbon dioxide is the result of cell respiration; Glucose is a raw material in cell respiration; Oxygen is a raw material in cell respiration; Water is a product of cell respiration; DNA is in the nucleus; Nuclei are found in all cells

Concepts	Connections	Junctions	Macro Structures	Micro Cell level	molecule	patterns	Hidden dimension
51	50	11	10	9	14	17	8

STH model	Examples from Eli concept map:
a) Junctions	Blood, capillaries, circulatory system, heart, arteries, energy, sugar concentration in blood, homeostasis, urea + water, insulin/glucagon.
b) Interaction between components	Energy – the circulatory system uses it to do its job; The brain directs the maintaining of homeostasis; The concentration of sugars in the blood is maintained by the secretion of insulin/glucagons; Energy is used to help the urinary system do its job.

c) Pattern Homeostasis	The concentration of urea and water in the body is regulated by homeostasis; Homeostasis is maintained by the concentration of sugars in the blood; Homeostasis of sugar concentration in the blood is powered by the secretion mechanism of insulin/Glucagon; The brain directs the maintaining of homeostasis; Insulin works through positive and negative feedback; The concentration of sugars in the blood is maintained by the secretion of insulin/glucagons; positive/ negative feedback is maintained by homeostasis.
d) Pattern Hierarchy	The circulatory system includes capillaries; The circulatory system includes veins; The circulatory system includes arteries; The circulatory system includes the heart.
e) Pattern Dynamism	Oxygen enters the body through the lungs; waste material is moved through the circulatory system—urea and water; urea and water are transferred from the circulatory system to the urinary system.
f) Hidden dimensions	Urea and water are transferred from the circulatory system to the urinary system; Glucogen is stored in the liver; insulin is secreted by the pancreas; glucose is a source of energy; glucose breaks down and becomes the concentration of sugars in the blood; oxygen is a source of energy.

Concepts	Connections	Junctions	Macro Structures	Micro Cell Level	Molecule	Patterns	Hidden Dimension
21	30	10	10	0	6	16	6

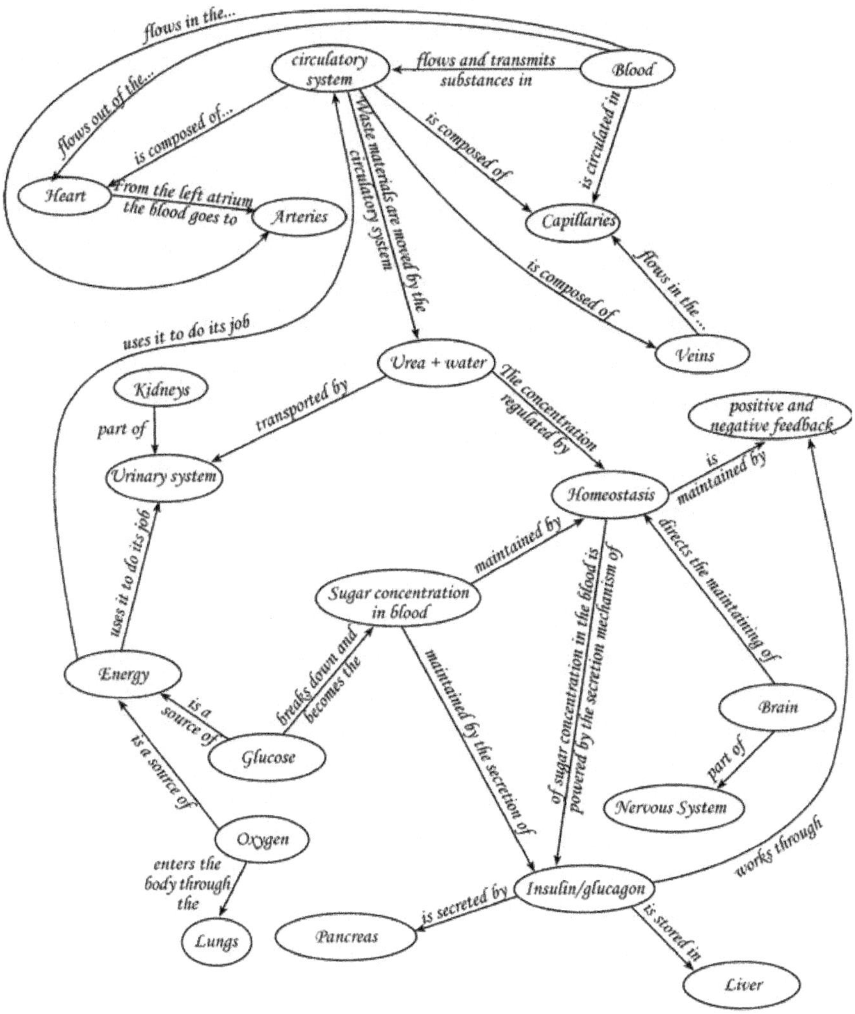

Eli's concept map.

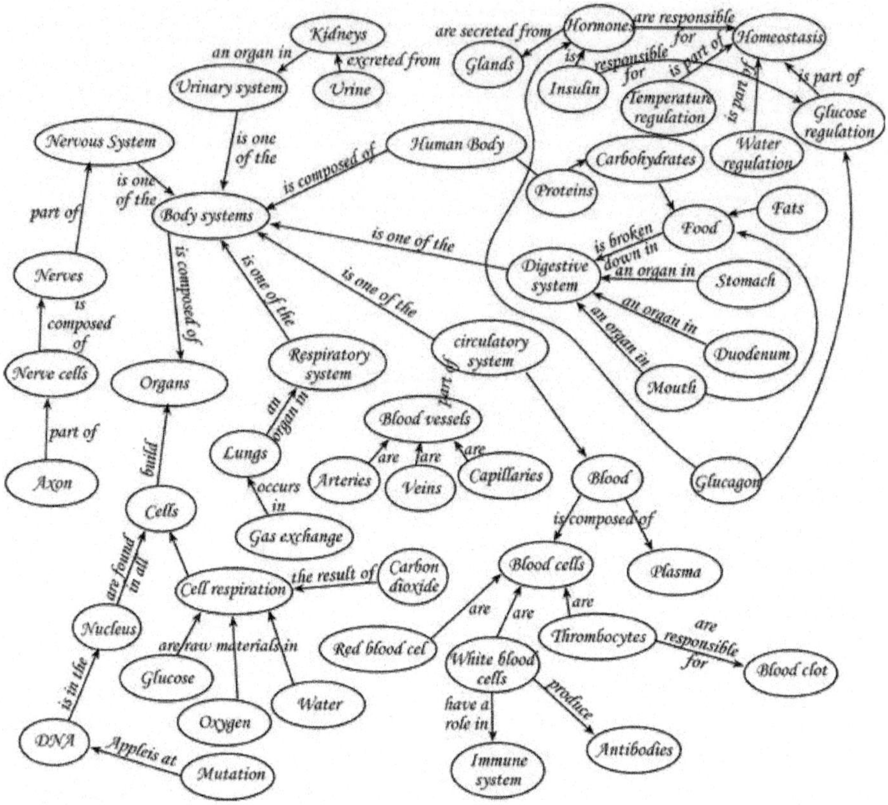

Tami's concept map.

Chapter 5

M/G/1 VACATION QUEUEING SYSTEMS WITH SERVER TIMEOUT

Oliver C. Ibe

Department of Electrical and Computer Engineering, University of Massachusetts, Lowell, USA

ABSTRACT

We consider a single-server vacation queueing system that operates in the following manner. When the server returns from a vacation, it observes the following rule. If there is at least one customer in the system, the server commences service and serves exhaustively before taking another vacation. If the server finds the system empty, it waits a fixed time c. At the expiration of this time, the server commences another vacation if no customer has arrived; otherwise, it serves exhaustively before commencing another vacation. Analytical results are derived for the mean waiting time in the system. The timeout scheme is shown to be a generalized scheme of which both the single vacation and multiple vacations schemes are special cases, with $c = \infty$ **and** $c = 0$, respectively. The model is extended to the N-policy vacation queueing system.

INTRODUCTION

A vacation queueing system is one in which a server may become unavailable for a random period of time from a primary service center. The time away from the primary service center is called a vacation, and customers who arrive while the server is on vacation will have to wait until he returns from vacation. A vacation can be the result of many factors. In some cases, the vacation can be the result of server breakdown, which means that the system must be repaired and brought back to service. It can also be a deliberate action taken to utilize the server in a secondary service center when there are no customers present at the primary service center. Thus, server vacations are useful for those systems in which the server's idle time is utilized for other purposes, and this makes the queueing model to be applicable to a variety of real world stochastic service systems.

One example of the application of vacation queueing systems is in token-passing local networks. In this case, a station can only transit its packets when it receives the token, which essentially corresponds to the server returning from a vacation. After transmitting the prescribed number of packets, the station passes the token to the next station in the transmitting order; this action is equivalent to the server taking a vacation with respect to that station. The station will transmit again when next it receives the token; or equivalently, when the server returns from vacation. In general, any system where service is provided in a round robin manner can be modeled by a vacation queueing system. Other applications that can be modeled by the vacation queueing system include machine breakdowns and maintenance in communication and computer systems. In these two cases, the server is unavailable to serve customers when the system breaks down or is intentionally taken out of service for maintenance. Thus, relative to the customers the server is on vacation.

There are different types of vacation queueing systems. In the single vacation scheme, the server takes a vacation of a random duration when the queue is empty. At the end of the vacation, the server returns to the queue. If there is at least one customer waiting when the server returns from vacation, the server performs one of the following actions depending on the service policy:

a) Under the exhaustive service policy, the server will serve all waiting customers as well as those that arrive while he is still serving at the station. He takes another vacation when the queue becomes empty.

b) Under the gated service policy, the server will serve only those customers that he finds at the queue upon his return from vacation. At the end of their service, the server will commence another vacation and any customers that arrive while the server is already serving at the station will be served when the server returns from the vacation.

c) Under the limited service policy, the server will serve only a predefined maximum number of customers and then will commence another vacation. The single-service scheme in which exactly one customer is served is a special type of this policy.

If the queue is empty on the server's return, the server waits to complete a busy period using one of the above service policies before taking another vacation.

In the multiple vacation scheme, if the server returns from a vacation and finds the queue empty, he immediately commences another vacation. If there is at least one waiting customer, then he will commence service according to the prevailing service policy.

Queueing systems with server vacations have attracted the attention of many researchers since the idea was first discussed in the paper of Levy and Yechiali [1] . Several excellent surveys on these vacation models have been done by Doshi [2] [3] , and the books by Takagi [4] and Tian and Zhang [5] are devoted to the subject. One of the most remarkable results for vacation queueing systems is the stochastic decomposition result, which is introduced by Fuhrmann and Cooper [6] and allows the system's behavior to be analyzed by considering separately the distribution of system (queue) size with no vacation and additional system size due to vacation.

Many new vacation queueing systems have been proposed. In the vacation queueing system, we have described the server completely stops service or is switched off when he is on vacation. Recently, Servi and Finn [7] introduced the working vacation scheme, in which the server worked at a different rate rather than completely stopping service during a vacation. The working vacation scheme has attracted a lot of research effort, and several authors have extended the original model. Wu and Takagi [8] generalized the model in [7] to an M/G/1 queue with general working vacations. Baba [9] studied a GI/M/1 queue with working vacations by using the matrix analytic method. Banik, Gupta and Pathak [10] analyzed the GI/ M/1/N queue with working vacations. Liu, Xu and Tian [11] established a stochastic decomposition result in the M/M/1 queue with working vacations. In [12] , Krishnamoorthy and Sreenivasan considered an M/M/2 queueing system with heterogeneous servers where one server is always available and the other goes on a working vacation when there are no customers in the system.

The differentiated multiple vacation queueing system is another vacation queueing model that is proposed by Ibe and Isijola [13] . The authors considered an M/M/1 multiple vacation queueing system with two types of vacations: Type 1 vacation is taken when the server returns from a vacation and finds at least one waiting customer; it takes the vacation after exhaustively serving all customers in the system. Type 2 vacation is taken when the server returns from a vacation and finds no customers in waiting to be served. The duration of one type of vacation is different from that of the other type.

Another model of the vacation queueing system is the system with vacation interruption in which the server can come back to the normal working level before the vacation ends. Li and Tian [14] analyzed an M/M/1 queue with working vacations and vacation interruption. The authors also extended the model to the GI/Geo/1 queue with working vacation and vacation interruptions [15] . The GI/M/1 queue with working vacations and vacation interruption was studied by Li, Tian and Ma [16] . Similarly, Zhang and Hou [17] discussed an M/G/1 queue with multiple working vacations and vacation interruption. In [18] , Sreenivasan, Chakravarthy and Krishnamoorthy considered a single server queueing model in which customers arrived according to a Markovian arrival process. They also introduces a threshold, N, where $1 \leq N < \infty$, such that the server offering services (at a lower rate) during a vacation will have the vacation interrupted the moment the queue size reaches N. In [19] Isijola-Adakeja and Ibe extended the model on [13] to include server vacation interruption.

In the vacation models that have been analyzed in the literature, server timeouts have not been considered. In this paper, we consider vacation queueing systems with server timeouts. Specifically, we consider a system that operates in the following manner. When the server has finished serving a customer and finds the system empty, it takes a vacation whose duration is independent of both the service time and the inter-arrival time. At the end of the vacation, the server returns to serve those customers, if any, who arrive during its vacation. It will commence another vacation when the system becomes empty. If no customer arrives during the vacation, the server waits for a fixed time c. If no customer arrives by the end of this period, the server commences another vacation. If a customer arrives before the period expires, the server commences service and serves exhaustively before commencing another vacation.

The paper is organized as follows. The model is formally defined in Section 2 and analyzed in Section 3. An extension of the model to the N-policy scheme is given in Section 4, and concluding remarks are made in Section 5.

MODEL DESCRIPTION

Figure 1 illustrates this scheme. Assume that the system has been operational for some time and the server returns from a vacation of duration V_1 at a time labeled t_1 and found two customers $C_1$and C_2 waiting. Before completing the service of these customers, another customer C_3 arrives. After serving the three customers the server takes another vacation of duration V_2. Upon returning from that vacation at a time t_2 it found no customer waiting. It waits for a time of duration c without any customer arriving. It thus leaves for another vacation of duration V_3 without serving a customer. It came back from the vacation at the time labeled t_3 and found two customers C_4 and C_5waiting. Before completing the service of these customers, another customer C_6 arrives. After serving these three customers, the server takes another vacation of duration V_4. Upon returning from the vacation at the time labeled t_4, it found no waiting customers. However, before the expiration of the timeout, customer C_7 arrives to initiate another busy period, and the process continues.

This vacation queueing is essentially a hybrid multiple and single vacation scheme that was introduced in Ibe and Trivedi [20] . The main objective of the model in [20] was to demonstrate how vacation queueing systems with exponentially distributed service times and finite population could be modeled by the stochastic Petri net. In this paper we assume that the population is infinite and service times have arbitrary distribution.

One important application of the vacation scheme with server timeout is to enhance resource utilization in the single vacation model. Specifically, if there is a problem in the arrival process that prevents customers from arriving for service, the server may be idle indefinitely after returning from a vacation in the single vacation. But when the idle time is bounded as described above, the server can be used to perform other functions at the expiration of the timeout rather than wait indefinitely.

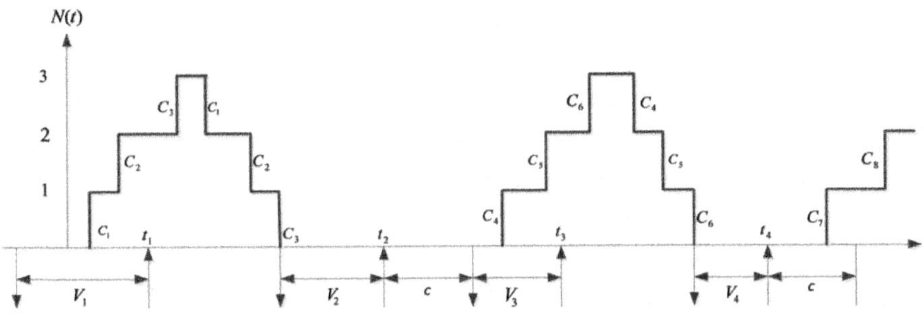

Figure 1: Vacation queueing scheme with timeout.

Thus, when the source is subject to breakdown, such as the disruption of the communication links along which messages arrive in a communication system, the server timeout scheme prevents the server from waiting indefinitely for customers to arrive after returning from a vacation.

Two other hybrid vacation schemes have been proposed. In Kella [21] , if the server returns from the $(i-1)$th vacation and finds the system empty, it takes another vacation with probability p_i and waits for the first customer to arrive with probability $1-p_i$. In the later case the server takes a vacation after serving exhaustively. In [4] if the server returns from a vacation and finds the system empty, it takes at most J vacations repeatedly until it finds at least one customer waiting in the system when it returns from a vacation. If no customer arrives by the Jth vacation, the server waits until a customer arrives.

ANALYSIS OF THE MODEL

We assume that customers arrive at the system according to a Poisson process with rate λ. The time, X, to serve a customer has a general distribution with cumulative distribution function (CDF) $F_X(x)$, mean $E[X]$ and second moment $E[X^2]$. The s-transform of the probability density function (PDF) of X, $f_X(x)$, is $M_X(s)$, which is defined by (see, for example, [22]):

$$M_X(s) = E\left[e^{-sX}\right] = \int_0^\infty e^{-sx} f_X(x)\,dx$$

where $f_X(x) = \frac{dF_X(x)}{dx}$.

The duration, V, of a vacation is also assumed to have a general distribution with CDF $F_V(v)$ and s-transform $M_V(s)$. The mean of V is $E[V]$ and its second moment is $E[V^2]$. X and V are mutually independent. Let the random variable A denote the number of customers in the system at the beginning of a busy period. The probability mass function (PMF) of A is $p_A(a) = P[A = a]$ whose z-transform, $G_A(z)$, is given by

$$G_A(z) = E\left[z^A\right] = \sum_{a=0}^{\infty} z^a p_A(a).$$

The mean of A is $E[A]$ and its second moment is $E[A^2]$. Let the random variable B denote the number of customers left behind by an arbitrary departing customer, and let L denote the number of customers in the system at an arbitrary point in time. The PMF of B is $p_B(b)$ whose z-transform is given by $G_B(z)$. The mean of B is $E[B]$ and its second moment is $E\left[B^2\right]$. Similarly, the PMF of L is $p_L(l)$ whose z-transform is given by $G_L(z)$. The mean of L is $E[L]$ and its second moment is $E\left[L^2\right]$. Let W_q denote the waiting time in the system, and let the utilization factor be defined by $\rho = \lambda E[X]$. The main result of the analysis is the following:

Theorem 1: The mean waiting time in the system is given by

$$E\left[W_q\right] = \frac{\lambda E\left[V^2\right]}{2\left\{\left[1 - e^{-\lambda c}\right] M_V(\lambda) + \lambda E[V]\right\}} + \frac{\lambda E\left[X^2\right]}{2(1-\rho)}.$$

The proof of this theorem is given in the Appendix 1.

We consider the limiting cases for c:

$$\lim_{c\to 0} E\left[W_q\right]=\frac{E\left[V^2\right]}{2E\left[V\right]}+\frac{\lambda E\left[X^2\right]}{2(1-\rho)}$$

$$\lim_{c\to 0} E\left[W_q\right]=\frac{\lambda E\left[V^2\right]}{2\left\{M_V(\lambda)+\lambda E\left[V\right]\right\}}+\frac{\lambda E\left[X^2\right]}{2(1-\rho)}.$$

These are the results obtained in [1] for the multiple vacation queueing system and the single vacation queueing system, respectively. Note that $E\left[W_q\right]$ monotonically decreases as c increases since

$$\frac{\mathrm{d}E\left[W_q\right]}{\mathrm{d}c}=\frac{-\lambda^2 E\left[V^2\right]M_V(\lambda)\mathrm{e}^{-\lambda c}}{2\left\{\left[1-\mathrm{e}^{-\lambda c}\right]M_V(\lambda)+\lambda E\left[V\right]\right\}^2}<0$$

which is consistent with the fact that the single vacation scheme $(c=\infty)$ has a smaller mean waiting time than the multiple vacation scheme $(c=0)$. Also, in [4] it is shown that the s-transform of the PDF of the waiting time is given by

$$M_W(s)=\frac{\lambda(1-\rho)\left\{1-G_A(1-s/\lambda)\right\}}{E\left[A\right]\left\{s-\lambda+\lambda M_X(s)\right\}}.$$

As shown in the Appendix,

$$G_A(z)=zp_2+\frac{M_V(\lambda-\lambda z)-M_V(\lambda)}{1-M_V(\lambda)}p_3$$

where

$$p_2=\frac{M_V(\lambda)\left\{1-\mathrm{e}^{-\lambda c}\right\}}{1-\mathrm{e}_V^{-\lambda c}M(\lambda)},\quad p_3=\frac{1-M_V(\lambda)}{1-\mathrm{e}_V^{-\lambda c}M(\lambda)}.$$

Applying this result to our model yields

$$M_W(s)=\frac{(1-\rho)\left\{\lambda\left[1-M_V(s)\right]+s\left(1-\mathrm{e}^{-\lambda c}\right)M_V(\lambda)\right\}}{\left\{s\left(1-\mathrm{e}^{-\lambda c}\right)M_V(\lambda)+\lambda E\left[V\right]\right\}\left\{s-\lambda+\lambda M_X(s)\right\}}$$

which, on differentiation and evaluating at $s=0$, gives the same value of $E\left[W_q\right]$ obtained earlier.

Although the model has been analyzed with fixed timeout c, the analysis can be extended to the case where the timeout is a random variable T with mean $E[T]$ and the s-transform of the PDF of T is $M_T(s)$. In this case only a slight modification is required in the results. Specifically, we replace the factor $e^{-\lambda c}$ with $M_T(\lambda)$.

Numerical Results

We assume that X is exponentially distributed with mean $E[X]=1 \Rightarrow \lambda=\rho$. Also, if we define the PDF of X by $f_X(x)=\mu e^{-\mu x}$, $x \geq 0$, then $E[X]=1/\mu=1$ and $E[X^2]=2/\mu^2=2$. Similarly, we assume that V has an exponential distribution with a mean of 5. This means that $f_V(v)=\gamma e^{-\gamma v}$, $v \geq 0$; $E[V]=1/\gamma=5 \Rightarrow \gamma=0.2$; $E[V^2]=2/\gamma^2=50$ and

$$M_V(s)=\frac{\gamma}{s+\gamma} \Rightarrow M_V(\lambda)=\frac{\gamma}{\lambda+\gamma}=\frac{0.2}{\rho+0.2}.$$

Thus we have that

$$E[T]=E[W_q]+E[X]=\frac{25\rho(\rho+0.2)}{0.2[1-e^{-\rho c}]+5\rho^2+\rho}+\frac{\rho}{1-\rho}+1.$$

Table 1 shows the values of $E[T]$ for different values of c.

From the table we observe that at low values of ρ (i.e., low offered load), the mean delay decreases as c increases. However, at high values of offered load, the mean delay becomes almost identical for all values of c. The reason for this is because at high values of offered load the server is likely to be serving customers more than he takes vacations.

Table 1: Values of $E[T]$ for different values of c

ρ	$c=0$	$c=1$	$c=2$	$c=5$	$c=10$	$c=15$	$c=\infty$
0.00	6.0000	5.1666	4.5714	3.5000	2.6667	2.2500	1.0000
0.10	6.1111	5.5481	5.1379	4.3906	3.8243	3.5671	3.2540
0.20	6.2500	5.8345	5.5424	5.0492	4.7408	4.6396	4.5833
0.30	6.4286	6.1053	5.8916	5.5705	5.4176	5.3852	5.3759
0.40	6.6667	6.4062	6.2464	6.0369	5.9636	5.9539	5.9524
0.50	7.0000	6.7848	6.6631	6.5253	6.4903	6.4874	6.4872
0.60	7.5000	7.3188	7.2249	7.1331	7.1163	7.1154	7.1154
0.70	8.3333	8.1785	8.1051	8.0433	8.0351	8.0348	8.0348
0.80	10.0000	9.8660	9.8081	9.7661	9.7620	9.7619	9.7619
0.90	15.0000	14.8829	14.8369	14.8079	14.8059	14.8058	14.8058
0.95	25.0000	24.8902	24.8490	24.8249	24.8234	24.8234	24.8234

THE N-POLICY WITH TIMEOUT

The N-policy was introduced by Yadin and Naor [23] and operates as follows. The server goes on vacation at the end of a busy period. The vacation ends with the arrival of the Nth customer since the end of the last busy period.

The N-policy scheme with timeout operates as follows. Consider cycles of busy periods and let T_{il} denote the arrival time of the ith customer in the lth cycle, where $i=1,2,\cdots,N$ and $l=1,2,\cdots$ Starting from the arrival of the first customer in a given cycle, if less than $N-1$ other customers arrive within the time interval c, then the server's vacation ends and service is performed exhaustively at the end which another vacation begins. Assume that the system is empty and, therefore, the server is on the lth vacation. The vacation ends at time T_l given by

$$T_l = \min\left(T_{1l} + c, T_{Nl}\right), \quad l = 1,2,\cdots$$

That is, the vacation ends at time T_{Nl} (when the Nth customer arrives) or (if less than $N-1$ other customers arrive within the timeout period c measured from when the first customer arrived), whichever comes first. This means that each busy period begins with either N customers, if $T_{Nl} \le T_{1l} + c$, or n customers, $1 \le n \le N-1$, otherwise. The server serves exhaustively and takes another vacation when the system becomes empty, whatever number of customers the busy period begins with.

The scheme is illustrated in Figure 2, where it is assumed that $N=3$. In the figure, during the server's first server vacation customer C_1 arrives and thus activates the timer. By the time the timeout expired only one other customer, C_2, arrived. Since the timeout has expired with less than 3 customers in the system, the server initiates a busy period with two customers at time t_1. Another customer, C_3, arrived during this busy period. After serving all 3 customers, the server takes a vacation. While the server is on vacation, customer C_4 arrives to activate the timer for the next cycle. The time expires at time t_2 and the server initiates a busy period at that time with only one customer. During this busy period, customers C_5 and C_6 arrive and are served before the server commences another vacation. While the server is on vacation, customer C_7 arrives and activates the time. Before the timer expires customer C_8 arrives. The timer expires at time t_3 and the server initiates a busy period at that time with two customers. During that busy period customer C_9 arrives and is served before the server commences another vacation. The process continues, as shown in the figure. The intervals labeled V_i, $i = 1,2,\cdots$, indicate the vacation intervals under the normal N-policy.

The analysis of the model is similar to that for the single vacation with server timeout and the mean waiting time in the system is given by the following theorem:

Theorem 2: The mean waiting time in the system is given by

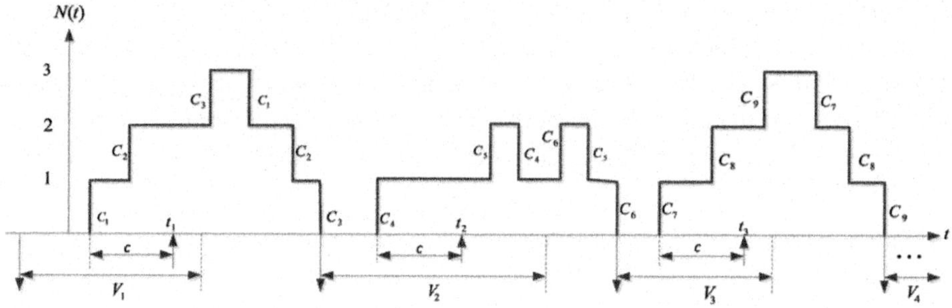

Figure 2: N-policy scheme with timeout.

$$E\left[W_q\right] = \frac{N-1}{2\lambda} + \frac{\lambda E\left[X^2\right]}{2(1-\rho)} - \sum_{n=1}^{N-1} \frac{(N-n)(\lambda c)^{n-1}}{(n-1)!} e^{-\lambda c}.$$

The proof of the theorem is given in Appendix 2 where it is also shown that

$$\lim_{c \to 0} E\left[W_q\right] = \frac{\lambda E\left[X^2\right]}{2(1-\rho)}$$

$$\lim_{c \to 0} E\left[W_q\right] = \frac{N-1}{2\lambda} + \frac{\lambda E\left[X^2\right]}{2(1-\rho)}$$

which are the results for a standard M/G/1 queue with no server vacation and a standard N-policy scheme without timeout, as shown in [23] , respectively. Also, we observe that

$$\frac{dE\left[W_q\right]}{dc} = \lambda e^{-\lambda c} \left\{ \sum_{n=1}^{N-1} \frac{(N-n)(\lambda c)^{n-1}}{(n-1)!} - \sum_{n=2}^{N-1} \frac{(N-n)(\lambda c)^{n-2}}{(n-2)!} \right\} > 0$$

which means that the expected waiting time monotonically increases with c. Therefore, the optimal value of c is zero.

Numerical Results

We use the same parameters as in Section 3.1 and assume that $N=3$. Then we have that

$$E[T]=E[W_q]+E[X]=\frac{1}{\rho}+\frac{\rho}{1-\rho}-\mathrm{e}^{-\rho c}\{2+\rho c\}+1.$$

Table 2 shows the values of $E[T]$ for different values of c.

Observe that when $c \neq 0$ the minimum value of $E[T]$ occurs around $\rho=0.5$. This can easily be verified for the case of $c=\infty$ by observing that

$$\lim_{c\to\infty} E[T]=\frac{1}{\rho}+\frac{\rho}{1-\rho}+1.$$

Differentiating this function with respect to ρ and setting the result to zero gives the optimal value at $\rho=0.5$. Since the second derivative is greater than zero, we conclude that $\rho=0.5$ gives the minimum mean delay. We can also rewrite the expression for $E[T]$ as follows:

$$E[T]=\frac{1}{\rho}+\frac{\rho}{1-\rho}-\mathrm{e}^{-\rho c}\{2+\rho c\}+1=E[T]_{c\to\infty}-\mathrm{e}^{-\rho c}\{2+\rho c\}.$$

Table 2: Values of $E[T]$ for different values of c at $N=3$.

ρ	$c=0$	$c=1$	$c=2$	$c=5$	$c=10$	$c=15$	$c=20$	$c=\infty$
0.05	1.0526	19.1026	19.1525	19.3003	19.5363	19.7536	19.9490	21.0526
0.10	1.1111	9.2109	9.3099	9.5948	10.0075	10.3302	10.5698	11.1111
0.20	1.2500	4.4488	4.6412	5.1464	5.7086	6.0010	6.1401	6.2500
0.30	1.4286	3.0580	3.3350	3.9809	4.5130	4.6897	4.7421	4.7619
0.40	1.6667	2.5579	2.9085	3.6253	4.0568	4.1468	4.1633	4.1667
0.50	2.0000	2.4837	2.8964	3.6306	3.9528	3.9947	3.9994	4.0000
0.60	2.5000	2.7397	3.2028	3.9177	4.1468	4.1653	4.1666	4.1667
0.70	3.3333	3.4211	3.9235	4.5958	4.7537	4.7615	4.7619	4.7619
0.80	5.0000	4.9919	5.5232	6.1401	6.2466	6.2499	6.2500	6.2500
0.90	10.0000	9.9320	10.4830	11.0389	11.1097	11.1111	11.1111	11.1111
0.95	20.0000	19.9117	20.4693	20.9942	21.0518	21.0526	21.0526	21.0526

Thus, the factor $\mathrm{e}^{-\rho c}\{2+\rho c\}$ has a damping effect on the mean delay at $c=\infty$. This means that as c increases, the mean delay tends to slightly increase, which agrees with the statement we made earlier that the optimal value of c is zero.

Observe also that at low offered load, the value of the expected delay when $c \neq 0$ is almost equal to the value at high offered load. This is due to the delay encountered by customers by waiting for the timeout to expire before they can be served.

CONCLUSIONS

We have derived expressions for the mean waiting time of a vacation queueing system in which the server does not immediately take another vacation upon returning from a vacation and finding the system empty, as in the multiple vacation scheme, or wait indefinitely for a customer to arrive, as in the single vacation scheme. In the proposed model, if the server returns from a vacation and finds the system empty, it waits for a fixed time c. If at the expiration of this time, no customer arrives, the server will take a vacation; otherwise it servers arriving customers exhaustively before taking another vacation. The results of the analysis are consistent with those of the multiple vacations scheme (where $c=0$), and the single vacation scheme (where $c=\infty$). Numerical results are also given that show the impact of the timeout period on the mean delay. A linear cost model was assumed to obtain the optimal value of for the assumed model. Numerical results are obtained under a set of parameter assumptions, and the results indicate that there is a very small range of values of the offered where meaning optimization can be applied.

The model is also extended to the N-policy scheme where the timeout is measured from the arrival of the first customer since the end of the last busy period. The results have also been shown to be consistent with earlier results for the case when $c=0$, which is the standard M/G/1 queue, and the case when $c=\infty$, which is the N-policy scheme without timeout. It is shown that the expected waiting time increases monotonically with c, which means that $c=0$ gives the minimum expected waiting time. Numerical results are also obtained that show that the mean delay has a minimum value around $\rho=0.5$.

REFERENCES

1. Levy, Y. and Yechiali, U. (1975) Utilization of Idle Time in an M/G/1 Queueing System. Management Science, 22, 202-211. http://dx.doi.org/10.1287/mnsc.22.2.202
2. Doshi, B.T. (1986) Queueing Systems with Vacations, a Survey. Queueing Systems, 1, 29-66. http://dx.doi.org/10.1007/BF01149327
3. Doshi, B.T. (1990) Single-Server Queues with Vacations. In: Takagi, H., Ed., Stochastic Analysis of Computer and Communications Systems, Elsevier, Amsterdam.

4. Takagi, H. (1991) Queueing Analysis: A Foundation of Performance Analysis, Volume 1: Vacation and Priority Systems, Part 1. Elsevier Science Publishers B.V., Amsterdam.
5. Tian, N. and Zhang, G. (2006) Vacation Queueing Models: Theory and Applications. Springer-Verlag, New York.
6. Fuhrmann, S.W. and Cooper, R.B. (1985) Stochastic Decomposition in M/G/1 Queue with Generalized Vacations. Operations Research, 33, 1117-1129.http://dx.doi.org/10.1287/opre.33.5.1117
7. Servi, L.D. and Finn, S.G. (2002) M/M/1 Queue with Working Vacations (M/M/1/WV). Performance Evaluation, 50, 41-52. http://dx.doi.org/10.1016/S0166-5316(02)00057-3
8. Wu, D. and Takagi, H. (2006) M/G/1 Queue with Multiple Working Vacation. Performance Evaluation, 63, 654-681. http://dx.doi.org/10.1016/j.peva.2005.05.005
9. Baba, Y. (2005) Analysis of a GI/M/1 Queue with Multiple Working Vacations. Operations Research Letters, 33, 201- 209. http://dx.doi.org/10.1016/j.orl.2004.05.006
10. Banik, A., Gupta, U. and Pathak, S. (2007) On the GI/M/1/N Queue with Multiple Working Vacations-Analytic Analysis and Computation. Applied Mathematical Modelling, 31, 1701-1710. http://dx.doi.org/10.1016/j.apm.2006.05.010
11. Liu, W., Xu, X. and Tian, N. (2007) Stochastic Decompositions in the M/M/1 Queue with Working Vacations. Operations Research Letters, 35, 595-600.http://dx.doi.org/10.1016/j.orl.2006.12.007
12. Krishnamoorthy, A. and Sreenivasan, C. (2012) An M/M/2 Queueing System with Heterogeneous Servers including One with Working Vacation. International Journal of Stochastic Analysis, 2012, Article ID: 145867. http://dx.doi.org/10.1155/2012/145867
13. Ibe, O.C. and Isijola, O.A. (2014) M/M/1 Multiple Vacation Queueing Systems with Differentiated Vacations. Modeling and Simulation in Engineering, 2014, Article 158247.http://dx.doi.org/10.1155/2014/158247
14. Li, J. and Tian, N. (2007) The M/M/1 Queue with Working Vacation and Vacation Interruption. Journal of Systems Science and Systems Engineering, 16, 121-127.http://dx.doi.org/10.1007/s11518-006-5030-6
15. Li, J. and Tian, N. (2007) The Discrete-Time GI/Geo/1 Queue with Working Vacations and Vacation Interruption. Applied Mathematics and Computation, 185, 1-10.http://dx.doi.org/10.1016/j.amc.2006.07.008

16. Li, J., Tian, N. and Ma, Z. (2008) Performance Analysis of GI/M/1 Queue with Working Vacations and Vacation Interruption. Applied Mathematical Modelling, 32, 2715-2730.http://dx.doi.org/10.1016/j.apm.2007.09.017
17. Zhang, M. and Hou, Z. (2010) Performance Analysis of M/G/1 Queue with Working Vacations and Vacation Interruption. Journal of Computational and Applied Mathematics, 234, 2977-2985. http://dx.doi.org/10.1016/j.cam.2010.04.010
18. Sreenivasan, C., Chakravarthy, S.R. and Krishnamoorthy, A. (2013) MAP/PH/1 Queue with Working Vacations, Vacation Interruptions and N-Policy. Applied Mathematical Modeling, 37, 3879-3893. http://dx.doi.org/10.1016/j.apm.2012.07.054
19. Isijola-Adakeja, O.A. and Ibe, O.C. (2014) M/M/1 Multiple Vacation Queueing Systems with Differentiated Vacations and Vacation Interruptions. IEEE Access, 2, 1384-1395.http://dx.doi.org/10.1109/ACCESS.2014.2372671
20. Ibe, O.C. and Trivedi, K.S. (1991) Stochastic Petri Net Analysis of Finite-Population Vacation Queueing Systems. Queueing Systems, 8, 111-128. http://dx.doi.org/10.1007/BF02412245
21. Kella, O. (1990) Optimal Control of the Vacation Scheme in an M/G/1 Queue. Operations Research Letters, 38, 724- 728. http://dx.doi.org/10.1287/opre.38.4.724
22. Ibe, O.C. (2014) Fundamentals of Applied Probability and Random Processes. 2nd Edition, Elsevier Academic Press, Waltham.
23. Yadin, M. and Naor, P. (1963) Queueing Systems with a Removable Service Station. Operational Research Quarterly, 14, 395-405.
24. Kleinrock, L. (1975) Queueing Systems, Volume 1: Theory. John Wiley & Sons, New York.
25. Little, J.D.C. (1961) A Proof of the Formula: L = λW. Operations Research, 9, 383-387.http://dx.doi.org/10.1287/opre.9.3.383

APPENDIX 1

In this section we provide a formal proof of theorem proposed in Section 3. Let $G_{L(M/G/1)}(z)$ denote the z-transform of the PMF of the number of customers in the system is a standard M/G/1 queue (i.e., one in which the server never takes a vacation). As defined earlier, let $G_L(z)$ denote the z-transform of the number of customers in the system at an arbitrary point in time and let $G_B(z)$ denote the z-transform of the PMF of the number of customers left behind by an arbitrary departing customer. From the stochastic decomposition result we have that

$$G_L(z) = G_B(z) G_{L(\mathrm{M/G/1})}(z).$$

It is well known (see Kleinrock [24] , for example) that

$$G_B(z) = \frac{1 - G_A(z)}{(1-z)E[A]},$$

$$G_{L(\mathrm{M/G/1})}(z) = \frac{(1-\rho)(1-z)M_X(\lambda - \lambda z)}{M_X(\lambda - \lambda z) - z}.$$

Let T denote the total time that a customer spends in the system. Thus, applying Little's law [25] we obtain the mean waiting time as

$$E[W_q] = \frac{1}{\lambda}E[T] - E[X] = \frac{1}{\lambda}\frac{\mathrm{d}G_L(z)}{\mathrm{d}z}\bigg|_{z=1} - E[X] = \frac{E[A^2] - E[A]}{2\lambda E[A]} + \frac{\lambda E[X^2]}{2(1-\rho)}.$$

Thus, once $G_A(z)$ is known, we can obtain $E[W_q]$. The remainder of the proof is devoted to deriving the expressions for $G_A(z)$ and $M_W(s)$, the s-transform of the PDF of the waiting time.

Consider the following mutually exclusive events associated with the server's experience after returning from a vacation:

1) The server returns from vacation, waits and commences another vacation without serving a customer; the probability of this event is $M_V(\lambda)e^{-\lambda c}$.
2) The server returns from vacation, waits and serves at least one customer before taking another vacation; the probability of this event is $M_V(\lambda)\{1 - e^{-\lambda c}\}$.
3) The server returns from vacation and finds at least one waiting customer; the probability of this event is $1 - M_V(\lambda)$.

Under event 2, a busy period is initiated with exactly one customer in the system. Similarly, under event 3, a busy period is initiated with at least one customer in the system. Therefore, if we define p_k as the probability of event k, given that a busy period was initiated before the server commences another vacation, where $k = 2,3$, then we obtain the following result:

$$G_A(z) = zp_2 + \frac{M_V(\lambda - \lambda z) - M_V(\lambda)}{1 - M_V(\lambda)} p_3$$

where

$$p_2 = \frac{M_V(\lambda)\{1-e^{-\lambda c}\}}{1-e^{-\lambda c}M_V(\lambda)},$$

$$p_3 = \frac{1-M_V(\lambda)}{1-e^{-\lambda c}M_V(\lambda)}.$$

From this we obtain the result

$$E[A] = p_2 + \frac{\lambda E[V]}{1-M_V(\lambda)} p_2,$$

$$E[A^2] = E[A] + \frac{\lambda^2 E[V^2]}{1-M_V(\lambda)} p_2,$$

$$E[W_q] = \frac{\lambda E[V^2]}{2E[A]\{1-e^{-\lambda c}M_V(\lambda)\}} + \frac{\lambda E[X^2]}{2(1-\rho)} = \frac{\lambda E[V^2]}{2\{(1-e^{-\lambda c})M_V(\lambda)+\lambda E[V]\}} + \frac{\lambda E[X^2]}{2(1-\rho)},$$

which completes the proof.

APPENDIX 2

In this appendix we prove Theorem 2 that was proposed in Section 4. For the current scheme, a busy period will commence with exactly N customers in the system with probability q_N, which is the probability of $N-1$ or more Poisson arrivals in an interval of length c and thus is given by

$$q_N = 1 - \sum_{n=0}^{N-2} \frac{(\lambda c)^n}{n!} e^{-\lambda c}.$$

Similarly, a busy period will commence with n customers, $1 \le n \le N-1$, with probability q_n, which is the probability of exactly $(n-1)$ Poisson arrivals in an interval of length c and is given by

$$q_n = \frac{(\lambda c)^{n-1}}{(n-1)!} e^{-\lambda c}, \quad 1 \le n \le N-1.$$

Therefore,

$$G_A(z)=z^N q_N+\sum_{n=0}^{N-1} z^n q_n = z^N\left\{1-\sum_{n=0}^{N-2}\frac{(\lambda c)^n}{n!}e^{-\lambda c}\right\}+\sum_{n=1}^{N-1} z^n \frac{(\lambda c)^{n-1}}{(n-1)!}e^{-\lambda c},$$

$$E[A]=\left.\frac{dG_A(z)}{dz}\right|_{z=1}=N\left\{1-\sum_{n=0}^{N-2}\frac{(\lambda c)^n}{n!}e^{-\lambda c}\right\}+\sum_{n=1}^{N-1}\frac{n(\lambda c)^{n-1}}{(n-1)!}e^{-\lambda c}=N-\sum_{n=0}^{N-2}\frac{(N-n-1)(\lambda c)^n}{n!}e^{-\lambda c}.$$

If we define D as the event that there are N customers at the beginning of a busy period and E as the event that there are less than N customers at the beginning of a busy period, then mimicking the method used in Takagi [4] we obtain the following conditional s-transforms of the PDF of the waiting time:

$$M_W(s|D)=\frac{(1-\rho)\left\{[\lambda/(s+\lambda)]^N-[M_X(s)]^N\right\}}{N\{[\lambda/(s+\lambda)]-M_X(s)\}}+\frac{\lambda(1-\rho)\left\{1-[M_X(s)]^N\right\}}{N\{s-\lambda+\lambda M_X(s)\}},$$

$$M_W(s|E,n)=\frac{(1-\rho)\left\{[\lambda/(s+\lambda)]^n-[M_X(s)]^n\right\}}{n\{[\lambda/(s+\lambda)]-M_X(s)\}}+\frac{\lambda(1-\rho)\left\{1-[M_X(s)]^n\right\}}{n\{s-\lambda+\lambda M_X(s)\}}.$$

Thus, the unconditional s-transform of the PDF of the waiting time is given by

$$\begin{aligned}M_W(s)&=M_W(s|D)q_N+\sum_{n=1}^{N-1}M_W(s|E,n)q_n\\&=M_W(s|D)q_N+\sum_{n=1}^{N-1}M_W(s|E,n)\frac{(\lambda c)^{n-1}}{(n-1)!}e^{-\lambda c}.\end{aligned}$$

From this we obtain the mean waiting time as

$$\begin{aligned}E[W_q]&=-\left.\frac{dW_W(s)}{ds}\right|_{s=0}=\frac{N-1}{2\lambda}q_N+\frac{\lambda E[X^2]}{2(1-\rho)}q_N+\sum_{n=1}^{N-1}\left\{\frac{\lambda E[X^2]}{2(1-\rho)}+\frac{n-1}{2\lambda}\right\}\frac{(\lambda c)^{n-1}}{(n-1)!}e^{-\lambda c}\\&=\frac{N-1}{2\lambda}+\frac{\lambda E[X^2]}{2(1-\rho)}-\sum_{n=1}^{N-1}\frac{(N-n)(\lambda c)^{n-1}}{(n-1)!}e^{-\lambda c}\\&=\frac{E[A]-1}{2\lambda}+\frac{\lambda E[X^2]}{2(1-\rho)}.\end{aligned}$$

Before considering the limiting values of c, recall that the expected value of A is given by

$$E[A]=N-\sum_{n=0}^{N-2}\frac{(N-n-1)(\lambda c)^n}{n!}e^{-\lambda c}=N-(N-1)e^{-\lambda c}-\sum_{n=1}^{N-2}\frac{(N-n-1)(\lambda c)^n}{n!}e^{-\lambda c}.$$

Thus, $\lim_{c\to 0} E[A]=1$ and $\lim_{c\to\infty} E[A]=N$. This implies that

$$\lim_{c\to 0} E\left[W_q\right]=\frac{\lambda E\left[X^2\right]}{2(1-\rho)},$$

$$\lim_{c\to\infty} E\left[W_q\right]=\frac{N-1}{2\lambda}+\frac{\lambda E\left[X^2\right]}{2(1-\rho)},$$

which are the results for a standard M/G/1 queue with no server vacation and a standard N-policy scheme without timeout, as shown in [23] , respectively. This completes the proof.

Chapter 6

OPTIMIZATION OF CRITICAL SYSTEMS FOR ROBUSTNESS IN A MULTISTATE WORLD

Edouard Kujawski
EJK Associates, Berkeley, USA

ABSTRACT

Critical systems are typically complex systems that are required to perform reliably over a wide range of scenarios, or multistate world. Seldom does a single system exist that performs best for all plausible scenarios. A robust solution, one that performs relatively well over a wide range of scenarios, is often the preferred choice for reduced risk at an acceptable cost. The alternative with the maximum expected utility may possess vulnerabilities that could be exploited. The best strategy is likely to be a hybrid solution. The von Neumann-Morgenstern Expected Utility Theory (EUT) would never select such a solution because, given its linear functional form, the expected utility of a hybrid solution cannot be greater than that of every constituent alternative. The continuity axiom and the independence axiom are assessed to be unrealistic for the problem of interest. Several well-known decision models are analyzed and demonstrated to be potentially misleading. The linear disappointment model modifies EUT by adding a term proportional to downside risk; however, it does not provide a mathematical basis for determining preferred hybrid solutions. The paper proposes a portfolio allocation model with stochastic optimization as a flexible and transparent method for defining choice problems and determining hybrid solutions for critical systems with desirable properties such as diversification and robustness.

INTRODUCTION

Broadly speaking, critical systems are systems necessary for mission success and whose failure poses a significant danger to life and property. They are typically complex systems required to perform reliably over a wide range of scenarios. For example, critical infrastructures are characterized as follows [1, p. 30]:

"Our critical infrastructures are particularly important because of the functions or services they provide to our country. Our critical infrastructures are also particularly important because they are complex systems: the effects of a terrorist attack can spread far beyond the direct target, and reverberate long after the immediate damage."

Seldom does a homogeneous system exist that performs best for all scenarios of interest. The focus of the paper is on choosing the "preferred" solution given a set of alternatives for scenarios with probabilities and consequences that have been assessed either objectively or subjectively. Following the risk-uncertainty classification of Luce and Raiffa [2], this is a problem of "decision making under risk".

Von Neumann and Morgenstern (vNM) [3] proved that the maximization of expected utility principle proposed by Daniel Bernoulli 200 years earlier is derivable from a set of axioms that appear to be reasonable. Savage [4] developed an axiomatic subjective expected utility theory with a focus on general decision problems rather than simply monetary lotteries. These seminal works resulted in the wide acceptance of what is referred to as Expected Utility Theory (EUT). The EUT is a normative theory that prescribes how the "Rational Individual"1 should make decisions under risk for any type of outcome2, money being a special case. But there is now a large body of evidence, mostly based on testing in the context of lotteries with monetary prizes, that the "rational individual"3 often systematically violate the EUT axioms. These deviations have been labeled "paradoxes". There are two plausible explanations for them: 1) they are caused by errors of reasoning, or 2) the EUT axioms, like all theories and models, have limitations as descriptive models of rational human behavior because the rational individual follows a different type of reasoning than the Rational Individual. The evidence and strength of the arguments are mounting in favor of the latter explanation. Fishburn [5, p. 86] writes:

"There are certain patterns of preference held by reasonable people for good reasons that simply do not agree with the axioms of expected utility theory and which suggest the need for serious reappraisal of the normative foundations of decision making under risk."

In recent years, several alternative theories of decision making under risk have been proposed to explain behavioral departures from EUT. The original prospect theory [6] and the newer cumulative prospect theory [7] have become the prominent descriptive theories of decision making under risk. They focus on biases in the human perception of probabilities and outcomes. Regret Theory (RT) [8-9] incorporates the concepts of regret and its counterpart, rejoicing, by adding a contribution to the EUT that accounts for these psychological influences. Disappointment Theory (DT) [10-11] models disappointment and its counterpart, elation, as complementary to both EUT and RT. There has also been much work to develop Generalized Expected Utility Theories (GEUT) by modifying the EUT axioms to account for preference patterns of rational individuals [12] and their perception catastrophic risks [13].

From the perspective of complex systems and Analysis of Alternatives (AoA), a key limitation of EUT is that it is a compensatory model. The poor performance of a system or bad outcome of a decision for one scenario can be mathematically compensated for by the other scenarios. Consequently, an alternative with the Maximum Expected Utility (MEU) may possess vulnerabilities for some scenarios that competitors in the commercial world and adversaries in the military world could exploit [14]. Depending on the context, a DM may prefer a robust solution, one that performs relatively well over a wide range of scenarios. The preferred choice for success and reduced risk often depends on FARness (Flexibility, Adaptiveness, and Robustness) rather than MEU [15]. Given its use of averages as decision criteria, EUT is susceptible to the "Flaw of Averages" [16] including insensitivity to low-probability catastrophic events. This limits its usefulness as a normative model for choosing critical systems [17].

Markowitz [18, p. 207] makes an interesting and convincing argument against maximizing the expected return from the field of investment portfolio theory:

"An investor who sought only to maximize the expected return would never prefer a diversified portfolio. If one security had greater return than any other, the investor would place all of his funds in this security."

The above argument readily generalizes to the selection of critical systems. Given a set of alternatives, the best strategy for FARness in a multistate world is likely to be a hybrid solution, or diversified portfolio. However, a Decision Maker (DM) who is an Expected Utility (EU) maximizer would never select a diversified portfolio; s(he) would select the MEU alternative.

This paper has three objectives:

1) Identify the differences between the EUT axioms and robust critical system requirements.
2) Investigate the validity of several well-known decision models as decision aids for choosing robust critical systems.
3) Develop a realistic and mathematically valid method for defining and selecting robust critical systems.

The remainder of the paper is structured as follows. Section 2 reviews the classical paradigm for AoA in a multistate world. A simple sensor selection problem is used to illustrate the limitations of EUT for critical systems. Section

3 critically analyzes the EUT axioms with a focus on the continuity and independence axioms. Section 4 presents a paradox for the selection hybrid systems. Section 5 discusses different notions of robustness. The minimax regret and the maximin criteria are shown to be potentially misleading criteria for robustness. Section 6 briefly reviews DT. The linear disappointment model is shown to provide a credible risk-based robustness metric. Section 7 proposes an approach based on portfolio allocation with stochastic optimization as a flexible and transparent method for defining choice problems and determining hybrid solutions with desirable properties such as diversification and robustness. Section 8 provides some concluding remarks.

ANALYSIS OF ALTERNATIVES

Classical Paradigm

The AoA problem is modeled with a decision matrix, **Table 1**. The m columns represent a set of m mutually exclusive scenarios $S=(S_1,S_2,\cdots,S_m)$ with probabilities $(p_1,p_2,\cdots,p_m)$ that sum to 1. The scenarios may be chance events or willful acts by adversaries. The n rows represent a set of n alternatives $A=(A_1,A_2,\cdots,A_n)$ with the performance across each scenario specified in terms of the associated outcomes, $u(A_i,S_j)=u_{i,j}$. Each alternative may then be thought of as a lottery; EUT is used to choose the preferred one. The AoA proceeds as follows:

1) The utility of a set of r mutually exclusive outcomes O_i with probabilities q_i is characterized by a vNM EU function4

Table 1: Generic decision matrix.

	Scenarios				
	S_1	--------	S_j	--------	S_m
Alternatives	p_1	--------	p_j	--------	p_m
A_1	$u_{1,1}$	--------	$u_{1,j}$	--------	$u_{1,m}$
-----------	--------	--------	--------	--------	--------
A_i	$u_{i,1}$	--------	$u_{i,1}$	--------	$u_{i,m}$
-----------	--------	--------	--------	--------	--------
A_n	$u_{n,1}$	--------	$u_{n,j}$	--------	$u_{n,m}$

$$U^{\text{vNM}}\left(\sum_{k=1}^{r} q_k O_k\right) = \sum_{k=1}^{r} q_k u(O_k), \sum_{k=1}^{r} q_k = 1. \tag{1}$$

2) The preference for each alternative A_i is represented by its EU,

$$\text{EU}^{\text{vNM}}(A_i) = \sum_{j=1}^{m} p_j u_{i,j}, \sum_{k=1}^{m} p_k = 1. \tag{2}$$

3) The Rational Individual should choose the alternative with the MEU,

$$A_i \cdot \ A_j \Leftrightarrow \sum_{k=1}^{m} p_k u_{i,k} \geq \sum_{k=1}^{m} p_k u_{j,k}. \tag{3}$$

The validity of the above equations requires additive independence. This is unlikely to be a realistic assumption for complex systems and Systems of Systems (SoS) because there are significant interactions among attributes with the result that the whole is greater the sum of the parts. The use of simulation is preferred for realism, especially for complex systems such as SoS [19].

The outlined process, as further discussed in the subsequent sections, may also mislead practitioners to overlook hybrid solutions with desirable properties such as FARness.

Single-Sensor Selection Example

As an example, consider a modified version of the decision problem in [20, p. 36]. To make it more concrete, it is assumed that the alternatives are 4 hypothetical sensors (A_1, A_2, A_3, A_4) for a border surveillance system. For the purpose of the paper, each sensor is approximated by a cookie-cutter model [21] with detection range R_n and probability of detection PoD(A_i, S_j) that depends on three states of nature, (S_1, S_2, S_3). The cookie-cutter data are specified in **Table 2** columns 1-45. It is assumed that potential intruders have no intelligence of the sensor performance; they try to infiltrate the border at random times and random locations.

The Rational Individual who equates PoD with utility is considered to be "risk neutral". S(he) would use Equation (1) with the values in **Table 2** columns 1-4 to compute each sensor EU. Based on the values in **Table 2** column 5, s(he) would select A_3. The Rational Individual who is either risk-averse or risk-seeking would argue that utility rather than PoD is the appropriate measure of sensor usefulness. In accordance with EUT, s(he) would base his/her preferences on the MEU.

Utility is associated with the performance of an alternative or the consequences of an act. The evaluation of a person's utility is a highly challenging task and for greater realism it should be experimentally solicited without invoking analytical mathematical functions [23]. The process is illustrated for a **Table 2** sensor.

1) PoD = 0.5 is identified as the reference level with u(0.5) = 0. Other values are assessed relative to it.
2) Outcomes associated with PoD > 0.5 are considered gains. The corresponding utilities have a diminishing characteristic of gain satiation. For ease of perception, a scale is established with u(1.0) = 10.
3) Outcomes associated with PoD < 0.5 are considered losses. In accordance with loss aversion, the shape of the utility function is steeper than for PoD > 0.5. A lower PoD threshold may be specified to screen out unacceptable alternatives. It is assumed that u(0) = –30.
4) The utilities of a few intermediate PoD values are determined by probing the DM about his/her indifference or willingness to bet a PoD_i sensor for a 50-50 gamble between a sensor with $PoD_h > PoD_i$ and a sensor with $PoD_l < PoD_i$. For example, consider a DM indifferent between a PoD = 0.06 sensor and a 50-50 gamble of a PoD = 0.5 sensor or nothing, i.e. sensor with PoD = 0. This equates to

$$u(0.06) = 0.5u(0.5) + 0.5u(0) = 0.5(-30.0) = -15.0.$$

Other intermediate points are determined the same way.

5) The resulting data are depicted in **Figure 1**. They can be fitted with the following utility function:

Table 2: Sensor cookie-cutter model data.

	PoD			
	S_1	S_2	S_3	
Alternatives	0.3	0.4	0.3	E(PoD)
A_1	0.89	0.40	0.34	0.53
A_2	0.70	0.50	0.40	0.53
A_3	0.80	0.70	0.20	0.58
A_4	0.50	0.50	0.50	0.50

$$u(x) = \lambda_1 (x-\gamma)^\alpha, x \geq \gamma$$
$$u(x) = \lambda_2 [\log(x+\delta) - \log(\gamma+\delta)], x < \gamma$$
$$\gamma = 0.5, \alpha = 0.2, \lambda_1 = 11.49, \lambda_2 = 17.57, \delta = 0.01. \quad (4)$$

The utility values for **Table 2** data are given in **Table 3**. Sensors A_1 and A_2 have unequal EUs. This reflects the fact that the utility function given by Equation (4) captures some aspects of risk aversion. The Rational Individual would select A_3. The rational individual, however, might disagree with this choice because of the poor performance of A_3 in state S_3. S(he) would argue that this choice is not robust because it is vulnerable to attacks by adversaries who may have acquired this information [14].

LIMITATIONS OF THE EUT AXIOMS

Problem Formulation

As with all theoretical models, EUT has limitations. Different but logically equivalent sets of axioms have been proposed. Most rely extensively on the concept of monetary lotteries. In a 1921 lecture entitled, "Geometry and Experience," Einstein [24, p. 233] states:

"As far as the propositions of mathematics refer to reality, they are not certain; as far as they are certain, they do not refer to reality."

The above comment is equally relevant with "decision theory" substituted for "mathematics".

This section follows Luce and Raiffa's formulation [2] because it provides greater visibility of the probabilistic aspects of the EUT axioms than recent more abstract mathematical formulations. Luce and Raiffa consider five basic assumptions: 1) ordering of alternatives including transitivity; 2) reduction of compound lotteries; 3) continuity; 4) substitutability6; and 5) monotonicity. The continuity and independence axioms are of special interest because of their key roles for the derivations of Equations (1)-(3). As a result, they are analyzed to systematically assess their validity for critical systems.

Consider a generic scenario, a set of n alternatives

Table 3: Sensor utility decision matrix.

	PoD			
	S_1	S_2	S_3	
Alternatives	0.3	0.4	0.3	EU
A_1	9.52	−1.67	−2.87	1.33
A_2	8.33	0.00	−1.67	2.00
A_3	9.03	8.33	−6.77	4.01
A_4	0.00	0.00	0.00	0.00

$A=(A_1,\cdots,A_n)$, and r basic consequences $(C_1,\cdots,C_r)$. In accordance with the ordering axiom, the consequences can be arranged in order of decreasing preference and numbered accordingly: C_1 贩C_2 $\cdots$? C_r. Every alternative can then be thought of as a lottery associated with an r-tuplet of basic consequences and probabilities q7:

$$A_i \sim \left(C_1, q_1^i; \cdots; C_r, q_r^i\right) \text{ with}$$

$$0 \le q_j^i \le 1 \text{ and } \sum_{j=1}^{r} q_j^i = 1. \tag{5}$$

The Continuity Axiom

For each outcome C_i, the Rational Individual8 is indifferent between C_i and an alternative involving just C_1 and C_r:

$$\forall C_1 \succ C_i \succ C_r, \exists p \in (0,1) \, s.t.$$

$$C_i \sim \left[(C_1, p; C_r, (1-p)\right] \equiv \tilde{C}_i. \tag{6}$$

C_i and $\tilde{C}_i$ are two different entities: $\tilde{C}_i$ can be any outcome of an equivalence class with certainty equivalent C_i [25].

Context is of the essence for validity of the continuity axiom. Taken literally, the Rational Individual should be indifferent between receiving an amount of money $Y < X$ and the lottery $\left[X, p; \text{Death}, (1-p)\right]$ for some probability p. Following this observation, Luce and Raiffa [2, p. 27] write:

"When put in such bold form, some, whom we would hesitate to charge with being "irrational", will say No... Even though the universality of the assumption is suspect, two thoughts are consoling. First, in few applications are such extreme alternatives as death present."

Critical systems and decisions often impact life and death issues. The DM who is a rational individual cannot invoke the above loophole. S(he) needs to mindfully assess low probability events with potentially dire consequences such as serious injuries, death, and heavy financial losses. The emotionally charged aspects of such decisions render the use of the continuity axiom impractical and unrealistic.

The Independence Axiom[9]

For every alternative in the set A, the Rational Individual is indifferent to the substitution of an equivalent outcome

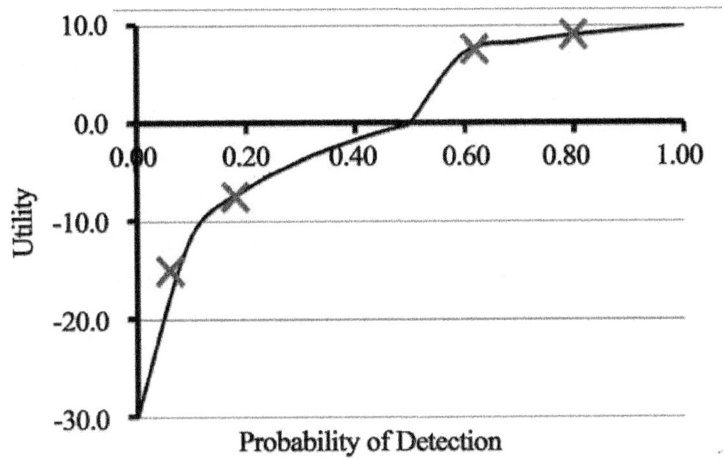

Figure 1: Evaluation of utility function.

$\tilde{C}_i$ for C_i:

$$\tilde{C}_i \sim C_i$$
$$\Leftrightarrow \left(C_1, p_1;\cdots;\tilde{C}_i, p_i;\cdots;C_r, p_r\right)$$
$$\sim \left(C_1, p_1;\cdots;C_i, p_i;\cdots;C_r, p_r\right). \tag{7a}$$

The independence condition given by Equation (7a) is more apparent if expressed in terms of the combination $Q \equiv \left(C_1, p_1;\cdots;\tilde{C}_i, 0;\cdots;C_r, p_r\right)$ in common with arbitrary alternatives G or H. The independence axiom then reduces to its common form,

$$G \sim H \Leftrightarrow \left[G, p; Q, (1-p)\right] \sim \left[H, p; Q, (1-p)\right]. \tag{7b}$$

It follows from the EUT continuity and transitivity axioms that

$$G \succ H \Leftrightarrow \left[G, p; Q, (1-p)\right] \succ \left[H, p; Q, (1-p)\right]. \tag{7c}$$

The probabilistic outcomes on the right-sides of Equations (7b) and (7c) are the mutually exclusive basic outcomes associated with Q, G and H. It would be wrong to think of them as mixtures of outcomes.

The independence axiom requires that when comparing probabilistic alternatives, the Rational Individual views the "common part" as irrelevant and preserves the original preference in accordance with Equations (7a)- (7c). This is a valid assumption if and only if there are no interaction effects including psychological influences between G and Q or H and Q. The use of decision trees is often presented as a graphical justification for the independence axiom. This argument is flawed because the folding back procedure substitutes the certainty equivalent for branches at nodes and thereby implicitly assumes that outcomes are independent. De Neuville [26, p. 366] writes:

"The axiom implies that the substitutions can occur regardless of the other opportunities in front of a person and, thus, regardless of how these substitutions alter the probabilistic distribution of the consequences."

In conclusion, the independence axiom is in conflict with probability calculus and the rational individual who makes decisions based on the complete risk curve rather than simply the EU [27].

The Chew Weighted Utility Theory

The Chew weighted utility theory [28] replaces the independence axiom with the weak substitution axiom:

$$\forall A_i, A_j, A_k \in A, A_i \sim A_j \Rightarrow \forall \beta \in (0,1), \exists \gamma \in (0,1)$$
$$s.t. \left[A_i, \beta; A_k, (1-\beta)\right] \sim \left[A_j, \gamma; A_k, (1-\gamma)\right]. \tag{8}$$

For $\beta = \gamma$, the weak substitution axiom reduces to the independence axiom, Equation (7b). The Chew weighted utility theory includes EUT as a special case. For $\beta \neq \gamma$ it admits the presence of complementarities of A_i and A_j with A_k. It, therefore, provides a resolution of the Allais paradox [29] as well as the two-sensor paradox of Section 4.

The Chew weighted utility function for a probabilistic mixture of two alternatives H and G has the following linear functional form:

$$U^{wut}\left(pH+(1-p)G\right)=\frac{pW(H)U(H)+(1-p)W(G)U(G)}{pW(H)+(1-p)W(G)}. \tag{9}$$

$U(\cdot)$ is a vNM utility function; $W(\cdot)$ is an additional weight function that enables it to accommodate the preference patterns of several key EUT paradoxes including the Allais paradox [29]. Equation (9) reduces to Equation (1) for $W(\cdot)=$ Constant. Otherwise, it significantly increases the complexity of EUT, which limits its applicability as a normative model for AoAs.

Choice Axioms for Catastrophic Risks

As discussed in Section 2, the EUT linear functional form causes the ranking of alternatives to be relatively insensitive to low probability scenarios. This is in conflict with the way rational individuals often choose options for a wide range of problems with low probability but catastrophic consequences. These include, but are not limited to, global environmental risks, national security, and critical infrastructures. Chichilnisky [13] proposed a set of 3 axioms to account for such choices. It results in a utility function $W(\cdot)$ that includes a term that assigns positive weights to low probability events:

$$\begin{aligned} &W(x)=\lambda U^{\mathrm{vNM}}(x)+(1-\lambda)\Phi(x)\\ &x \succ y \Leftrightarrow W(x)>W(y), \end{aligned} \tag{10}$$

where $0<\lambda<1$ and $\Phi(\cdot)$ is an additive measure with a functional form yet to be specified. Chichilnisky [13] characterizes the status of her work as follows: "The new axioms require a new calculus of variation.... Some results already exist, but much work is still needed."

TWO-SENSOR PARADOX

The Section 2.2 sensors are considered for a border surveillance system of length $4R_0$. The problem is which two sensors to select to maximize the detection probability of potential intruders. It is assumed that the latter possess no intelligence of sensor performance; they try to infiltrate the border at random locations and times.

EUT Analysis

For the Rational Individual,

$$A_i \succ A_j \Rightarrow pA_i+(1-p)A_i \succ pA_i+(1-p)A_j. \tag{11}$$

S(he) would then conclude that the homogeneous combination $A_i - A_i$ is preferable to the hybrid combination $A_i - A_j$. By extending this reasoning to all sensor combinations, the Rational Individual concludes that the best combination consists only of the highest utility sensor, i.e. A_3. The rational individual would disagree with this solution because it lacks diversity and it is highly vulnerable to infiltration given the low PoD in scenario S_3 (see Table 2). There are several other examples of EUT paradoxes with anomalous risk behaviors [30-31].

Two-Dimensional Preference Analysis

Given 4 alternatives, there are 10 possible two-sensor combinations to consider. Each combination (A_i, A_j) may be represented by a two-dimensional vector with the PoD utility as elements. A two-dimensional utility function is needed to represent preferences over the combinations such that

$$(A_i, A_j) \succ (A_k, A_l) \Leftrightarrow U(x_i, x_j) > U(x_k, x_l). \tag{12}$$

By analogy to the determination of the sensor utility function in Section 2.2, it is assumed that $U(x, y)$ is given by treating Equation (4) as a function of the joint variable xy:

$$\begin{aligned} &U^D(x, y) = \lambda_1 \left(\sqrt{xy} - \gamma\right)^\alpha, x > \gamma \\ &U^D(x, y) = \lambda_2 \left[\log\left(\sqrt{xy} + \delta\right) - \log(\gamma + \delta)\right], x < \gamma \\ &\gamma = 0.5, \alpha = 0.2, \lambda_1 = 11.49, \lambda_2 = 17.57, \delta = 0.01. \end{aligned} \tag{13}$$

The multiplicative form xy has advantages over the linear form $(x + y)$. It more realistically accounts for two-sensor vulnerabilities. Furthermore, the $U^D(x, y)$ indifference or isoutility curves represent simple convex preferences1⁰.

The utilities for the 10 two-sensor combination are computed using Equation (13) and averaged over the 3 scenarios. The results are presented in **Table 4**.

Table 4: Two-dimensional preference analysis.

	Utility				
	*S*1	*S*2	*S*3		
Combination	0.3	0.4	0.3	EU	Rank
A_1-A_1	9.52	−1.67	−2.87	1.33	9
A_2-A_2	8.33	0.00	−1.67	2.00	6
A_3-A_3	9.03	8.33	−6.77	4.01	4
A_4-A_4	0.00	0.00	0.00	0.00	10
A_1-A_2	8.97	−0.79	−2.25	1.70	8
A_1-A_3	9.26	5.70	−4.58	3.68	5
A_1-A_4	8.06	−0.79	−1.30	1.71	7
A_2-A_3	8.71	7.10	−3.80	4.31	3
A_2-A_4	7.10	7.10	−0.79	4.73	1
A_3-A_4	7.64	7.10	−2.66	4.33	2

1) $A_2 - A_4 \succ A_3 - A_3$. The MEU solution, $A_2 - A_4$ does not consist of the two MEU alternatives. It draws on the synergistic interaction between the two alternatives.

2) $U^D(\cdot)$ exhibits diversification [32]. For some combinations of $A_i \prec A_k$ and $A_j \prec A_k$,

$$U^D\left(pA_i,(1-p)A_j\right) > U^D\left(A_k\right). \tag{14}$$

3) The robust combination A_4-A_4 ranks last. This result combined with the above two demonstrates that convex preferences exhibit diversification but that diversification is not equivalent to robustness.

A resolution of the two-sensor paradox is that EUT needs to be modified to be compatible with preference for robust and hybrid solutions. Options include 1) replacing the independence axiom with a modification such as the weak substitution axiom, or 2) introducing corrections to EUT such as DT (see Section 6).

ROBUST SOLUTIONS

Various Notions

In semi-layman parlance, robust solutions perform reasonably well and have acceptable outcomes over a wide range of plausible scenarios without assuming that "everything goes right". The importance of robustness as a criterion for good decisions is reflected by the fact that DMs often purposefully forgo the MEU option for ones that perform well over a wide range of scenarios and are relatively insensitive to uncertainties. The professional literature contains different notions of robustness with variations that depend on the application domain.

The following three notions of robustness are of interest for critical systems consideration.

1) Lempert et al. [33] define a robust strategy as "one with relatively small regret compared to the alternatives across a wide range of plausible futures." They evaluate the regret of long-term policies using an approach based on Savage's minimax regret rule1[1].
2) Krokhmal et al. [34] advocate generating robust decision by optimizing an upper percentile of the risk probability distribution1[2]. They note that "in this regard, risk management in military applications is similar to other fields such as finance, nuclear safety, etc., where decisions targeted at achieving the maximal expected performance may lead to an excessive risk exposure."
3) Ullman [35] advocates developing designs with low sensitivity to noise or uncertainty in accordance with the Taguchi method. He considers performance, risk and cost among the factors that need to be addressed [35, p. 36]: "The result of a robust decision is an option chosen with known and acceptable satisfaction and risk... A major goal in making a robust decision is to eliminate all the uncertainty possible within the scope of available resources."

This section examines Savage's minimax regret rule as a regret-based robustness criterion. Risk-based robustness is addressed in Section 6.

The Minimax Regret Rule

Savage's minimax regret rule is a recommendation that under uncertainty a person should choose the alternative that minimizes the maximum difference from the highest achievable utility in each scenario. The anticipated regret for choice A_i in scenario S_j depends on the other alternatives:

$$R(A_i,S_j)=\underset{A_i}{\text{Max}}\,U(A_i,S_j)-U(A_i,S_j), \tag{15a}$$

where $U(A_i,S_j)$ is a vNM utility. Equation (15a) represents the largest possible loss for A_i in state S_i relative to the best alternative. The minimax regret choice is the alternative A^* with the smallest regret over all of the scenarios:

$$\underset{S_j}{\text{Max}}\,R(A^*,S_j)=\underset{A_i}{\text{Min}}\,\underset{S_j}{\text{Max}}\,R(A_i,S_j),\,A^*\in(A_1,\cdots,A_n). \tag{15b}$$

The minimax regret rule leads to the comparison of the **Table 3** sensors in **Table 5**. Sensor A_3 is the recommended alternative; but, it performs poorly in S_3. The rational individual who seeks a robust solution is likely to reject it. The well balanced A_4 ranks 3rd. These results raise serious doubts about the usefulness of the minimax regret rule as a robustness criterion1[3].

The maximin criterion, also known as Wald's minimax for losses, recommends selecting the alternative with the maximum minimum utility, A_4 (see **Table 3**). The maximin criterion and minimax regret criterion have different characteristics. The maximin criterion captures aversion to the worst possible outcome for each alternative. The minimax regret criterion is concerned with missing beneficial opportunities. Numerous analysts believe that the minimax regret criterion is preferable to the maximin criterion; this is in conflict with the above results.

DISAPPOINTMENT THEORY FOR AOA

The term "disappointment" has psychological connotetions that are not relevant in the context of the paper. Nevertheless, it is used in the paper because it is entrenched in the decision theory literature.

General Overview

Bell [10] and Loomes and Sugden [11] proposed DT to account for the fact that rational individuals may anticipate disappointment when they counterfactually compare the outcome of an action in one scenario relative to the better outcomes in the other scenarios. The rational individual may account ex-ante for these negative feelings in decision making. Similarly, feelings of elation are associated with the better outcomes. DT, unlike RT, does not perform comparisons across alternatives. This avoids potentially misleading results of pairwise comparisons [36]. The disappointment level and utility of an alternative are therefore independent of the other alternatives.

Several DT models have been proposed as corrections to EUT. Bell [10] and Loomes and Sugden [11] assumed that for a given alternative the strength of the disappointment associated with a probabilistic outcome is measured relative to the alternative's EU. Delquié and Cillo [37] noted that rational individuals are more likely to compare the outcome obtained against the outcomes that they failed to get had other scenarios realized.

Table 5. The minimax regret decision matrix corresponding to **Table 3**.

	S_1	S_2	S_3		
Alternative		Regret		Max Regret	Selected *A*
A_1	0.00	10.00	2.87	10.00	
A_2	1.19	8.33	1.67	8.33	
A_3	0.49	0.00	6.77	6.77	A_3
A_4	9.52	8.33	0.00	9.52	
			Minimax Regret	6.77	

They hypothesized that the level of disappointment associated with outcome O_i depends explicitly on the probabilities and utilities of the outcomes that are preferred, i.e. the outcomes $O_k, k=1,\cdots,(i-1)$14. The disappointment associated with alternative A_j given the realization of scenario S_i, is modeled as follows:

$$DT(A_j,S_i)=\sum_{k=1}^{i} p_k f_D\left[U^{\text{vNM}}(A_j,S_k)-U^{\text{vNM}}(A_j,S_i)\right], \tag{16}$$

where $f_D(\cdot)$ is a positive function of $U^{\text{vNM}}(\cdot)$ with $f_D(0)=0$; p_k is the probability of scenario S_k.

Disappointment Correction to EUT

For the purpose of this paper, $f_D(\cdot)$ in Equation (16) is assumed to be a linear function. The overall disappointment is then obtained by averaging Equation (16) over all the m scenarios. This simplified version of the Delquié and Cillo Linear Disappointment Model (LDM) [37] modifies the vNM EU, Equation (2), as follows:

$$\text{EU}^{\text{LDM}}(A_i)=\sum_{j=1}^{m} p_j\left[u_{i,j}-\text{d}\sum_{k=1}^{j} p_k\left(u_{i,k}-u_{i,j}\right)\right], \tag{17}$$

where d is a positive constant that captures the rational individual's sensitivity to disappointment.

Equation (17) has several notable properties:

1) The functional form is similar to the vNM EU but with rank-dependent weighted probabilities.
2) The second term represents downside risk1[5] relative to the best possible outcome. It provides a credible measure of risk-based robustness.
3) The parameter d can be varied to reflect varying degrees of risk tolerance and thereby identify alternatives with unacceptable consequences.

Two-Sensor Selection Example

The two-sensor paradox of Section 4 is analyzed using the LDM. Given that disappointment is explicitly accounted for, the analysis uses the **Table 2** physical data. The results for d = 0.8 are presented in **Table 6. Figure 2** compares several two-sensor combinations versus d.

The results are consistent with a rational preference for diversification and the notion of risk-based robustness. The two-sensor combinations change rank as d varies. For $d > 0.7$, A_4-A_4 rank 1st and A_2-A_4 ranks 2nd. A_3-A_3 changes rank from 1st for $d = 0$ to last for $d > 0.6$. To reach a final decision, a risk-averse DM would peruse **Figure 2** and chooses between A_2-A_4 or A_4-A_4 depending on preferences for risk tolerance and diversification.

The LDM1[6] analysis is simpler and more intuitive than the utility theory two-dimensional preference analysis of Section 4.2. It can readily be extended to complex combinations of multiple constituents. However, all available DT models lack a mathematical basis for developing complex hybrid solutions or portfolios. Such an approach is presented in the next section.

PORTFOLIO OPTIMIZATION FOR ROBUSTNESS

General Formulation

As discussed in Sections 4-6, a risk-adverse DM is likely to prefer a hybrid solution that performs relatively well over all scenarios of interest rather than a homogeneous solution consisting solely of the MEU alternative. Consider a situation of n alternatives $(A_i, i = 1, \cdots, n)$ and m mutually exclusive probabilistic scenarios $(S_j, p_j, j = 1, \cdots, m)$. A hybrid solution may be thought of as a portfolio characterized by a multidimensional probability distribution function [39]

Table 6: Results using the linear disappointment model.

Combination	EU	DT d = 0.8	U^{LDM}	Rank
A_1-A_1	0.53	0.09	0.44	10
A_2-A_2	0.53	0.05	0.48	4
A_3-A_3	0.58	0.10	0.48	6
A_4-A_4	0.50	0.00	0.50	1
A_1-A_2	0.53	0.07	0.46	8
A_1-A_3	0.56	0.10	0.46	9
A_1-A_4	0.52	0.05	0.47	7
A_2-A_3	0.56	0.08	0.48	5
A_2-A_4	0.52	0.03	0.49	2
A_3-A_4	0.54	0.05	0.49	3

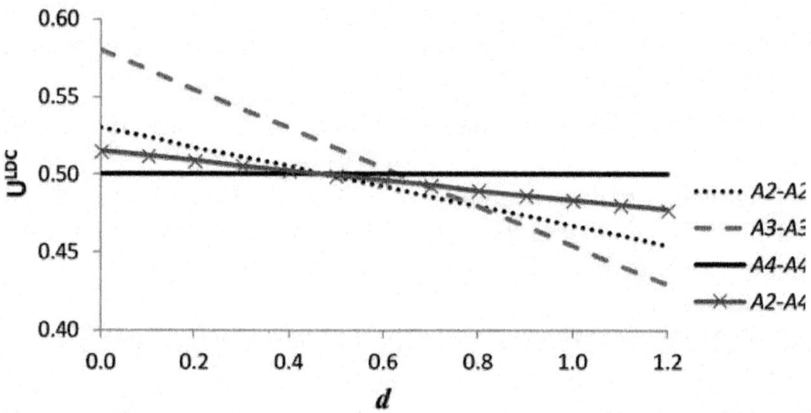

Figure 2: Comparison of selected two-sensor combinations versus disappointment parameter d using the linear disappointment model.

$$P_{HS}(x) = \sum_{i=1}^{n} x_i F\left[\sum_{j=1}^{m} f_{pd}(A_i, S_j), p_j\right], \tag{19a}$$

where $x \equiv (x_1, x_2, \cdots, x_n)$ is the decision vector of allocation variables with x_i being the A_i allocation fraction. The multidimensional function $F[\cdot]$ is a generalized discrete probability distribution with values being probability distributions functions (pdf) rather than point estimates. $f_{pd}(A_i,S_j)$ is a pdf that models the A_i outcome[17] for scenario S_j.

The determination of the decision vector x may be specified in terms of the following stochastic optimization problem:

Maximize the zth percentile of $P_{HS}(x)$, Equation (19a), subject to the normalization, scenario performance, and affordability constraints, Equations (19b)-(19d):

$$\sum_{i}^{n} x_i = 1 \tag{19b}$$

$$\phi^{th}\text{percentile of } R(x,S_j) = \sum_{i=1}^{n} x_i f_{pd}(A_i,S_j) \geq \alpha_j \forall S_j \tag{19c}$$

$$C_T(x) = N_0 \sum_{i=1}^{n} x_i C_i \leq B_0. \tag{19d}$$

The values z and f are percentiles[18] that a DM specifies in accordance with his/her risk tolerance; a_j is his/her lowest acceptable value for the fth percentile for performance in scenario S_j; C_i is the A_i cost; N_0 is the number of required systems; and B_0 is the available budget. Equation (19d) can accommodate pdfs for C_i with an agreed-to percentile for the cost constraint [27].

The above formulation provides a flexible and transparent way for the analyst and DM to define the choice problem and generate mixed solutions with desirable properties such as diversification and robustness[19].

Hybrid Solution Example

The above approach is applied to the design of a robust border surveillance system of N_0 sensors using the four **Table 2** sensors. The problem is to determine the combination of sensors (x_1,x_2,x_3,x_4), Equation (19a), that maximizes the expected PoD over the three scenarios (S_1,S_2,S_3) subject to the normalization constraint, Equation (19b), and $\text{PoD}(S_3) \geq 0.43$ for the performance constraint, Equation (19c). The budget is assumed not to be a constraint. The hybrid solution data are summarized in **Table 7** and its cumulative distribution [41] and several of interest are depicted in **Figure 3**.

Table 7: Sensor hybrid solution optimized for robustness.

Objective	Constraints	Decision Variables			
Maximize E(PoD)	PoD(S_3) $\geq$ 0.43	A_1	A_2	A_3	A_4
0.52	0.43	0.0	0.7	0.0	0.3

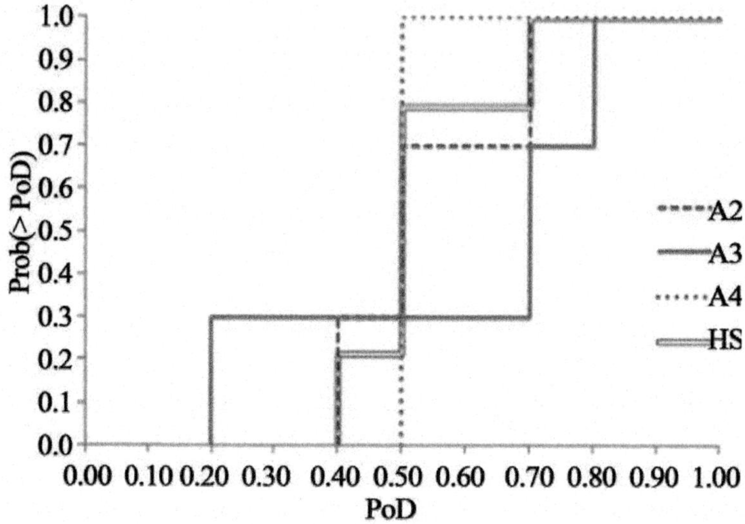

Figure 3: Cumulative distributions for the hybrid sensor solution (HS) optimized for robustness, its constituent sensors A2 and A4, and the MEU alternative A3.

The hybrid solution is robust: 1) the associated cumulative distribution is narrower than those of the homogeneous alternatives, 2) the minimum PoD = 0.4, and 3) there is a 80% probability that the PoD ^{3} 0.5 for the identified scenarios. Given the **Table 2** alternatives, this is the solution that a DM who desires some level of robustness at an acceptable performance cost is likely to intuitively prefer. Diversity provides additional benefits such as defense against common-cause failures [42] and increased deterrence against attackers [14].

The solution to this example is obvious. Nevertheless, it illustrates the significant advantages and benefits of portfolio allocation with stochastic optimization for developing robust critical systems. Solving complex realworld problems will require tools with integrated Monte Carlo simulation and stochastic optimization capabilities. They are available in commercial and proprietary versions. For information purposes, Crystal Ball® with

OptQuest® was applied to the illustrative example. OptQuest® is a general purpose optimizer developed to effectively solve complex stochastic, nonlinear, and combinatorial optimization problems [43].

CONCLUSIONS

The paper begins by analyzing the suitability of EUT for AoAs of critical systems in a multistate world. The concept of robust solutions is discussed within the context of multiple scenarios. Several well-known models are compared for a simplified example of choosing sensors for a hypothetical border surveillance system. Portfolio allocation with stochastic optimization is proposed and demonstrated to be a valid and realistic method for defining robust critical systems.

Some of the key conclusions are:

1) The EUT is inadequate for critical systems AoA because MEU alternatives may possess vulnerabilities that could be exploited and/or are susceptible to common-cause failures. The continuity axiom is unrealistic given the emotionally charged aspects of critical decisions. The independence axiom is in conflict with the rational individual who is not an expected utility maximizer and prefers to make decisions using risk curves.
2) The two-sensor paradox raises serious doubts about the validity of the minimax regret rule as a robustness criterion for decision making under risk and the belief that it is superior to the maximin rule. Both have serious problems and are potential misleading for choosing critical systems.
3) If the utility is linear in the probabilities, the EU of a hybrid solution cannot be greater than the EU of every constituent alternative. A GEUT and nonlinear utility functions are required to exhibit robustness and diversification.
4) The LDM simply modifies the vNM EU by adding a term proportional to downside risk. The results of the two-sensor selection example confirm that it is a promising model for choosing robust homogeneous solutions. However, the LDM and other available DT models do not provide a mathematical basis for developing hybrid solutions.
5) Portfolio allocation with stochastic optimization provides a flexible and transparent approach for defining the choice problem and determining hybrid solutions for critical systems with desirable properties such as diversification and robustness. The best combination does not necessarily consist of the alternatives with the highest EUs; it draws

on the strengths and complementarities that any of the alternatives can provide.

6) The composition of the hybrid solution is determined by optimizing a specified percentile of the system performance subject to robustness and cost constraints. Performance and cost uncertainties can be modeled using realistic pdf's.

Ultimately, the selection of robust solutions depends on the identified alternatives. Critical systems require analysis that is mindful of the following considerations: 1) the limitations of EUT and GEUT models; 2) risk-mitigation properties such as FARness; and 3) the importance of hybrid solutions as options. Given the availability of commercial and proprietary tools with integrated Monte Carlo simulation and stochastic optimization capabilities, the future direction is the implementation of the proposed portfolio allocation with stochastic optimization as a practical design and analysis tool for complex systems and SoS on actual projects.

ACKNOWLEDGMENTS

This work was stimulated by a Naval Postgraduate School capstone project prepared by the 2008 Cohort from the Naval Surface Warfare Center Dahlgren Division (NSWCDD) and the Naval Undersea Warfare Center Division Newport [44]. The author thanks Neil Baron, Distinguished Scientist NSWCDD, for insightful comments on hybrid solutions.

REFERENCES

1. US Office of Homeland Security, "The National Strategy for Homeland Security," 2002. http://www.ncs.gov/library/policy_docs/nat_strat_hls.pdf
2. R. D. Luce and H. Raiffa, "Games and Decisions: Introduction and Critical Survey," Dover Publications, New York, 1957.
3. J. von Neumann and O. Morgenstern, "Theory of Games and Economic Behavior," 2nd Edition, Princeton University Press, Princeton, 1947.
4. L. J. Savage, "The Foundations of Statistics," 2nd Revised Edition, Dover Publications, New York, 1972.
5. P. C. Fishburn, "Normative Theories of Decision Making under Risk and Uncertainty," In: D. Bell, H. Raiffa and A. Tversky, Eds., Decision Making: Descriptive, Normative, and Prescriptive Interactions, Cambridge University Press, Cambridge, 1988, pp. 78-98.doi:10.1017/CBO9780511598951.006

6. D. Kahneman and A. Tversky, "Prospect Theory: An Analysis of Decisions under Risk," Econometrica, Vol. 47, No. 3. 1979, pp. 263-291. doi:10.2307/1914185

7. Tversky and D. Kahneman, "Advances in Prospect Theory: Cumulative Representation of Uncertainty," Journal of Risk and Uncertainty, Vol. 5, No. 4, 1992, pp. 297-323.

8. D. E. Bell, "Regret in Decision Making under Uncertainty," Operations Research, Vol. 30, No. 5, 1982, pp. 961-981. doi:10.1287/opre.30.5.961

9. G. Loomes and R. Sugden, "Regret Theory: An Alternative Theory of Rational Choice under Uncertainty," Economic Journal, Vol. 92, No. 368, 1982, pp. 805-824. doi:10.2307/2232669

10. D. E. Bell, "Disappointment in Decision Making under Uncertainty," Operations Research, Vol. 33, No. 1, 1985, pp. 1-27. doi:10.1287/opre.33.1.1

11. G. Loomes and R. Sugden, "Disappointment and Dynamic Consistency in Choice under Uncertainty," Review of Economic Studies, Vol. 53, No. 2, 1986, pp. 271-282.doi:10.2307/2297651

12. E. Dekel and B. L. Lipman, "How (Not) to Do Decision Theory," Annual Review of Economics, Vol. 2, No. 1, 2010, pp. 257-282. doi:10.1146/annurev.economics.102308.124328

13. G. Chichilnisky, "An Axiomatic Approach to Choice under Uncertainty with Catastrophic Risks," Resource and Energy Economics, Vol. 22, No. 3, 2000, pp. 221- 231.doi:10.2139/ssrn.1522307

14. G. G. Brown and L. A. Cox Jr., "Making Terrorism Risk Analysis Less Harmful and More Useful: Another Try," Risk Analysis, Vol. 31, No. 2, 2011, pp. 193-195. doi:10.1111/j.1539-6924.2010.01563.x

15. P. K. Davis, R. D. Shaver and J. Beck, "Portfolio-Analysis Methods for Assessing Capability Options," RAND, Santa Monica, 2008.

16. S. Savage, "The Flaw of Averages: Why We Underestimate Risk in the Face of Uncertainty," John Wiley & Sons, Hoboken, 2009.

17. Y. Y. Haimes, "Risk Modeling, Assessment, and Management," 3rd Edition, John Wiley & Sons, Hoboken, 2009.

18. H. M. Markowitz, "Portfolio Selection: Efficient Diversification of Investments," 2nd Edition, Blackwell, Cambridge, 1997.

19. Office of the Deputy under Secretary of Defense for Acquisition and Technology, Systems and Software Engineering, "Systems Engineering Guide for Systems of Systems," Version 1.0, ODUSD(A&T)SSE, Washington, 2008.

20. D. H. Wagner, W. C. Mylander and T. J. Sanders, "Naval Operations Analysis," 3rd Edition, Naval Institute Press, Annapolis, 1999.
21. Washburn and M. Kress, "Combat Modeling," Springer, New York, 2009.doi:10.1007/978-1-4419-0790-5
22. 2011 Naval Surface Warfare Center Crane Cohort, "System Engineering Approach to Improving Arizona Border Patrol C4ISR Mission Operations," MSSE Capstone Project, Naval Postgraduate School, Monterey, 2011.
23. R. Hastie and R. M. Dawes, "Rational Choice in an Uncertain World: The Psychology of Judgment and Decision Making," Sage Publications, Thousand Oaks, 2001.
24. Einstein, "Ideas and Opinions," Bonanza Books, New York, 1954.
25. E. Malinvaud, "Note on von Neumann-Morgenstern's Strong Independence Axiom," Econometrica, Vol. 20, No. 4, 1952, p. 679. doi:10.2307/1907650
26. R. de Neuville, "Applied Systems Analysis: Engineering Planning and Technology Management," McGraw-Hill, New York, 1999.
27. E. Kujawski, M. Alvaro and W. Edwards, "Incorporating Psychological Influences in Probabilistic Cost Analysis," Systems Engineering, Vol. 7, No. 3, 2004, pp. 195-216.doi:10.1002/sys.20004
28. S. H. Chew, "Axiomatic Utility Theories with the Betweenness Property," Annals of Operations Research, Vol. 19, No. 1, 1989, pp. 273-298. doi:10.1007/BF02283525
29. M. Allais, "Le Comportement de l'Homme Rationnel devant le Risque: Critique des Postulats et Axiomes de l'Ecole Américaine," Econometrica, Vol. 21, No. 4, 1953, pp. 503-546.doi:10.2307/1907921
30. P. Samuelson, "Risk and Uncertainty: A Fallacy of Large Numbers," Scientia, Vol. 98, No. 4, 1963, pp. 108-113.
31. M. Rabin and R. H. Thaler, "Anomalies: Risk Aversion," The Journal of Economic Perspectives, Vol. 15, No. 1, 2001, pp. 219-232. doi:10.1257/jep.15.1.219
32. E. Dekel, "Asset Demands without the Independence Axiom," Econometrica, Vol. 57, No. 1, 1989, pp. 163-169. doi:10.2307/1912577
33. R. J. Lempert, D. G. Groves, S. W. Popper and S. C. Bankes, "A General, Analytic Method for Generating Robust Strategies and Narrative Scenarios," Management Science, Vol. 52, No. 4, 2006, pp. 51-528. doi:10.1287/mnsc.1050.0472

34. P. Krokhmal, R. Murphey, P. Pardalos, S. Uryasev and G. Zrazhevsky, "Robust Decision Making: Addressing Uncertainties in Distributions," In: S. Butenko, et al., Eds., Cooperative Control: Models, Applications and Algorithms, Kluwer Academic Publishers, Dordrecht, 2003, pp. 165-185.

35. D. G. Ullman, "Making Robust Decisions: Decision Management for Technical, Business, & Service," Trafford Publishing, Victoria, 2006.

36. G. Friedman, "The Intransitivity of Pairwise Comparisons Even with a Single Rational Decision Maker or: Homomorphisms from Allegedly Paradoxical Dice to Decision-Making in the Military, Business and Sports World," Presentation to the NSF Decision-Based Design Workshop, Long Beach, 1999. http://dbd.eng.buffalo.edu/papers/friedman.html

37. P. Delquié and A. Cillo, "Disappointment without Prior Expectation: A Unifying Perspective on Decision under Risk," Journal of Risk and Uncertainty, Vol. 33, No. 3, pp. 197-215.doi:10.1007/s11166-006-0499-4

38. D. Nawrocki, "A Brief History of Downside Risk Measures," Journal of Investing, Vol. 8, No. 3, 1999, pp. 9-25. doi:10.3905/joi.1999.319365

39. E. Kujawski and G. A. Miller, "Quantitative Risk-Based Analysis for Military Counterterrorism Systems," Systems Engineering, Vol. 10, No. 4, 2007, pp. 273-289. doi:10.1002/sys.20075

40. D Bertsimas, D. B. Brown and C. Caramanis, "The Theory and Applications of Robust Optimization," SIAM Review, Vol. 53, No. 3, 2011, pp. 464-501. doi:10.1137/080734510

41. M. Levy, "Almost Stochastic Dominance and Efficient Investment Sets," American Journal of Operations Research, Vol. 2, No. 3, 2012, pp. 313-321. doi:10.4236/ajor.2012.23038

42. E. Kujawski, I. M. Jacobs and A. M. Smith, "An Evaluation of the Use of Signal Validation Techniques as a Defense against Common-Cause Failures," Electric Power Research Institute, Palo Alto, 1987.

43. F. Glover, M. Laguna and R. Martí, "Scatter Search," In: A. Ghosh and S. Tsutsui, Eds., Advances in Evolutionary Computation: Theory and Applications, Springer-Verlag, New York, 2003, pp. 519-537.

44. 2008 Cohort from the Naval Surface Warfare Center Dahlgren Division and the Naval Undersea Warfare Center Division Newport, "Enhanced Vessel Awareness Capability," MSSE Capstone Project, Naval Postgraduate School, Monterey, 2011.

Chapter 7

BI-CRITERIA OPTIMIZATION TECHNIQUE IN STOCHASTIC SYSTEM MAINTENANCE ALLOCATION PROBLEM

Irfan Ali, S. Suhaib Hasan

Department of Statistics & Operations Research, Aligarh Muslim University, Aligarh, India

ABSTRACT

In this paper, the problem of optimum allocation of repairable and replaceable components in a system is formulated as a Bi-objective stochastic nonlinear programming problem. The system maintenance time and cost are random variable and has gamma and normal distribution respectively. A Bi-criteria optimization technique, weighted Tchebycheff is used to obtain the optimum allocation for a system. A numerical example is also presented to illustrate the computational details.

INTRODUCTION

We consider a system which requires performing a sequence of identical production runs after every given (fixed) period. A production run in the system consists of several subsystems where each subsystem can work properly if at least one of its components is operational. The following assumptions are also made:

1) all the components can be repaired if deteriorated or failed;
2) all component states are independent.

We assume that the system comprises two types of subsystem. One is the type of subsystems in which the components are very sensitive to the functioning of the whole system and, therefore, on deterioration these should be replaced by new ones. Let these subsystems range from 1 to S. The other type of subsystems is those in which the components after deterioration can be repaired and then replaced. Let such subsystems range from $s+1$ to m. In **Figure 1** the Group X consists of the S subsystems with sensitive components which on failure are replaced by new ones and Y the remaining $(m-s)$ subsystems in which the components can be repaired (see Ali et al. [1]).

Ideally, all the failed components in all the subsystem of Group X are replaced by new ones prior to the beginning of the next mission/ run. In a similar way, ideally all the failed components in subsystem of Group Y are repaired and then replaced prior to the beginning of the next mission/run. However, due to the constraints on the cost and time it may not be possible to repair and replace all the failed components in Groups X and Y. For this a mathematical programming frame-work is established for assisting decision-makers in determining the optimal subset of maintenance activities to perform prior to beginning of the next mission. This decision-making process is referred to as selective maintenance.

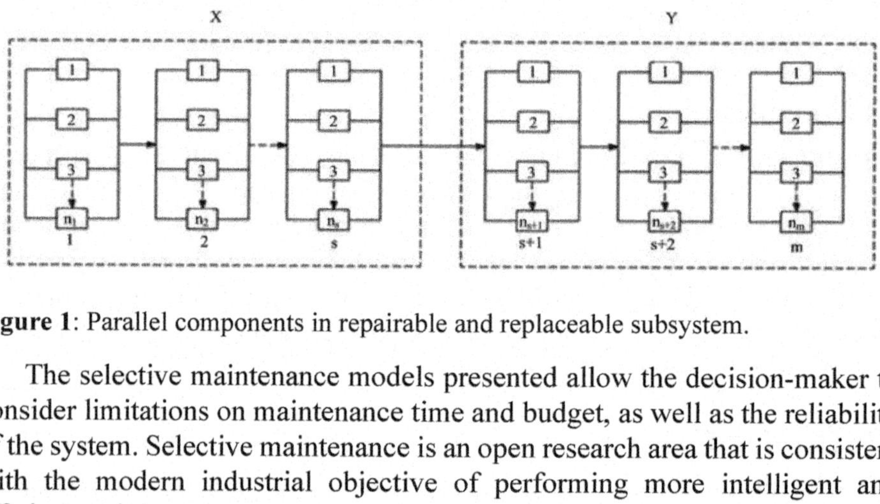

Figure 1: Parallel components in repairable and replaceable subsystem.

The selective maintenance models presented allow the decision-maker to consider limitations on maintenance time and budget, as well as the reliability of the system. Selective maintenance is an open research area that is consistent with the modern industrial objective of performing more intelligent and efficient maintenance.

For this let us suppose a_i be the total failed components in the subsystems and d_i be the number of components in the ith subsystem, which can be repaired and replaced prior to the beginning of the next mission (See Rice et al. [2]). Thus under the selective maintenance the number of components available for the next mission in the ith subsystem will be

$$(n_i - k_i) + d_i, i = 1, 2, \cdots, m \quad (1)$$

Therefore the reliability of the subsystems range from 1 to S for a production run is given by

$$R(d_i) = \left\{ \prod_{i=1}^{s} \left[1 - (1 - r_i)^{n_i - a_i + d_i} \right] \right\} \tag{2}$$

and the reliability of the subsystems range from $s+1$ to m for a production run is given by

$$R(d_i) = \left\{ \prod_{i=S+1}^{m} \left[1 - (1 - r_i)^{n_i - a_i + d_i} \right] \right\} \tag{3}$$

The maintenance time constraint for the system is given as

$$\sum_{i=1}^{m} t_i d_i \leq T_0 \tag{4}$$

and the maintenance cost constraint for the system is given as

$$\sum_{i=1}^{m} c_i d_i \leq C_0 \tag{5}$$

However, in the event the reliability of the subsystems of Groups X and Y time are of equally serious concern. Let us consider, for instance, the following multi-objective problem (please see the Equation (6) below).

Secondly, a Bi-objective programming problem in which time and the cost spent on system maintenance is minimized simultaneously for the required reliability $R_i^*(d_i)$ (say). The mathematical model of the problem is given as Equation (7) below.

Recently many authors have discussed the allocation problem of repairable components. Among them are Rice et al. [2], Schneider and Cassady [3], Rajaopalan and Cassady [4], Schneider et al. [5], Iyoob et al. [6], Ali et al. ([1,7-10]), Faisal and Ali [11] and many others.

In this paper, we have formulated stochastic system maintenance problem as a multi-objective programming problem.

$$
\left.
\begin{array}{ll}
\text{Maximize } R(d_i) = \left\{ \prod_{i=1}^{s} \left[1 - (1 - r_i)^{n_i - a_i + d_i} \right] \right\} & \text{(i)} \\
\text{and Maximize } R(d_i) = \left\{ \prod_{i=S+1}^{m} \left[1 - (1 - r_i)^{n_i - a_i + d_i} \right] \right\} & \text{(ii)} \\
\text{Subject to } \sum_{i=1}^{m} t_i d_i \le T_0 & \text{(iii)} \\
\sum_{i=1}^{m} c_i d_i \le C_0 & \text{(iv)} \\
0 \le d_i \le a_i, \forall d_i \text{ are integer} & \text{(v)} \\
n_i \ge a_i, i = 1, 2, \cdots, m & \text{(vi)}
\end{array}
\right\} \qquad (6)
$$

$$
\left.
\begin{array}{ll}
\text{Minimize } T = \sum_{i=1}^{m} t_i d_i & \text{(i)} \\
\text{and Minimize } C = \sum_{i=1}^{m} c_i d_i C_0 & \text{(ii)} \\
\text{Subject to } \left\{ \prod_{i=1}^{s} \left[1 - (1 - r_i)^{n_i - a_i + d_i} \right] \right\} \ge R_1^*(d_i) & \text{(iii)} \\
\left\{ \prod_{i=S+1}^{m} \left[1 - (1 - r_i)^{n_i - a_i + d_i} \right] \right\} \ge R_2^*(d_i) & \text{(iv)} \\
0 \le d_i \le a_i, \forall d_i \text{ are integer} & \text{(v)} \\
n_i \ge a_i, i = 1, 2, \cdots, m & \text{(vi)}
\end{array}
\right\} \qquad (7)
$$

We have discussed components repairable and replaceable time and cost as a random variable in the constraint and has Gamma and Normal distribution respectively. The Probabilistic constraints function is then converted into an equivalent deterministic non-linear programming form by using chance constrained programming.

THE CHANCE CONSTRAINED PROGRAMMING

In many practical situations the constraint Equations (iii) and (iv) are not fixed and taken as probabilistic. Thus the above problem (6) can be written in the following chance constrained programming form as Equation (8) below, where $p_0, 0 \le p_0 \le 1$ is a specified probability.

In the above problem (8), let us assume that t_i and c_i are independently gamma and normally distributed random variables.

Let us assume that $t_i, i = 1, \cdots, m$ are independent Gamma distributed random variables in the constraint 8 (iii), i.e., $t_i \sim G(\alpha_i, \beta_i)$.

Then the,

$$\text{Mean}(t_i) = \frac{\beta_i}{\alpha_i}, \quad \text{Variance}(t_i) = \frac{\beta_i}{\alpha_i^2}. \tag{8}$$

Now let

$$f(d) = \sum_{i=1}^{m} t_i d_i$$

Then mean is

$$E\{f(d)\} = \sum_{i=1}^{m} d_i E(t_i) = \sum_{i=1}^{m} \frac{\beta_i}{\alpha_i} d_i$$

Further, as t_i are independently distributed, we have

$$V\{f(d)\} = \sum_{i=1}^{m} t_i^2 V(d_i) = \sum_{i=1}^{m} \frac{\beta_i}{\alpha_i^2} d_i$$

Now the constraints 8 (iii) can be written as

$$P(f(d) \le T_0) \ge p_0, \tag{9}$$

Since number of components within the system are assumed to be large we have from Liapounoff's central limit theorem

$$f(d) \sim N(E\{f(d)\}, V\{f(d)\}).$$

Thus (9) is equivalent to

$$P\left\{\frac{f(d) - E\{f(d)\}}{\sqrt{V\{f(d)\}}} \le \frac{T_0 - E\{f(d)\}}{\sqrt{V\{f(d)\}}}\right\} \ge p_0,$$

where $\left[\frac{f(d) - E\{f(d)\}}{\sqrt{V\{f(d)\}}}\right]$ is a standard normal variate with mean zero and variance one. Thus the probability of realizing $\{f(d)\}$ less than or equal to T_0 can be written as

$$P(f(d) \le T_0) = \phi\left[\frac{T_0 - E\{f(d)\}}{\sqrt{V\{f(d)\}}}\right], \tag{10}$$

where $\phi(z)$ represents the cumulative density function of the standard normal variable evaluated at Z. If K_α represents the value of the standard normal variable at which $\phi(K_\alpha) = p_0$, then the constraint (10) can be written as

$$\phi\left[\frac{T_0 - E\{f(d)\}}{\sqrt{V\{f(d)\}}}\right] \geq \phi(K_\alpha).$$

(11)

The inequality will be satisfied only if

$$\left[\frac{T_0 - E\{f(d)\}}{\sqrt{V\{f(d)\}}}\right] \geq K_\alpha,$$

$$\left.\begin{array}{lr}
\text{Maximize } R(d_i) = \left\{\prod_{i=1}^{s}\left[1-(1-r_i)^{n_i-a_i+d_i}\right]\right\} & \text{(i)} \\
\text{and Maximize } R(d_i) = \left\{\prod_{i=S+1}^{m}\left[1-(1-r_i)^{n_i-a_i+d_i}\right]\right\} & \text{(ii)} \\
\text{Subject to } P\left(\sum_{i=1}^{m} t_i d_i \leq T_0\right) \geq p_0 & \text{(iii)} \\
P\left(\sum_{i=1}^{m} c_i d_i \leq C_0\right) \geq p_0 & \text{(iv)} \\
0 \leq d_i \leq a_i, \forall d_i \text{ are int eger} & \text{(v)} \\
n_i \geq a_i, i = 1, 2, \cdots, m & \text{(vi)}
\end{array}\right\} \quad (8)$$

or equivalently $\dfrac{T_0 - \sum_{i=1}^{m} \frac{\beta_i}{\alpha_i} d_i}{\sqrt{\sum_{i=1}^{m} \frac{\beta_i}{\alpha_i^2} d_i^2}} \geq K_\alpha,$

Thus an equivalent deterministic constraint to the stochastic constraint is given by

$$\left(\sum_{i=1}^{m} \frac{\beta_i}{\alpha_i} d_i\right) + K_\alpha \sqrt{\sum_{i=1}^{m} \frac{\beta_i}{\alpha_i^2} d_i^2} \leq T_0$$

(12)

Now we consider the case when c_i are independently normally distributed random variables in the constraint 8 (iv), i.e. $c_i \sim N(\mu_i, \sigma_i^2)$.

The constraint $P(f(c) \leq C_0) \geq p_0$, is equivalent to

$$P\left\{\frac{f(c)-E\{f(c)\}}{\sqrt{V\{f(c)\}}} \le \frac{C_0-E\{f(c)\}}{\sqrt{V\{f(c)\}}}\right\} \ge p_0, \quad (13)$$

Now in this case

$$E\{f(c)\} = \sum_{i=1}^{m} d_i E(c_i) = \sum_{i=1}^{m} \mu_i d_i$$

and

$$V\{f(c)\} = \sum_{i=1}^{m} d_i^2 V(c_h) = \sum_{i=1}^{m} \sigma_i^2 d_i^2$$

Therefore, the deterministic equivalent of 8 (iv) in this case is

$$\left(\sum_{i=1}^{m} \mu_i d_i\right) + K_\alpha \sqrt{\sum_{i=1}^{m} \sigma_i^2 d_i^2} \le C_0 \quad (14)$$

The equivalent deterministic non-linear programming problem (8) to the stochastic programming problem is given by

$$\left.\begin{array}{lr}
\text{Maximize } R(d_i) = \left\{\prod_{i=1}^{s}\left[1-(1-r_i)^{n_i-a_i+d_i}\right]\right\} & \text{(i)}\\
\text{and Maximize } R(d_i) = \left\{\prod_{i=S+1}^{m}\left[1-(1-r_i)^{n_i-a_i+d_i}\right]\right\} & \text{(ii)}\\
\text{Subject to} \left(\sum_{i=1}^{m} \frac{\beta_i}{\alpha_i} d_i\right) + K_\alpha \sqrt{\sum_{i=1}^{m} \frac{\beta_i}{\alpha_i^2} d_i^2} \le T_0 & \text{(iii)}\\
\left(\sum_{i=1}^{m} \mu_i d_i\right) + K_\alpha \sqrt{\sum_{i=1}^{m} \sigma_i^2 d_i^2} \le C_0 & \text{(iv)}\\
0 \le d_i \le a_i, \forall d_i \text{ are int eger} & \text{(v)}\\
n_i \ge a_i, i = 1, 2, \cdots, m & \text{(vi)}
\end{array}\right\} \quad (15)$$

MODIFIED E-MODEL

Consider the situations in which the time taken and cost spent on maintenance are not fixed and taken as probabilistic in the objective function in Equations (i) and (ii). Thus the above problem (7) can be written in the following probabilistic objective function form as:

$$\left.\begin{array}{ll}\text{Minimize } T = p\left(\sum_{i=1}^{m} t_i d_i\right) & \text{(i)}\\ \text{and Minimize } C = p\left(\sum_{i=1}^{m} c_i d_i\right) & \text{(ii)}\\ \text{Subject to}\left\{\prod_{i=1}^{s}\left[1-(1-r_i)^{n_i-a_i+d_i}\right]\right\}\times\left\{\prod_{i=S+1}^{m}\left[1-(1-r_i)^{n_i-a_i+d_i}\right]\right\}\geq R^*(d_i) & \text{(iii)}\\ 0\leq d_i\leq a_i, \forall d_i \text{ are int eger} & \text{(v)}\\ n_i\geq a_i, i=1,2,\cdots,m & \text{(vi)}\end{array}\right\} \quad (16)$$

Using Modified E-model technique, the problem (16) is formulated as

$$\left.\begin{array}{ll}\text{Minimize } T = k_1\left(\sum_{i=1}^{m}\frac{\beta_i}{\alpha_i}d_i\right)+k_2\sqrt{\sum_{i=1}^{m}\frac{\beta_i}{\alpha_i^2}d_i^2} & \text{(i)}\\ \text{and Minimize } C = k_1\left(\sum_{i=1}^{m}\mu_i d_i\right)+k_2\sqrt{\sum_{i=1}^{m}\sigma_i^2 d_i^2} & \text{(ii)}\\ \text{Subject to}\left\{\prod_{i=1}^{s}\left[1-(1-r_i)^{n_i-a_i+d_i}\right]\right\}\times\left\{\prod_{i=S+1}^{m}\left[1-(1-r_i)^{n_i-a_i+d_i}\right]\right\}\geq R^*(d_i) & \text{(iii)}\\ 0\leq d_i\leq a_i, \forall d_i \text{ are int eger} & \text{(iv)}\\ n_i\geq a_i, i=1,2,\cdots,m & \text{(v)}\end{array}\right\}$$

(17)

where k_1 and k_2 are non-negative constants, and their values show the relative importance of the expectation and the variance. Some authors suggest that $k_1+k_2=1$see Rao [12].

The two others Bi-objective programming models in different prospects for the decision-makers are

Model 1:

$$\left.\begin{array}{ll}\text{Minimize } T = k_1\left(\sum_{i=1}^{m}\frac{\beta_i}{\alpha_i}d_i\right)+k_2\sqrt{\sum_{i=1}^{m}\frac{\beta_i}{\alpha_i^2}d_i^2} & \text{(i)}\\ \text{and Maximize } R_2(d_i)=\left\{\prod_{i=S+1}^{m}\left[1-(1-r_i)^{n_i-a_i+d_i}\right]\right\} & \text{(ii)}\\ \text{Subject to}\left(\sum_{i=1}^{m}\mu_i d_i\right)+K_\alpha\sqrt{\sum_{i=1}^{m}\sigma_i^2 d_i^2}\leq C_0 & \text{(iii)}\\ \left\{\prod_{i=1}^{s}\left[1-(1-r_i)^{n_i-a_i+d_i}\right]\right\}\times\left\{\prod_{i=S+1}^{m}\left[1-(1-r_i)^{n_i-a_i+d_i}\right]\right\}\geq R^*(d_i) & \text{(iv)}\\ 0\leq d_i\leq a_i, \forall d_i \text{ are int eger} & \text{(v)}\\ n_i\geq a_i, i=1,2,\cdots,m & \text{(vi)}\end{array}\right\} \quad (18)$$

Model 2:

$$\left.\begin{array}{ll}\text{Minimize } C = k_1\left(\sum_{i=1}^{m}\mu_i d_i\right)+k_2\sqrt{\sum_{i=1}^{m}\sigma_i^2 d_i^2} & \text{(i)}\\ \text{and Maximize } R_1(d_i)=\left\{\prod_{i=1}^{s}\left[1-(1-r_i)^{n_i-a_i+d_i}\right]\right\} & \text{(ii)}\\ \text{Subject to}\left(\sum_{i=1}^{m}\frac{\beta_i}{\alpha_i}d_i\right)+K_\alpha\sqrt{\sum_{i=1}^{m}\frac{\beta_i}{\alpha_i^2}d_i^2}\le T_0 & \text{(iii)}\\ \left\{\prod_{i=1}^{s}\left[1-(1-r_i)^{n_i-a_i+d_i}\right]\right\}\times\left\{\prod_{i=S+1}^{m}\left[1-(1-r_i)^{n_i-a_i+d_i}\right]\right\}\ge R^*(d_i) & \text{(iv)}\\ 0\le d_i\le a_i, \forall d_i \text{ are int eger} & \text{(v)}\\ n_i\ge a_i, i=1,2,\cdots,m & \text{(vi)}\end{array}\right\} \quad (19)$$

A MULTI-CRITERIA WEIGHTED TCHEBYCHEFF OPTIMIZATION TECHNIQUE

Let us consider a multi-objective programming problem

$$\left.\begin{array}{l}\text{Min } f(x)=f_1(x),f_2(x),\cdots,f_k(x)\\ \text{Subject to } x\in s\end{array}\right\} \quad (20)$$

assumed to have $k(k\ge 2)$ competing objective functions $(f_i:\Re^n\to\Re)$ that are to be minimized simultaneously. The following definitions illustrate the concepts of efficient and weakly efficient decision vectors.

Definition: A decision vector $x^*\in X$ is efficient (Pareto optimal) for multi-objective programming problem if there does not exist a $x\in X, x\ne x^*$ such that $f_i(x)\le f_i(x^*)$ for $i=1,2,\cdots,k$ with strict inequality holding for at least one index i. ($x^*\in X$ is efficient, $f(x^*)$ is non-dominated).

Definition: A decision vector $x^*\in X$ is weakly efficient (weakly Pareto optimal) for multi-objective programming problem if there does not exist a $x\in X,\ x\ne x^*$ such that $f_i(x)<f_i(x^*)$ for $i=1,2,\cdots,k$. ($x^*\in X$ is weakly efficient, $f(x^*)$ is weakly nondominated).

There are several metrics that are found in the literature related to multi-objective programming problem. If ξ_i is the reference point and the ideal point, $\xi_i=\text{Min}_{x\in X}f_i(x)$ is used as the reference point, the general weighted L_p-metric$(1\le p\le\infty)$ is defined as

$$\left.\begin{array}{l}\text{Min}\left(\sum_{i=1}^{k} w_i \left|f_i(x)-\xi_i\right|^p\right)^{1/p} \\ \text{Subject to } x \in X\end{array}\right\} \tag{21}$$

We assume that $w_i \geq 0, \forall i = 1, 2, \cdots, k$ and $\sum_{i=1}^{k} w_i = 1$ where the w_i's are weighting coefficients provided by the decision maker reflecting the relative importance. If $p = \infty$, problem (21) reduces to a "weighted Tchebycheff Technique" (see Bowman [13]).

$$\left.\begin{array}{l}\text{Min} \underset{i=1,2,\cdots,k}{\text{Max}}\left[w_i \left|f_i(x)-\xi_i\right|\right] \\ \text{Subject to } x \in X\end{array}\right\} \tag{22}$$

If the reference point is the global optimal solution of $f_i(x)$, then the absolute value signs in problem (22) can be removed (see Miettinen [14]) yielding

$$\left.\begin{array}{l}\text{Min} \underset{i=1,2,\cdots,k}{\text{Max}}\left[w_i \left(f_i(x)-\xi_i\right)\right] \\ \text{Subject to } x \in X\end{array}\right\} \tag{23}$$

Miettinen [14] also showed that if the objectives and constraints are differentiable form of problem (23) can be defined as

$$\left.\begin{array}{l}\text{Min } \delta \\ \text{Subject to } \delta \geq \left\{w_i \left(f_i(x)-\xi_i\right)\right\}, \forall i, x \in X\end{array}\right\} \tag{24}$$

The solution of problem (24) is guaranteed weekly non-dominated for positive weights and at least one non-dominated solution is also guaranteed. If the solution is unique, then it is non-dominated, however if it is not unique, then it might be weakly non-dominated (see Wierzbicki [15].

The two objective functions in the Equation (15) are to be maximizing the total reliability of replaceable components of Group X and the reliability of repairable components of Group Y. We have convert the following maximizing problem into minimization problem using the property $\text{Max}Z = \text{Min}(-Z)$. Therefore the problem defined in Equation (15) is converted into a two criterion minimization problem: $\text{Min} - \{Z_1, Z_2\}$ subject to the constraints (see Khasawneh et al. [16]). Now the efficient solution is obtained by using the weighted Tchebycheff technique

$$\left.\begin{array}{ll}
\text{Minimize } \delta & \text{(i)} \\
\text{Subject to } \delta \geq w_1\left\{\left\{-\prod_{i=1}^{s}\left[1-(1-r_i)^{n_i-a_i+d_i}\right]\right\}-\xi_1\right\} & \text{(ii)} \\
\delta \geq w_2\left\{\left\{\prod_{i=S+1}^{m}\left[1-(1-r_i)^{n_i-a_i+d_i}\right]\right\}-\xi_2\right\} & \text{(iii)} \\
\left(\sum_{i=1}^{m}\frac{\beta_i}{\alpha_i}t_i\right)+K_\alpha\sqrt{\sum_{i=1}^{m}\frac{\beta_i}{\alpha_i^2}d_i^2} \leq T_0 & \text{(iv)} \\
\left(\sum_{i=1}^{m}\mu_i d_i\right)+K_\alpha\sqrt{\sum_{i=1}^{m}\sigma_i^2 d_i^2} \leq C_0 & \text{(v)} \\
w_1+w_2=1, w_1, w_2 \geq 0 & \text{(vi)} \\
0 \leq d_i \leq a_i, \forall d_i \text{ are int eger} & \text{(vii)} \\
n_i \geq a_i, i=1,2,\cdots,m & \text{(viii)}
\end{array}\right\} \tag{25}$$

In practice, ξ_i can be defined as the minimum individual values of the following problems:

$$\left.\begin{array}{ll}
\xi_1 = \text{Min}\left(-R(d_i)\right) = -\left\{\prod_{i=1}^{s}\left[1-(1-r_i)^{n_i-a_i+d_i}\right]\right\} & \text{(i)} \\
\text{Subject to } \left(\sum_{i=1}^{m}\frac{\beta_i}{\alpha_i}t_i\right)+K_\alpha\sqrt{\sum_{i=1}^{m}\frac{\beta_i}{\alpha_i^2}d_i^2} \leq T_0 & \text{(ii)} \\
\left(\sum_{i=1}^{m}\mu_i d_i\right)+K_\alpha\sqrt{\sum_{i=1}^{m}\sigma_i^2 d_i^2} \leq C_0 & \text{(iii)} \\
0 \leq d_i \leq a_i, \forall d_i \text{ are int eger} & \text{(iv)} \\
n_i \geq a_i, i=1,2,\cdots,m & \text{(v)}
\end{array}\right\} \tag{26}$$

and similarly

$$\left.\begin{array}{ll}
\xi_2 = \text{Min}\left(-R(d_i)\right) = -\left\{\prod_{i=s+1}^{m}\left[1-(1-r_i)^{n_i-a_i+d_i}\right]\right\} & \text{(i)}\\
\text{Subject to}\left(\sum_{i=1}^{m}\frac{\beta_i}{\alpha_i}t_i\right)+K_\alpha\sqrt{\sum_{i=1}^{m}\frac{\beta_i}{\alpha_i^2}d_i^2}\le T_0 & \text{(ii)}\\
\left(\sum_{i=1}^{m}\mu_i d_i\right)+K_\alpha\sqrt{\sum_{i=1}^{m}\sigma_i^2 d_i^2}\le C_0 & \text{(iii)}\\
0\le d_i\le a_i, \forall d_i \text{ are int eger} & \text{(iv)}\\
n_i\ge a_i, i=1,2,\cdots,m & \text{(v)}
\end{array}\right\} \tag{27}$$

In similar way, the problem defined in Equation (17) is also a two criterion minimization problem: $\text{Min}\{T,C\}$ subject to the constraints. Now the efficient solution is obtained by using the weighted Tchebycheff technique

$$\left.\begin{array}{ll}
\text{Minimize } \delta & \text{(i)}\\
\text{Subject to } \delta\ge w_1\left\{\left\{k_1\left(\sum_{i=1}^{m}\frac{\beta_i}{\alpha_i}t_i\right)+k_2\sqrt{\sum_{i=1}^{m}\frac{\beta_i}{\alpha_i^2}d_i^2}\right\}-\xi_1\right\} & \text{(ii)}\\
\delta\ge w_2\left\{k_1\left(\sum_{i=1}^{m}\mu_i d_i\right)+k_2\sqrt{\sum_{i=1}^{m}\sigma_i^2 d_i^2}-\xi_2\right\} & \text{(iii)}\\
\left\{\prod_{i=1}^{s}\left[1-(1-r_i)^{n_i-a_i+d_i}\right]\right\}\times\left\{\prod_{i=S+1}^{m}\left[1-(1-r_i)^{n_i-a_i+d_i}\right]\right\}\ge R^*(d_i) & \text{(iv)}\\
w_1+w_2=1, w_1, w_2\ge 0 & \text{(v)}\\
0\le d_i\le a_i, \forall d_i \text{ are int eger} & \text{(vi)}\\
n_i\ge a_i, i=1,2,\cdots,m & \text{(vii)}
\end{array}\right\} \tag{28}$$

The values of ξ_i can be defined as the minimum individual values of the following problems:

$$\left.\begin{aligned}
&\xi_1 = \operatorname{Min} T = k_1\left(\sum_{i=1}^{m}\frac{\beta_i}{\alpha_i}d_i\right)+k_2\sqrt{\sum_{i=1}^{m}\frac{\beta_i}{\alpha_i^2}d_i^2} && \text{(i)}\\
&\text{Subject to}\left\{\prod_{i=1}^{s}\left[1-(1-r_i)^{n_i-a_i+d_i}\right]\right\}\times\left\{\prod_{i=S+1}^{m}\left[1-(1-r_i)^{n_i-a_i+d_i}\right]\right\}\geq R^*(d_i) && \text{(ii)}\\
&0\leq d_i\leq a_i, \forall d_i \text{ are int eger} && \text{(iii)}\\
&n_i\geq a_i, i=1,2,\cdots,m && \text{(iv)}
\end{aligned}\right\} \quad (29)$$

and similarly

$$\left.\begin{aligned}
&\xi_2 = \operatorname{Min} C = k_1\left(\sum_{i=1}^{m}\mu_i d_i\right)+k_2\sqrt{\sum_{i=1}^{m}\sigma_i^2 d_i^2} && \text{(i)}\\
&\text{Subject to}\left\{\prod_{i=1}^{s}\left[1-(1-r_i)^{n_i-a_i+d_i}\right]\right\}\times\left\{\prod_{i=S+1}^{m}\left[1-(1-r_i)^{n_i-a_i+d_i}\right]\right\}\geq R^*(d_i) && \text{(ii)}\\
&0\leq d_i\leq a_i, \forall d_i \text{ are int eger} && \text{(iii)}\\
&n_i\geq a_i, i=1,2,\cdots,m && \text{(iv)}
\end{aligned}\right\} \quad (30)$$

Now the problem defined in Model 1; Equation (18) is also a two criterion minimization problem:

$\operatorname{Min}\{T - R_s(d_i)\}$ subject to the constraints. Now the efficient solution is obtained by using the weighted Tchebycheff technique

$$\left.\begin{aligned}
&\text{Minimize } \delta && \text{(i)}\\
&\text{Subject to } \delta\geq w_1\left\{\left\{k_1\left(\sum_{i=1}^{m}\frac{\beta_i}{\alpha_i}t_i\right)+k_2\sqrt{\sum_{i=1}^{m}\frac{\beta_i}{\alpha_i^2}d_i^2}\right\}-\xi_1\right\} && \text{(ii)}\\
&\delta\geq w_2\left\{\left\{\prod_{i=S+1}^{m}\left[1-(1-r_i)^{n_i-a_i+d_i}\right]\right\}-\xi_2\right\} && \text{(iii)}\\
&\left\{\prod_{i=1}^{s}\left[1-(1-r_i)^{n_i-a_i+d_i}\right]\right\}\times\left\{\prod_{i=S+1}^{m}\left[1-(1-r_i)^{n_i-a_i+d_i}\right]\right\}\geq R^*(d_i) && \text{(iv)}\\
&\left(\sum_{i=1}^{m}\mu_i d_i\right)+K_\alpha\sqrt{\sum_{i=1}^{m}\sigma_i^2 d_i^2}\leq C_0 && \text{(v)}\\
&w_1+w_2=1, w_1, w_2\geq 0 && \text{(vi)}\\
&0\leq d_i\leq a_i, \forall d_i \text{ are int eger} && \text{(vii)}\\
&n_i\geq a_i, i=1,2,\cdots,m && \text{(viii)}
\end{aligned}\right\} \quad (31)$$

The values of ξ_1 and ξ_2 is the minimum individual values obtained as

$$\xi_1 = \text{Minimize } T$$

$$= k_1\left(\sum_{i=1}^{m}\frac{\beta_i}{\alpha_i}d_i\right)+k_2\sqrt{\sum_{i=1}^{m}\frac{\beta_i}{\alpha_i^2}d_i^2}$$

subject to the (iii) to (vi) of Equation (18)

$$\xi_2 = \text{Min}R_2(d_i) = -\left\{\prod_{i=S+1}^{m}\left[1-(1-r_i)^{n_i-a_i+d_i}\right]\right\}$$

subject to the (iii) to (vi) of Equation (18).

In Model 2, Equation (19) is also a two criterion minimization problem: $\text{Min}\{C - R_1(d_i)\}$ subject to the constraints. Now the efficient solution is obtained by using the weighted Tchebycheff technique

$$\left.\begin{array}{ll}
\text{Minimize } \delta & \text{(i)}\\
\text{Subject to } \delta \geq w_1\left\{k_1\left(\sum_{i=1}^{m}\mu_i d_i\right)+k_2\sqrt{\sum_{i=1}^{m}\sigma_i^2 d_i^2}-\xi_1\right\} & \text{(ii)}\\
\delta \geq w_2\left\{\left\{\prod_{i=1}^{s}\left[1-(1-r_i)^{n_i-a_i+d_i}\right]\right\}-\xi_2\right\} & \text{(iii)}\\
\left\{\prod_{i=1}^{s}\left[1-(1-r_i)^{n_i-a_i+d_i}\right]\right\}\times\left\{\prod_{i=S+1}^{m}\left[1-(1-r_i)^{n_i-a_i+d_i}\right]\right\}\geq R^*(d_i) & \text{(iv)}\\
\left(\sum_{i=1}^{m}\frac{\beta_i}{\alpha_i}t_i\right)+K_\alpha\sqrt{\sum_{i=1}^{m}\frac{\beta_i}{\alpha_i^2}d_i^2}\leq T_0 & \text{(v)}\\
w_1+w_2=1, w_1, w_2\geq 0 & \text{(vi)}\\
0\leq d_i\leq a_i, \forall d_i \text{ are int eger} & \text{(vii)}\\
n_i\geq a_i, i=1,2,\cdots,m & \text{(viii)}
\end{array}\right\} \quad (32)$$

where the values of ξ_1 and ξ_2 is the minimum individual values obtained as

$$\xi_1 = \text{Minimize } C = k_1\left(\sum_{i=1}^{m}\mu_i d_i\right)+k_2\sqrt{\sum_{i=1}^{m}\sigma_i^2 d_i^2}$$

subject to the (iii) to (vi) of Equation (19)

$$\xi_2 = \text{Min } R_1(d_i) = -\left\{\prod_{i=1}^{s}\left[1-(1-r_i)^{n_i-a_i+d_i}\right]\right\}$$

subject to the (iii) to (vi) of Equation (19).

NUMERICAL ILLUSTRATIONS

Consider a system having the Group X consisting of 3 subsystems and also the Group Y consisting of 4 subsystems. The available time between two missions for repairing and replacing is 150 time units. The available cost of maintenance for repairing and replacing for the next mission is 860 units. For simplicity we have considered in the above numerical illustration: the reliability of each component in a subsystem is same, cost spent and time taken on replacing and repairing each component within a subsystem are same. The remaining parameters for the various subsystems are given in **Table 1**.

SOLUTION OF CHANCE CONSTRAINED PROGRAMMING BY USING WEIGHTED TCHEBYCHEFF TECHNIQUE

Before applying the Weighted Tchebycheff Technique firstly we find the individual optimum values $\xi_i = (\xi_1, \xi_2)$.

For the values given in **Table 1**, the SNLPP (26) for the first optimum value is

$$\left.\begin{array}{l} \xi_1 = \text{Min}\left(-R_1(d_i)\right) = -\left\{\left[1-(1-0.8)^{(3+d_1)}\right]\times\left[1-(1-0.75)^{(2+d_2)}\right]\times\left[1-(1-0.8)^{(4+d_3)}\right]\right\}, \\ \text{Subject to}\left[2d_1+3d_2+d_3+20d_4+28d_5+22d_6+22d_7\right] \\ +2.99\sqrt{0.33d_1^2+0.60d_2^2+0.10d_3^2+2.86d_4^2+3.11d_5^2+1.83d_6^2+2.2d_7^2} \le 150 \\ \left[120d_1+110d_2+120d_3+40d_4+30d_5+45d_6+65d_7\right] \\ +2.99\sqrt{13d_1^2+10d_2^2+15d_3^2+4d_4^2+3d_5^2+5d_6^2+6d_7^2} \le 860 \\ 0 \le d_i \le a_i, \forall d_i \text{ are int eger } n_i \ge a_i, i = 1,2,\cdots,7. \end{array}\right\} \quad (33)$$

The optimal solution of (33) provided by LINGO is

$$d_1^* = 2, d_2^* = 3, d_3^* = 2, d_4^* = 0, d_5^* = 0, d_6^* = 0, d_7^* = 0$$

with the value of objective function as

$$R_1\left(d_{ij}\right) = 0.9986398.$$

Similarly using (27) the SNLPP for the second optimum values

Table 1: The number of failed components and the respective cost and time etc. in the various subsystems.

	Group X (Replaced)			Group Y (Repair)			
Subsystem	**1**	**2**	**3**	**4**	**5**	**6**	**7**
n_i	6	5	10	7	9	12	10
r_i	0.8	0.75	0.8	0.8	0.75	0.8	0.7
β_i	12	15	10	140	252	264	220
α_i	6	5	10	7	9	12	10
$E(t_i)=\frac{\beta_i}{\alpha_i}$	2	3	1	20	28	22	22
$V(t_i)=\frac{\beta_i}{\alpha_i^2}$	0.33	0.60	0.10	2.86	3.11	1.83	2.2
$\overline{c}_i$	120	110	120	40	30	45	65
$\sigma_{c_i}^2$	13	10	15	4	3	5	6
a_i	3	3	6	5	7	9	7

$$\left.\begin{array}{l}
\xi_2=\text{Min}\left(-R_2\left(d_i\right)\right)=-\left\{\left[1-(1-0.8)^{(2+d_4)}\right]\times\left[1-(1-0.75)^{(2+d_5)}\right]\times\left[1-(1-0.8)^{(3+d_6)}\right]\times\left[1-(1-0.70)^{(3+d_7)}\right]\right\}, \\
\text{Subject to}\left[2d_1+3d_2+d_3+20d_4+28d_5+22d_6+22d_7\right] \\
+2.99\sqrt{0.33d_1^2+0.60d_2^2+0.10d_3^2+2.86d_4^2+3.11d_5^2+1.83d_6^2+2.2d_7^2}\le 150 \\
\left[120d_1+110d_2+120d_3+40d_4+30d_5+45d_6+65d_7\right]+2.99\sqrt{13d_1^2+10d_2^2+15d_3^2+4d_4^2+3d_5^2+5d_6^2+6d_7^2}\le 860 \\
0\le d_i\le a_i, \forall d_i \text{ are int eger } n_i\ge a_i, i=1,2,\cdots,7.
\end{array}\right\} \quad (34)$$

The optimal solution of (34) provided by LINGO is

$d_1^*=0, d_2^*=0, d_3^*=0, d_4^*=2, d_5^*=1, d_6^*=1, d_7^*=2$

with the value of objective function as

$R_2\left(d_{ij}\right)=0.9788431$

From the Equations (33) and (34) the optimum values $\xi=(-0.9986398,-0.9788431)$. For simplicity we assumed that the reliability of both the Groups X and Y subsystems are equally important, that is $w_1=w_2=0.5$.

For the values given in **Table 1**, the SNLPP (25) efficient solution is obtained by using the weighted Tchebycheff Technique

$$\left.\begin{array}{l}\text{Minimize } \delta \\ \text{Subject to } \delta \ge w_1\left\{-\left\{\left[1-(1-0.8)^{(3+d_1)}\right]\times\left[1-(1-0.75)^{(2+d_2)}\right]\times\left[1-(1-0.8)^{(4+d_3)}\right]\right\}-(-0.9983842)\right\} \\ \delta \ge w_2\left\{-\left\{\left[1-(1-0.8)^{(2+d_4)}\right]\times\left[1-(1-0.75)^{(2+d_5)}\right]\times\left[1-(1-0.8)^{(3+d_6)}\right]\times\left[1-(1-0.7)^{(3+d_7)}\right]\right\}-(-0.9994604)\right\} \\ \left[2d_1+3d_2+d_3+20d_4+28d_5+22d_6+22d_7\right] \\ +2.99\sqrt{0.15d_1^2+0.18d_2^2+0.10d_3^2+0.35d_4^2+0.40d_5^2+0.50d_6^2+0.60d_7^2} \le 150 \\ \left[120d_1+110d_2+120d_3+40d_4+30d_5+45d_6+65d_7\right]+2.99\sqrt{10d_1^2+8d_2^2+15d_3^2+8d_4^2+5d_5^2+7d_6^2+9d_7^2} \le 860 \\ 0 \le d_i \le a_i, \forall d_i \text{ are int eger}, w_1+w_2=1, w_1, w_2 \ge 0, n_i \ge a_i, i=1,2,\cdots,7.\end{array}\right\} \quad (35)$$

The optimum allocation under the Weighted Tchebycheff Technique

$$d^*_{\text{Tcheb}} = \left(\underline{d}_1^*, d_2^*, \underline{d}_3^*, \underline{d}_4^*, \underline{d}_5^*, \underline{d}_6^*, \underline{d}_7^*\right)$$

is obtained as

$$d_1^* = 2, d_2^* = 3, d_3^* = 0, d_4^* = 2, d_5^* = 2, d_6^* = 0, d_7^* = 1.$$

The corresponding value of objective function is 0.00076.

Solution of Modified E-Model by Using Weighted Tchebycheff Technique

The individual optimum values $\xi_i = (\xi_1, \xi_2)$. For the values given in **Table 1** and for simplicity take $k_1 = k_2 = 0.5$ then the SNLPP (29) for the first optimum value is

$$\left.\begin{array}{l}\xi_1 = \text{Min}T = k_1\left[2d_1+3d_2+d_3+20d_4+28d_5+22d_6+22d_7\right] \\ \quad +k_2\sqrt{0.33d_1^2+0.60d_2^2+0.10d_3^2+2.86d_4^2+3.11d_5^2+1.83d_6^2+2.2d_7^2} \\ \text{Subject to } \left\{\left[1-(1-0.8)^{(3+d_1)}\right]\times\left[1-(1-0.75)^{(2+d_2)}\right]\times\left[1-(1-0.8)^{(4+d_3)}\right]\right\} \\ \qquad \times\left\{\left[1-(1-0.8)^{(2+d_1)}\right]\times\left[1-(1-0.75)^{(2+d_2)}\right]\times\left[1-(1-0.8)^{(3+d_3)}\right]\times\left[1-(1-0.70)^{(3+d_2)}\right]\right\} \ge 0.99 \\ 0 \le d_i \le a_i, \forall d_i \text{ are int eger}, n_i \ge a_i, i=1,2,\cdots,7.\end{array}\right\} \quad (36)$$

And the SNLPP (30) for the second optimum value

$$\left.\begin{array}{l}\xi_2 = \text{MinC} = k_1\left[120d_1+110d_2+120d_3+40d_4+30d_5+45d_6+65d_7\right] \\ \quad +k_2\sqrt{10d_1^2+8d_2^2+15d_3^2+8d_4^2+5d_5^2+7d_6^2+9d_7^2} \\ \text{Subject to } \left\{\left[1-(1-0.8)^{(3+d_1)}\right]\times\left[1-(1-0.75)^{(2+d_2)}\right]\times\left[1-(1-0.8)^{(4+d_3)}\right]\right\} \\ \qquad \times\left\{\left[1-(1-0.8)^{(2+d_1)}\right]\times\left[1-(1-0.75)^{(2+d_2)}\right]\times\left[1-(1-0.8)^{(3+d_3)}\right]\times\left[1-(1-0.70)^{(3+d_2)}\right]\right\} \ge 0.99 \\ 0 \le d_i \le a_i, \forall d_i \text{ areint eger}, n_i \ge a_i, i=1,2,\cdots,7.\end{array}\right\} \quad (37)$$

From the Equations (36) and (37) the optimum values $\xi = (101.73, 418.40)$.

For simplicity we assumed that the maintenance time taken and cost spent for both the Groups X and Y subsystems are equally important, that is $w_1 = w_2 = 0.5$. For the values given in **Table 1**, the SNLPP (28) efficient solution is obtained by using the weighted Tchebycheff Technique

$$\left.\begin{array}{l}
\text{Minimize } \delta \\
\text{Subject to } \delta \ge w_1\left\{\left\{0.5\left[2d_1+3d_2+d_3+20d_4+28d_5+22d_6+22d_7\right]\right.\right. \\
\qquad\qquad \left.\left.+0.5\sqrt{0.33d_1^2+0.60d_2^2+0.10d_3^2+2.86d_4^2+3.11d_5^2+1.83d_6^2+2.2d_7^2}\right\}-101.73\right\} \\
\qquad \delta \ge w_2\left\{-\left\{0.5\left[120d_1+110d_2+120d_3+40d_4+30d_5+45d_6+65d_7\right]\right.\right. \\
\qquad\qquad \left.\left.+0.5\sqrt{10d_1^2+8d_2^2+15d_3^2+8d_4^2+5d_5^2+7d_6^2+9d_7^2}\right\}-418.40\right\} \\
\text{Subject to } \left\{\left[1-(1-0.8)^{(3+d_1)}\right]\times\left[1-(1-0.75)^{(2+d_2)}\right]\times\left[1-(1-0.8)^{(4+d_3)}\right]\right\} \\
\qquad \times\left\{\left[1-(1-0.8)^{(2+d_1)}\right]\times\left[1-(1-0.75)^{(2+d_2)}\right]\times\left[1-(1-0.8)^{(3+d_3)}\right]\times\left[1-(1-0.70)^{(3+d_2)}\right]\right\}\ge 0.99 \\
0 \le d_i \le a_i, \forall d_i \text{ are int eger}, n_i \ge a_i, i = 1,2,\cdots,7.
\end{array}\right\} \quad (38)$$

The optimum allocation under the Weighted Tchebycheff Technique

$$d^*_{\text{Tcheb}} = \left(\underline{d}_1^*, \underline{d}_2^*, \underline{d}_3^*, \underline{d}_4^*, \underline{d}_5^*, \underline{d}_6^*, \underline{d}_7^*\right)$$

is obtained as

$$d_1^* = 1, d_2^* = 3, d_3^* = 0, d_4^* = 3, d_5^* = 3, d_6^* = 1, d_7^* = 2.$$

The corresponding value of objective function is 6.48.

The individual optimum values of Model 1 are $\xi = (64.47, -0.9994)$. Therefore, the optimum allocation under the Weighted Tchebycheff Technique of SNLPP (31) is obtained as

$$d_1^* = 1, d_2^* = 3, d_3^* = 1, d_4^* = 2, d_5^* = 1, d_6^* = 1, d_7^* = 1.$$

The corresponding value of objective function is 0.01306.

In the same manner, we obtained the individual optimum values for Model 2 are $\xi = (277.57, -0.9989)$. Therefore, the optimum allocation under the Weighted Tchebycheff Technique of SNLPP (32) is obtained as

$$d_1^* = 1, d_2^* = 2, d_3^* = 0, d_4^* = 2, d_5^* = 2, d_6^* = 0, d_7^* = 1.$$

The corresponding value of objective function is 0.00299.

The optimum allocations obtained corresponding to the various Bi-criteria models are summarized as given below in **Table 2**.

CONCLUSIONS

The multi-objective problem of allocation of repairable and replaceable components becomes complicated because an allocation that is optimal for one objective is usually far from optimal for other objectives. In such situations, we need a compromise criterion that gives an allocation which is optimum for all objectives in some sense. This paper is an attempt to utilize weighted Tchebycheff approach to the solution of optimum compromise allocation of repairable and replaceable components in a system.

The allocation problem of repairable and replaceable components for a parallel-series system considered as a Bi-objective stochastic optimization problem and discussed the four different situations. In the first situation, the reliabilities of Groups X and Y are considered as two different objectives. While in the next three situations

Table 2: Optimum allocation of replaceable and repairable components under various Bi-criteria models.

S. No.	Bi-criteria models	d_1	d_2	d_3	d_4	d_5	d_6	d_7
1	$-\text{Min}\{R_1, R_2\}$	2	3	0	2	1	0	2
2	$\text{Min}\{T, C\}$	2	1	0	2	1	0	3
3	$\text{Min}\{C, -R_1\}$	1	2	0	2	2	0	1
4	$\text{Min}\{T, -R_2\}$	1	3	1	2	1	1	1

The maintenance cost and time, maintenance cost and Group X subsystem reliabilities, maintenance time and Group Y subsystem reliabilities respectively are considered as two different objectives. Selective maintenance policy is used to select the repairable and replaceable components.

An equivalent deterministic model of these Bi-objective stochastic optimization problems is established by using Chance Constrained programming method. The following four different stochastic problems are then solved by using the Bi-criteria optimization technique, weighted Tchebycheff. The weighted Tchebycheff technique provides compromise allocations of repairable and replaceable components which are optimum for both the objectives function (see **Table 2**). Since a compromise criteria differ from method to method, so comparison can not be made.

REFERENCES

1. I. Ali, M. F. Khan, Y. S. Raghav and A. Bari, "Allocating Repairable and Replaceable Components for a System Availability Using Selective Maintenance with Probabilistic Constraints," American Journal of Operations Research, Vol. 1, No. 3, 2011, pp. 147-154.doi:10.4236/ajor.2011.13016
2. W. F. Rice, C. R. Cassady and J. A. Nachlas, "Optimal Maintenance Plans under Limited Maintenance Time," Proceedings of the 7th Industrial Engineering Research Conference, Banff, 9-10 May 1998.
3. C. R. Cassady, E. A. Pohl and J. Song, "Managing Availability Improvement Efforts with Importance Measures and Optimization," IMA Journal of Management Mathematics, Vol. 15, No. 2, 2004, pp. 161-174. doi:10.1093/imaman/15.2.161
4. R. Rajagopalan and C. R. Cassady, "An Improved Selective Maintenance Approach," Journal of Quality in Maintenance Engineering, Vol. 12, No. 2, 2006, pp. 172-185.doi:10.1108/13552510610667183
5. K. Schneider, L. M. Maillart, C. R. Cassady and C. Rainwater, "Selective Maintenance Decision-Making over Extended Planning Horizons," IEEE Transactions on Reliability, Vol. 58, No. 3, 2009, pp. 462-469. doi:10.1109/TR.2009.2026689
6. I. M. Iyoob, C. R. Cassady and E. A. Pohl., "Establishing Maintenance Resource Levels Using Selective Maintenance," The Engineering Economist, Vol. 51, No. 2, 2006, pp. 99-114. doi:10.1080/00137910600695627
7. I. Ali, Y. S. Raghav and A. Bari, "Allocating Repairable and Replaceable Components for a System Availability Using Selective Maintenance: An Integer Solution," Safety and Reliability Society, Vol. 31, No. 2, 2011, pp. 9-18.
8. I. Ali, Y. S. Raghav and A. Bari, "Integer Goal Programming Approach for Finding a Compromise Allocation of Repairable Components," International Journal of Engineering Science and Technology, Vol. 3, No. 6, 2011, pp. 184-195.
9. I. Ali, Y. S. Raghav and A. Bari, "Compromise Allocation in Multivariate Stratified Surveys with Stochastic Quadratic Cost Function," Journal of Statistical Computation and Simulation, 2011. doi:10.1080/00949655.2011.643890
10. I. Ali, M. F. Khan, Y. S. Raghav and A. Bari, "Selective Maintenance in System Reliability with Random Costs of Repairing and

Replacing the Components," Communications in Statistics—Simulation and Computation, 2012. http://www.tandfonline.com/doi/abs/10.1080/03610918.2012

11. M. F. Khan and I. Ali, "Allocation Problem of Repairable and Replaceable Components for a System Reliability using Selective Maintenance: An Integer Solution," International Journal of Scientific and Engineering Research, Vol. 3, No. 5, 2012, pp. 1-4.
12. S. S. Rao, "Optimization Theory and Applications," Wiley Eastern Limited, New Delhi, 1979.
13. V. J. Bowman Jr., "On the Relationship of the Tchebycheff Norm and the Efficient Frontier of Multiple-Criteria Objectives," In: H. Thiriez and S. Zionts, Eds., "Multiple Criteria Decision Making," Lecture Notes in Economics and Mathematical Systems, Springer-Verlag, Berlin, 1976, pp. 76-85.
14. K. M. Miettinen, "Non Linear Multiobjective Optimization," Kluwer Academic Publishers, Boston, 1999.
15. A. P. Wierzbicki, "On the Completeness and Constructiveness of Parametric Characterizations to Vector Optimization Problem," Operations-Research-Spektrum, Vol. 8, No. 2, 1986, pp. 73-87. doi:10.1007/BF01719738
16. M. T. Khasawneh, S. R. Bowling, S. Kaewkuekool and B. R. Cho, "A Cost-Effective Strength-Stress Reliability Modeling and Optimization in Engineering Design," Proceeding of the Industrial Engineering Research Conference, Orlando, 19-21 May 2002.

Chapter 8

A PARTICLE SWARM OPTIMIZATION ALGORITHM FOR A 2-D IRREGULAR STRIP PACKING PROBLEM

Mohamed A. Shalaby, Mohamed Kashkoush

Mechanical Design and Production Department, Faculty of Engineering, Cairo University, Giza, Egypt

ABSTRACT

Two-Dimensional Irregular Strip Packing Problem is a classical cutting/packing problem. The problem is to assign, a set of 2-D irregular-shaped items to a rectangular sheet. The width of the sheet is fixed, while its length is extendable and has to be minimized. A sequence-based approach is developed and tested. The approach involves two phases; optimization phase and placement phase. The optimization phase searches for the packing sequence that would lead to an optimal (or best) solution when translated to an actual pattern through the placement phase. A Particle Swarm Optimization algorithm is applied in this optimization phase. Regarding the placement phase, a combined algorithm based on traditional placement methods is developed. Competitive results are obtained, where the best solutions are found to be better than, or at least equal to, the best known solutions for 10 out of 31 benchmark data sets. A Statistical Design of Experiments and a random generator of test problems are also used to characterize the performance of the entire algorithm.

INTRODUCTION

The problem of packing or cutting out shapes with minimal waste is a classical problem that exists in many industrial fields. Common applications are found in sheet metal, leather, furniture, shipbuilding, and textile Industries. Cutting stock, trim loss, bin packing, strip packing, pallet loading, nesting, and knapsack problems, are all related names for cutting and packing problems. The first general classification for cutting and packing problems is introduced in Dychkoff [1]. Dychkoff developed a special classification for cutting and packing problems in which he called it a typology. Recently, an improved typology was developed by Wäscher [2], which is partially based on Dyckhoff's original one, but it adopts new categorization criteria.

The scope of this research is the two-dimensional Irregular Strip Packing Problem which is classified as an Open Dimension Problem (ODP), according to Wäscher's improved typology. The problem is to assign (cut or pack), a set of two-dimensional irregular-shaped items to a rectangular object (sheet for cutting or container for packing). The width of the object is fixed, while its length is extendable and has to be minimized. A typical application for this problem is found in the textile Industry under the name "Marker-Making". Another popular name is "Nesting" which has been, and still, used by many authors for the Irregular Strip Packing Problems, and even for other irregular cutting and packing problems in general. The problem is schematically demonstrated in **Figure 1**, where the best arrangement for the given items into the strip to minimize the pattern length is sought.

The 2-D rectangular packing problem has been shown by Fowler et al. [3] to be NP-complete. As the irregular case is more complex, hence it could also be considered as NP-complete, and as a result they are practically tackled through heuristic approaches. Two main heuristic approaches are applied; sequence-based approach and direct approach. The first approach utilizes an optimization technique to find the packing sequence of items that would lead to the best possible pattern when employed under a certain placement method (Babu and Babu [4], Fischer and Dagli [5], Takahara et al. [6], and Burke et al. [7]). While the second approach heuristically builds an initial pattern that may not be a feasible one, and an optimization technique is directly applied to the items positions seeking the best possible feasible pattern (Bennell and Dowsland [8], Gomes and Oliveira [9], Immamichi et al. [10], and Umetani et al. [11]).

Other solution approaches are found in the literature that could not be categorized under any of the two main approaches mentioned above. These approaches involve; mathematical programming approach followed by Grinde and Daniels [12] and Fischetti and Luzzi [13], Constraint Logic Programming (CLP) approach followed by Carravilla et al. [14], and dynamic sequence-based approach followed by Oliveira et al. [15], and Bennell and Song [16].

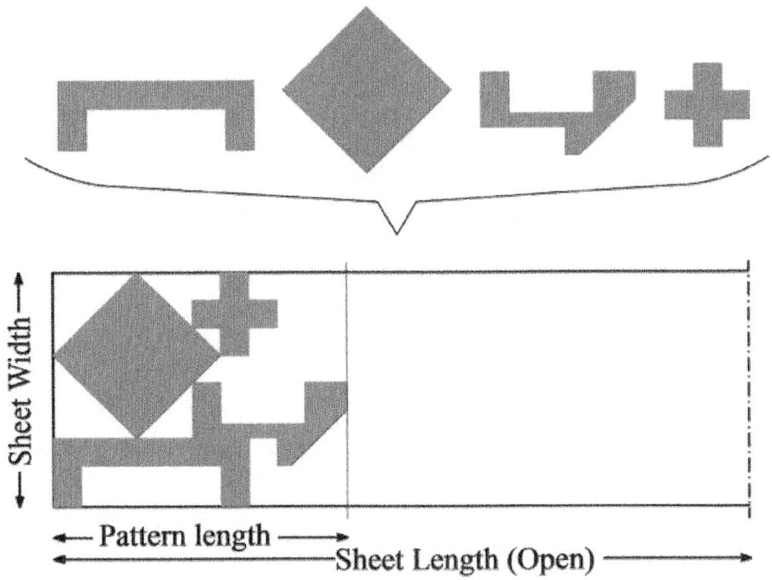

Figure 1: A schematic illustration for the 2-D irregular strip packing problem.

Developing a geometrical technique for handling the graphical aspects of irregular shapes is used to be the main obstacle among extending the research in this problem, compared to the regular cutting and packing one. Three geometry handling techniques are widely used in the literature. Pixel or Raster Method is an approximate method that is employed by Jain and Gea [17], Babu and Babu [4], and Fischer and Dagli [5]. Direct trigonometry techniques are used by Hifi and M'Hallah [18], Takahara et al. [6], and Burke et al. [7]. A more sophisticated technique is the no-fit polygon which is applied by Oliveira et al. [15], Bennell and Song [16], and Imamichi et al. [10].

Many authors used to incorporate meta-heuristic search techniques in their solution algorithms. Three main meta-heuristics are commonly used with irregular cutting and packing problems, they are; Genetic Algorithm (Jakobs [19], Hifi and M'Hallah [18], and Hu-yao and Yuan-jun [20]), Simulated Annealing (Oliveira and Ferreira [21], Han and Na [22], Heckmann and Lengauer [23], and Gomes and Oliveira [9]), and Tabu Search (Blazewicz et al. [24], Bennell and Dowsland [8], and Burke et al. [7]).

From the above review, much interest could be noticed in the Irregular Strip Packing Problem for the last few years. This could be attributed to two main reasons; firstly, until now, no optimal solution could be found for a real-sized problem, a very small improvement in a given arrangement in a cutting or packing operation could save a considerably large amount of money. Secondly, is the significant increase in the capabilities of modern computational engines, which made it available to apply techniques that would be impractical to apply 15 years ago. Several heuristic and meta-heuristic optimization techniques have been applied and investigated with the problem. Particle Swarm Optimization (PSO) algorithm, originally introduced by Kennedy and Eberhart [25], is a lately developed stochastic optimization technique that has proved good performance in various optimization fields, but has not been applied to a Strip Packing Problem or any cutting or packing problem. This research investigates the performance of PSO to find a competitive (or even better) performance to the existing approaches.

PROPOSED SOLUTION APPROACH

The solution approach proposed in this paper, which is given the name "SwarmNest", is a sequence-based approach that consists of two phases; optimization phase and placement phase. SwarmNest assumes only hole-free polygons that are allowed to rotate by increments of 90°.

For the optimization phase, a Particle Swarm Optimization (PSO) algorithm is employed to search for the optimal packing sequence. Up to our knowledge, this is the first time Particle Swarm is utilized in tackling an irregular packing problem. A proposed Local Search routine is also used to enhance the algorithm performance. As to the placement phase, a combined placement algorithm is developed and employs two different geometry handling techniques; pixel method and direct trigonometry. The algorithm involves two stages; the first is an initial placement stage that applies a pixelbased placement procedure, which alternatively could be a D-function-based one. The second is a refinement placement stage that incorporates a heuristic compaction routine. A general flowchart for the entire approach (SwarmNest) is shown in **Figure 2**.

The remaining of this paper is organized as follows. In the next Section, the combined placement algorithm along with the employed geometry handling techniques is explained. The proposed Particle Swarm Optimization algorithm is described in Section 4. Then, the experimental results of the entire approach (SwarmNest) are presented in Section 5. SwarmNest performance is then characterized in Section 5. Finally, the summary and main conclusions are presented in Section 6.

COMBINED PLACEMENT ALGORITHM

As mentioned in the previous section, the proposed solution approach is a sequence-based approach that employs a placement method for allocating items into an open ended sheet to translate a given sequence of items into a feasible pattern. A combined placement algorithm is developed that involves two stages; initial placement and refinement placement. The algorithm makes use of two common geometry handling techniques; the pixel (or raster) method, and direct trigonometry.

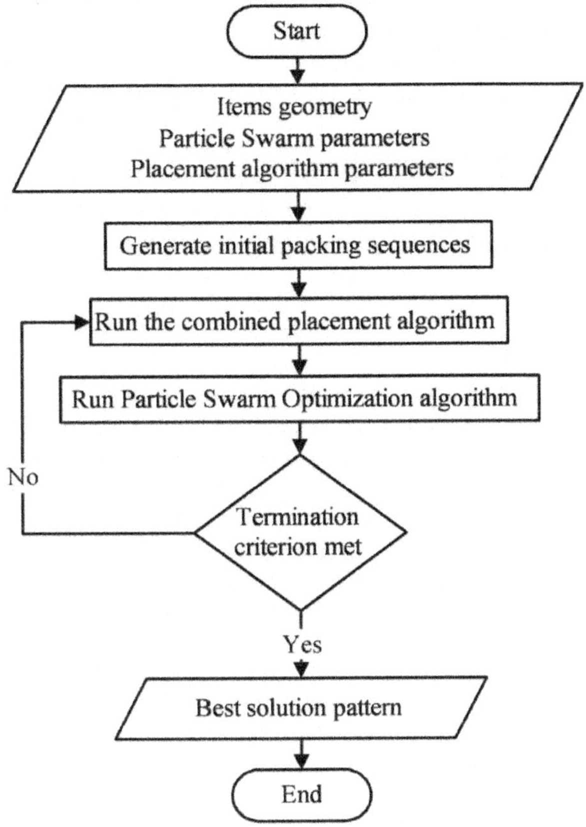

Figure 2: A general flowchart for the entire approach (SwarmNest).

Pixel method is employed under a pixel-based placement procedure during the initial placement stage; while direct trigonometry is applied through a heuristic compaction routine during the refinement placement stage.

Stage I: Initial Placement

Pixel Method

Pixel method is an approximate geometry handling technique. The geometry of an item through this method is first approximated by a set of square cells that form a rectilinear enclosure. Such an approximated form is then encoded by a 2-D array (item array). An item array involves two types of cells: interior and exterior cells. An interior cell is the one corresponding to one of the rectilinear enclosure cells, while other remaining cells within the item array are exterior cells. **Figure 3**illustrates the idea through the simplest encoding scheme; the binary encoding scheme used by Oliveira and Ferreira [21]. In the same manner, the placement sheet is encoded by a 2-D array (sheet array). This array also involves two types of cells; occupied and unoccupied cells. Occupied cells are those cells utilized via former placement operations, while unoccupied cells are the remaining free cells. Assigning a certain location inside the sheet to a given item is done by adding the corresponding item array, to the sheet array at that place. Hence, for a feasible placement (overlap-free placement), any of the interior cells of the item array should not be added to an occupied cell within the sheet array.

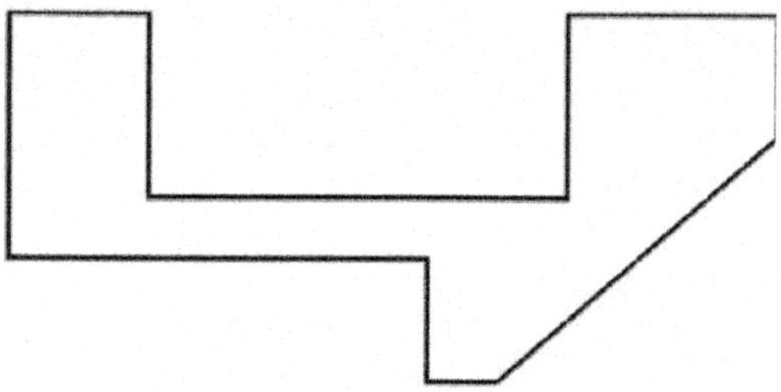

Original geometry

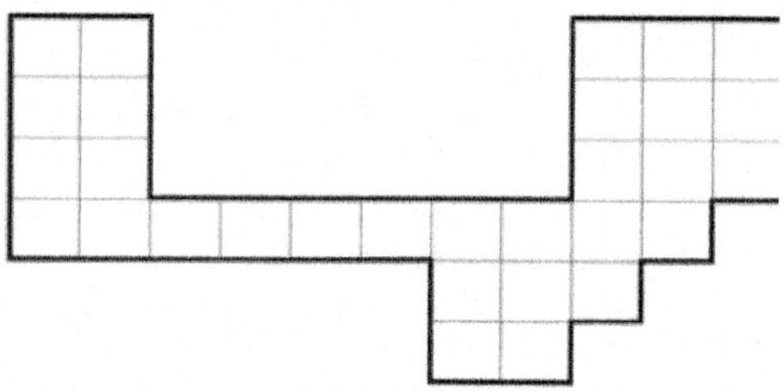

Rectilinear enclosure

1	1	0	0	0	0	0	0	1	1	1
1	1	0	0	0	0	0	0	1	1	1
1	1	0	0	0	0	0	0	1	1	1
1	1	1	1	1	1	1	1	1	1	0
0	0	0	0	0	0	1	1	1	0	0
0	0	0	0	0	0	1	1	0	0	0

Encoded item array

Figure 3. Geometry encoding through binary encoding scheme (Oliveira and Ferreira (1993)).

Initial Placement Algorithm

The strategy by which the algorithm works is a scanning strategy. Each time a new item is required to be placed or packed, the sheet is scanned through a predefined path for the first feasible available positioning for that item. The path by which the sheet array is scanned is exhibited in**Figure 4**.

Accordingly, the scanning is carried out along the sheet length direction from left to right, where each column is scanned from bottom to top; usually called bottom-left placement policy. The default starting position is the sheet origin, while in order to reduce the scanning time, an item starts its scanning path from the farthest position a smaller item, or even the same item, has reached. A smaller item is the one that has a smaller area, length, and width than the item being placed. In this way many potentially infeasible positions are neglected.

During the placement procedure, feasibility check, at each of the scanned position is carried out for each of the allowed orientations (0°, 90°, 180°, and/or 270°). The selected orientation is the one with the highest contacting condition; contact with other already placed items and with the sheet borders.

Stage II: Refinement Placement

Direct Trigonometry Method (D-Function)

Through the heuristic compaction routine, to be described in the next subsection, the items are represented by their actual geometries (hole-free polygons), where the direct trigonometry method described by Bennell and Oliveira [26] is employed as a geometry handling technique.

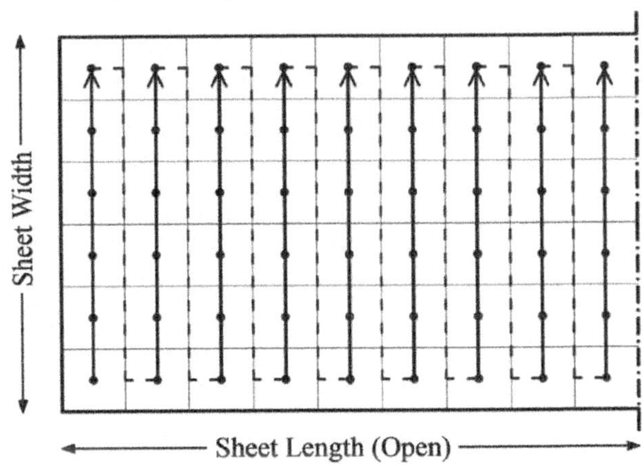

Figure 4: Proposed scanning path.

Accordingly, an overlap between any two items is detected through three consecutive levels of trigonometric tests. The core calculations are carried out through the third level. This level detects intersection between any two lines that have an overlap potential (according to the former levels). A mathematical function known as Dfunction is employed through this level. It was first proposed by Konopasek [27], and the D-function is given by the following expression:

$$D_{ABP} = (X_A - X_B)(Y_A - Y_P) - (Y_A - Y_B)(X_A - X_P) \quad (1)$$

Upon calculating the D-function value for a point P and an oriented edge AB, the relative position of P with respect to AB, as well the extension of AB in both directions (an infinite line), could be defined according to the sign of D_{ABP}. If $D_{ABP} > 0$, then P is on the left side of AB and its extensions. If $D_{ABP} < 0$, then P is on the right side of AB and its extension. While If $D_{ABP} = 0$, then P lies on AB or its extension. **Figure 5** presents the case where P is on the left side of $AB(D_{ABP} > 0)$.

Refinement Placement Algorithm

The second stage of the combined placement algorithm applies a heuristic compaction routine on the pattern obtained from the first stage. A similar routine to the proposed one is found in Hopper [28] as "iteration routine". Another one is introduced by Jakobs [19] as "shrinking algorithm". Both authors primarily employ a rectangular enclosure representation for the items.

Through the proposed routine, the compaction process is carried out on a number of successive identical passes. A compaction pass sequentially considers the items one by one. Each of the considered items is tried for compaction in eight directions; leftward firstly then upward, downward, up-left, and down-left. Compacting an item in a given direction is performed by incremental translational movements carried out in that direction.

After each movement, the item's new position is checked for being a feasible one (no overlaps). If the new position is feasible, then it is being approved and the compaction process is continued in that direction, otherwise the item retrieves its last feasible position before continuing to the next compaction direction. The routine is performed via more than one pass so as to utilize the probable in between gaps, resulted after each pass.

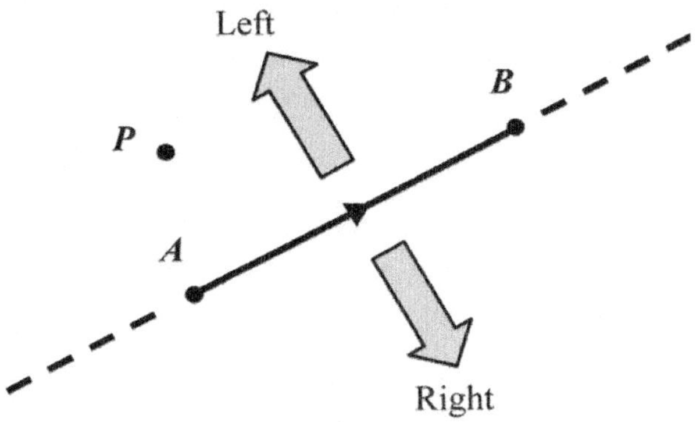

Figure 5: The case at which the value of D_{ABP} is greater than zero.

PARTICLE SWARM OPTIMIZATION

Particle swarm optimization (PSO) is a population-based stochastic optimization technique devised to model the social behavior of animals, birds, or insects. It was originally developed by Kennedy and Eberhart [25]. PSO has been successfully applied to several classical optimization problems such as; scheduling, traveling salesman, network deployment, task assignment, and neural network training problems (Blum and Merkle [29]). PSO is used here during the optimization phase of SwarmNest to search for a better placement sequence, where the feasibility and value yielded by any sequence is assessed by applying the combined placement algorithm.

Basics of Continuous Particle Swarm Optimization

In PSO, individual particles represent possible solutions, which spread through the problem search space looking for an optimal, or even a good enough solution, and each particle transmits its current state to other particles in the population. In an evolutionary manner, the position of each particle is updated according to its current position, the particle's best position so far (personal best position), and the global best position ever found by the whole population. As the search continues, the whole swarm is supposed to be directed more and more towards the position corresponding to the global best solution for the whole population. The textbook of Blum and Merkle [29] is a good introduction for PSO.

The algorithm is initialized by randomly generating a population of particles; where each particle is associated with a position vector and velocity vector (i.e. rate of change of a particle's position per iteration). Elements of the position vector represent the set of continuous decision variables. Then the fitness (objective value) of each particle is evaluated as a function of its current position so as to find the particle of the highest fitness. The first iteration is then started by updating both the position and the velocity vectors of the entire population according to Equations (2) and (3).

$$\overline{x}_{n+1}^{i} = \overline{x}_{n}^{i} + \overline{v}_{n+1}^{i} \quad (2)$$

$$\overline{v}_{n+1}^{i} = \overline{v}_{n}^{i} + c_1 r_1 \left(\overline{p}_{n}^{i} - \overline{x}_{n}^{i}\right) + c_2 r_2 \left(\overline{p}_{n}^{g} - \overline{x}_{n}^{i}\right) \quad (3)$$

where:

i: Refers to the ith particle of the swarm n, n + 1: Denote iterations, n being the current iteration.

r_1, r_2: Pseudo-random numbers in the interval [0, 1], generated for each variable within each particle at each iteration

$\overline{x}$: Position vector whose elements are decision variables $x^1, x^2, \ldots, x^m$, where m is the number of decision variables.

$\overline{v}$: Velocity vector.

$\overline{p}^i$: Personal best position vector found by particle $i\left(\text{initially} = \overline{x}_{o}^{i}\right)$.

$\overline{p}^g$: Global best position vector through the whole swarm.

c_1, c_2: Adjustable optimization parameters used to control the swarm's behavior, both values are kept fixed for the entire algorithm.

New fitness values are evaluated at the updated particle's positions, and the best position found so far by each particle, as well as the global best position, are updated. The procedure is then continued until a predefined termination criterion is met. An additional adjustment can be applied to Equation (3), whereby velocity is updated, by inserting what is known as inertia reduction weight(ω), hence the velocity equation is to be written as shown in Equation (4):

$$\overline{v}_{n+1}^{i} = \omega \overline{v}_{n}^{i} + c_1 r_1 \left(\overline{p}_{n}^{i} - \overline{x}_{n}^{i}\right) + c_2 r_2 \left(\overline{p}_{n}^{g} - \overline{x}_{n}^{i}\right) \quad (4)$$

The schematic vector diagram in **Figure 6** illustrates the updating process of a position vector according to Equations (2) and (4).

Proposed PSO Algorithm

Based on the approach followed by Tasgetiren et al. [30] in tackling the flow shop scheduling problem, our proposed PSO algorithm for finding the optimal packing sequence is a continuous one that employs a proposed criterion in extracting the actual solutions. Through the proposed algorithm, the position of a particle, denotes a packing sequence, is represented by a vector with the same length of the total number of items. The value of each element in the position vector always imply an item number, thus it initially takes a positive integer number from 1 to total number of items.

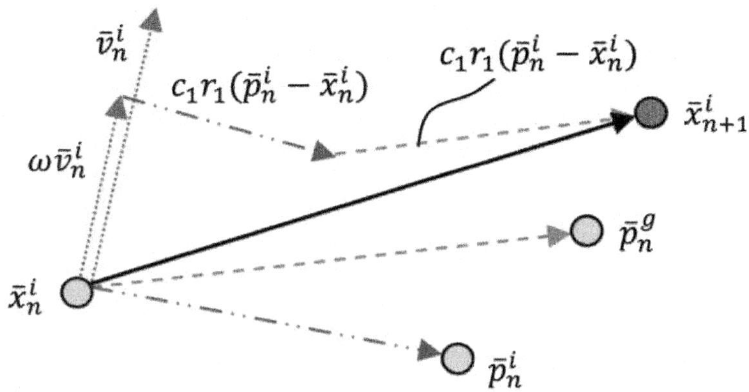

Figure 6: A schematic vector diagram for updating position vector $\bar{x}_n^i$.

Thus, the initial position vectors represent a set of feasible packing sequences. Hence, the velocity value of a given particle is equivalent to a difference between some two items numbers. Velocity values are initially set to zero. The initial population is made up of two sets of sequences. The first set involves the result of applying 10 heuristic ordering rules (e.g. by area, by length, and by width), while the other set is completely randomly generated sequences.

The velocity vectors are updated using Equation (4) that incorporates inertia reduction parameter (ω), but additionally a decrement factor (β) is multiplied by (ω) during each iteration as in Tasgetiren et al. [30]. This would results in a gradual increase in the inertia reduction with consecutive iterations; hence, a premature convergence is to be avoided in early iterations, while an intensive search behavior could be realized in later ones. A minimum value for (ω) is preserved to prevent the complete elimination of the inertia term in the velocity equation.

Moreover, the applied velocity limits are always set to be half the total number of items in both directions (positive and negative); this would force the velocity and position values to be always in magnitude of the initial position values that signify items numbers. A proposed criterion is applied for obtaining the actual discrete sequence from a given continuous position vector. Each element i in the position vector contains a real value (− or +). The order of the ith element, when all elements are ascendingly ordered, becomes the ith element in the actual sequence (see **Figure** 7).

In addition, an item number is originally assigned based on its area, relative to other larger or smaller items, that is, the number 1 is given to the item with the smallest area and so on. Thus, the value taken by each position element implies a physical meaning. A pseudocode for the proposed PSO is outlined below.

Input $c_1, c_2, \omega, \omega_{min}, N_p, v_{min}, v_{max}, T_{max}$ //parameters Set $g = 1$ //initially set global best particle to be the first one

Variable:	1	2	3	4	5	6
Pos. vector:	1.80	−0.99	3.01	−0.72	−1.20	2.15
Act. seq.:	4	2	6	3	1	5

Figure 7: Proposed solution extraction criterion.

For i = 1 to N_p //N_p is the population size Initialize $\bar{x}^i, \bar{v}^i$

$Z(i) = \textbf{Evaluate}(\bar{x}^i)$ //call the placement algorithm Set $\bar{p}^i = \bar{x}^i, \textbf{Best}(i) = Z(i)$

If $Z(i) \le Z(g)$

g = i //global best particle

End if

Next i While $t < T_{max}$ //T_{max} is the maximum no. of iterations t = t +1 For i = 1 to N_p

For j = 1 to Total no. of items

$r_1 = \textbf{rand}(0,1), r_2 = \textbf{rand}(0,1)$

Update v^{ij} //Equation 4 (velocity equation)

If $v^{ij} < v_{\min}$,

$v^{ij} = v_{\min}$

Else if $v^{ij} > v_{\max}$,

$v^{ij} = v_{\max}$

End if

Update x^{ij} //Equation 2 Next j

$\bar{s}^{i} = \textbf{Extract}\left(\bar{x}^{i}\right)$ //proposed extraction criterion

$Z(i) = \textbf{Evaluate}\left(\bar{s}^{i}\right)$

If $Z(i) \leq \text{Best}(i)$

Set $\bar{p}^{i} = \bar{x}^{i}, \text{Best}(i) = Z(i)$ //update personal best

End if

If $\text{Best}(i) \leq \text{Best}(g)$,

$g = i$ //update global best

End if

Next i

End While

Local Search Routine

The performance of PSO algorithms are highly improved when local or neighborhood search is integrated to it. The proposed Local Search routine applies three Local Search operations; interchange between two randomly selected items (**Figure 8**(a)), interchange between a randomly selected item and its consecutive one (**Figure 8**(b)), and Insertion of a randomly selected item into a randomly selected place (**Figure 8**(c)).

An interchange operation is completed if only the items to be interchanged represent different items. Those operations are applied during each PSO iteration.

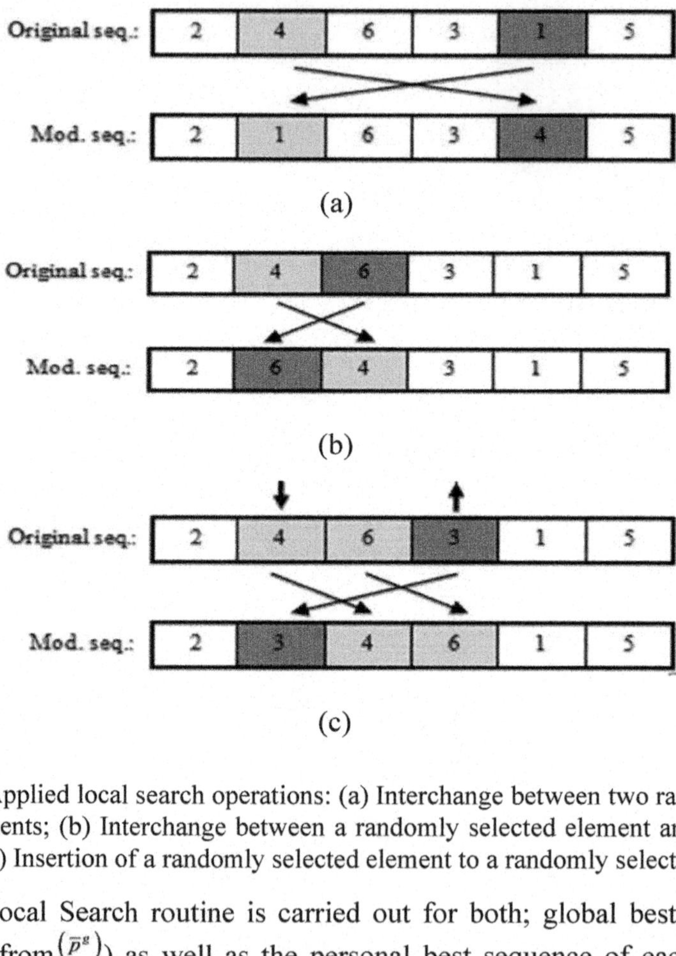

Figure 8: Applied local search operations: (a) Interchange between two randomly selected elements; (b) Interchange between a randomly selected element and it's consecutive; (c) Insertion of a randomly selected element to a randomly selected place.

The Local Search routine is carried out for both; global best sequence (obtained from $(\bar{p}^g)$) as well as the personal best sequence of each particle (obtained from $(\bar{p}^i)$). The entire routine is repeated up to a predefined number of local search iterations.

EXPERIMENTAL RESULTS

In this section, SwarmNest is benchmarked and characterized. Firstly, the results obtained by SwarmNest for 31 benchmark data sets available in the literature are compared with the published results. Then, 25 factorial design of experiments is conducted to characterize SwarmNest performance. Experiments are conducted on a laptop of 1.5 GB RAM supported by an Intel® Core Duo processor of 1.66 GHz CPU speed and 2 MB cache size, operated with a Microsoft Windows XP Professional O/S, and algorithms are written in Visual Basic.

Benchmarking of SwarmNest

The work of Hopper [28] presents the first considerable attempt for collecting and organizing benchmark data sets for the 2-D Irregular Strip Packing Problem. Hopper collected 18 data sets from the literature. Most of the shapes of these data sets simulate the shapes involved in the textile industry. Two data sets (Dighe1 and Dighe2) are jigsaw problems (or puzzle problems) with known optimum of 100% utilization. Additional 9 problems (from Poly1a to Poly5b) were also randomly generated and documented for testing purposes. All the data sets in Hopper [28] are collections of hole-free polygons that do not include any internal holes or arcs. Hopper's data sets were then made available by the EURO Special Interest Group on Cutting and Packing (ESICUP) (http://paginas.fe.up.pt/~esicup/).

In Burke et al. [7], 10 new data sets (Profile1, ..., Profile10) were introduced for benchmarking purposes. Only three data sets are included, and two of which are jigsaw problems (Profile7 and Profile8). Others are not included as they contain holes and/or arcs in the shapes of their items. Furthermore, we extracted a single data set (Grinde) from Hifi and M'Hallah [18].

Seven sets of published results were available for our comparison; the first set is definitely the results obtained by Hopper [28]. Few years later, Gomes and Oliveira [9] introduced the first published results to consider Hopper's data sets. They also conducted a comparison between the best ever results published for Hopper's data sets and the corresponding results obtained by their own algorithms; SAHA and GLSHA. Gomes and Oliveira reported improvements over the best known results for 15 data sets, but they considered only 15 out of the 18 data sets, and did not consider any of the data sets Poly1a to Poly5b. Gomes and Oliveira [9] results and comparisons were then used as the basis for the subsequent literature comparisons. Such comparisons are reported in Egeblad et al. [31], Bennell and Song [16], Imamichi et al. [10] and Umetani et al. [11]. Also Burke et al. [7] introduced 10 new data sets, in addition to considering almost all Hopper's data sets in their reported comparison.

SwarmNest allows Particle Swarm to be used with either a combined pixel-based, or alternatively, a combined D-function-based placement algorithm.

The alternative placement algorithm applies a D-function-based placement procedure during its first stage, instead of the pixel-based one described in the second section. Both procedures are conceptually similar to a great extent, in which both adapt the same scanning strategy. The combined pixel-based algorithm is the main one, but some data sets would be better tackled with the combined D-function-based algorithm, especially those that are either jigsaw saw or randomly generated data sets. Hence, 14 data sets (Han, Dighe1, Dighe2, Profile7, Profile8, and Poly1a to Poly5b) were solved with the SwarmNest that employs a D-function-based algorithm.

For every data set, SwarmNest is allowed to run for 20 times and maximum time allowed for each run is set to 1 hour. The best results obtained by SwarmNest along with the best results reported by each of the 7 publications are shown in **Table 1**, in which a typical utilization% measure (total items area/(sheet width x used sheet length)) is used as a unified efficiency measure. The best ever solution for each problem is recognized with a grey background.

SwarmNest produced better solutions, or at least equal to, the best known solutions for 10 out of the 31 benchmark data sets. Moreover, the overall average of SwarmNest best results over the 31 data sets (78.37%) is better than the average of all the published results (77.47%). A sample from the best solution patterns obtained by SwarmNest is shown in **Figure 9**.

Characterization of SwarmNest

In order to characterize the better performance of SwarmNest, a Statistical Design of Experiments approach is adapted. Accordingly, five key factors are defined, and two levels are set for each factor, and thus a (2^5) factorial design with 32 different experiments is used. The five factors are defined, provided that the value for each factor, excluding the fourth one, is given by the mean value among an entire data set:

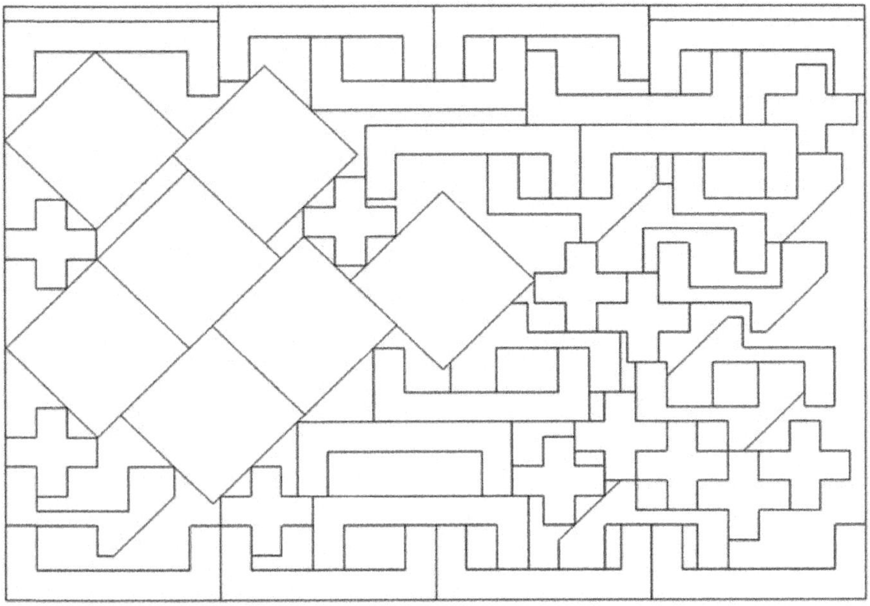

Shapes1 (length = 56, util. = 71.25%)

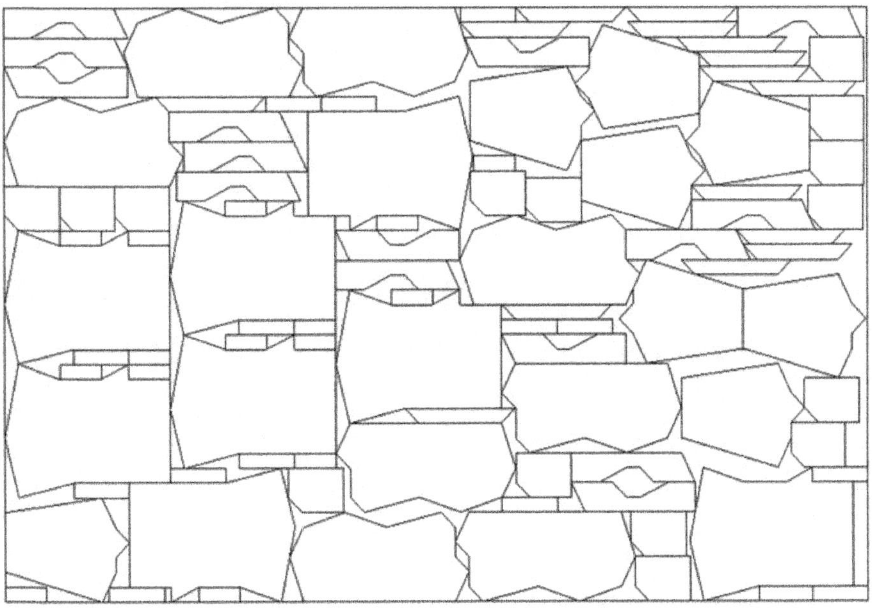

Shirts (length = 62.6, util. = 86.26%)

Figure 9: A sample from the best solution patterns for SwarmNest (length, utilization %).

Table 1: Best published results and best results obtained by SwarmNest for the 31 benchmark data sets.

Data set No.	Label	Hopper [28] (2000)	Burke *et al.* [7] (2006)	Gomes & Oliveira [9] (2006)	Egeblad *et al.* [31] (2007)	Bennell & Song [16] (2010)	Imamichi *et al.* [10] (2009)	Umetani *et al.* [11] (2009)	SwarmNest
		Hopper's	BLF	SAHA	2DNest	BS	ILSQN	FITS	
1	Blaz1	70.42	79.41	83.60	81.59	81.29	84.25	81.68	80.9
2	Blaz2	63.97	74.5	-	-	-	-	-	74.74
3	Dagli	77.1	83.7	87.15	87.05	87.99	87.40	86.35	83.97
4	Fu	83.82	86.9	90.96	92.03	90.82	90.67	91.23	89.06
5	Jakobs1	74.25	82.6	81.67*	89.07	85.96	86..89	89.09	81.67
6	Jakobs2	68.4	74.8	77.28	81.07	80.40	82.51	80.84	77.2
7	Marques	82.73	86.47	88.14	89.82	88.92	89.03	89.21	86.47
8	Shapes	63.33	67.6	-	-	-	-	-	71.25
9	Shapes0	57.83	61.4	66.50	67.09	65.41	68.44	66.50	65.41
10	Shapes1	62.25	68.3	71.25	73.84	72.55	73.84	73.88	71.25
11	Shirts	79.65	85.7	86.80*	87.38	89.69	88.78	86.92	86.26
12	Swim	67.42	68.4	74.37	72.49	75.04	75.29	74.54	68.25
13	Trousers	79.12	89.6	89.96	90.46	90.38	89.79	89.40	88.36
14	Albano	84.09	84.6	87.43	87.88	87.88	88.16	87.56	83.36
15	Dighe1	71.01	77.4	100	99.86	100	99.89	99.89	100
16	Dighe2	70.23	79.4	100	99.95	100	99.99	100	100
17	Mao	68.65	79.5	82.54	85.15	84.07	83.44	83.73	78.4
18	Han	68.9	-	-	-	-	-	-	77.19
19	Poly1a*	69.7	73.2	-	-	-	-	-	73.73
20	Poly2a*	68.1	72.8	-	-	-	-	-	75.33
21	Poly3a*	76.11	76.2	-	-	-	-	-	73.87
22	Poly4a*	72.05	74.6	-	-	-	-	-	73.06
23	Poly5a*	71.57	73.9	-	-	-	-	-	72.86
24	Poly2b*	68.34	75.4	-	-	-	-	-	73.41
25	Poly3b*	73	74.9	-	-	-	-	-	73.43
26	Poly4b*	73.15	74.8	-	-	-	-	-	73.66
27	Poly5b*	72.38	75.8	-	-	79.51	-	-	72.25
28	Profile7	-	77.1	-	-	-	-	-	81.93
29	Profile8	-	75.8	-	-	-	-	-	79.98
30	Profile10	-	66.2	-	-	-	-	-	63.1
31	Grinde	Only Hifi and M'Hallah [18]: 78.6%							78.97

[a]All the items shall initially be rotated to their minimum bounding rectangle orientations; [b]Another algorithm simpler but faster than SAHA is used, GL-SHA.

1) No. of items (NI): Total no. of items;
2) Size ratio (SR): Length + width of the item, divided by twice the sheet width;
3) No. of lines (NL): No. of lines or vertices per item or shape;
4) Coeff. of var. (CV): Standard deviation of the items areas, divided by mean area;
5) Complexity (C): No. of interior angles which are greater than 180°, divided by No. of lines.

The first factor (NI) reflects the dimensionality of a given problem, as the problem of finding the optimal packing sequence becomes more complicated with larger no. of items. The second factor (SR) expresses the relative size of an average item size with respect to the size of the placement sheet. The third factor (NL) represents an aspect of geometrical complexity of a given item. The fourth factor, Coefficient of Variation, (CV) signifies the extent of variability between different item sizes within the same data set. The last factor (C) is another aspect of item geometrical complexity; higher complexity factor means higher concavities in an item's shape, while zero complexity means convex-shaped items.

Two levels, high and low, are selected for each factor. Since generating a data set of random shaped items with an exact value for each factor is very difficult, a range of values is used to specify each factor level. The benchmark data sets were a good guide for selecting those ranges. The applied range for each factor at each level is shown in **Table 2**.

Thirty two (2^5) different experimental runs are conducted, where each run consisted of 10 replications, and each replication requires one data set of N Items. Hence, the 320 (32×10) data sets are randomly generated and solved by SwarmNest with the combined pixel-compaction placement algorithm, as being the main placement algorithm. The sheet width is always set to 60 units and all the items are allowed to rotate by increments of 90°. Main and interaction effects of the five factors on the response are exhibited in **Table 3**.

Table 2: Applied range for each factor level.

No.	Factor	Low level (–)		High level (+)	
		Value	Range	Value	Range
1	No. of items (NI)	20	(10, 30)	50	(40, 60)
2	Size ratio (SR)	0.15	(0.1, 0.2)	0.35	(0.3, 0.4)
3	No. of lines (NL)	4.5	(3, 6)	8.5	(7, 10)
4	Coeff. of var. (CV)	0.45	(0.3, 0.6)	0.95	(0.8, 1.1)
5	Complexity (C @ low NL)	0.025	(0, 0.05)	0.125	(0.1, 0.15)

Table 3: Main and interaction effects.

No.	Factors	Effect	No.	Factors	Effect
1	NI	0.456	17	NI*SR*CV	0.102
2	SR	3.992	18	NI*SR*C	–0.758
3	NL	0.067	19	NI*NL*CV	–0.309
4	CV	2.633	20	NI*NL*C	0.645
5	C	–3.555	21	NI*CV*C	0.462
6	NI*SR	–1.508	22	SR*NL*CV	0.34
7	NI*NL	–0.28	23	SR*NL*C	0.147
8	NI*CV	0.306	24	SR*CV*C	0.07
9	NI*C	0.295	25	NL*CV*C	–0.416
10	SR*NL	–0.91	26	NI*SR*NL*CV	0.194
11	SR*CV	0.656	27	NI*SR*NL*C	–0.112
12	SR*C	0.541	28	NI*SR*CV*C	0.048
13	NL*CV	0.069	29	NI*NL*CV*C	0.322
14	NL*C	–1.637	30	SR*NL*CV*C	–0.089
15	CV*C	–0.039	31	NI*SR*NL*CV*C	–0.179
16	NI*SR*NL	0.258			

The response is the sheet utilization, or the algorithm efficiency. A positive sign associated with an effect value means that the response value increases with increasing the corresponding factor value, and vice versa, and the magnitude of the effect represents its relative weight. Statistical package of Minitab 15 is used for calculations and analysis of results.

Clearly, many of the calculated effects appear to be not significant enough, which indicates that SwarmNest performance is of a low sensitivity to a problem attributes. The best performance for SwarmNest is obtained with problems of large relative item sizes, low geometrical complexity or irregularity, and high variability between item sizes.

CONCLUSIONS

In this research, the common industrial problem "2-D Irregular Strip Packing Problem" is addressed, and a heuristic sequence-based approach is developed for solving it. A Particle Swarm Optimization algorithm (PSO) is applied during the optimization phase of the approach, where it is the first time a PSO algorithm is incorporated in solving this problem. The introduced solution approach is given the name "SwarmNest".

Using 31 benchmark data sets, the best solutions of SwarmNest are compared with the best published results for seven different recent publications. Competitive results were obtained and the best solutions of SwarmNest are found to be better than, or at least equal to, the best known solutions for 10 out of the 31 benchmark data sets.

Furthermore, the performance of SwarmNest is characterized through a Statistical Design of Experiments that incorporates a specially designed random problem generator. The results showed that the best performance for SwarmNest is obtained at large item sizes (with respect to the sheet), low geometrical complexity, and high variability between item sizes.

REFERENCES

1. H. Dyckhoff, "A Typology of Cutting and Packing Problems," European Journal of Operational Research, Vol. 44, No. 2, 1990, pp. 145-159. doi:10.1016/0377-2217(90)90350-K
2. G. Wäscher, H. Haußner, and H. Schumann, "An Improved Typology of Cutting and Packing Problems," European Journal of Operational Research, Vol. 183, No. 3, 2007, pp. 1109-1130. doi:10.1016/j.ejor.2005.12.047
3. R. J. Fowler, M. S. Paterson and S. L. Tanimoto, "Optimal Packing and Covering in the Plane Are NP-Complete," Information Processing Letters,

Vol. 12, No. 3, 1981, pp. 133-137. doi:10.1016/0020-0190(81)90111-3

4. A. R. Babu and N. R. Babu, "A Generic Approach for Nesting of 2-D Parts in 2-D Sheets Using Genetic and Heuristic Algorithms," Computer-Aided Design, Vol. 33, No. 12, 2001, pp. 879-891. doi:10.1016/S0010-4485(00)00112-3
5. A. D. Fischer and C. H. Dagli, "Employing Subgroup Evolution for Irregular-Shape Nesting," Journal of Intelligent Manufacturing, Vol. 15, No. 2, 2004, pp. 187-199.doi:10.1023/B:JIMS.0000018032.38317.f3
6. S. Takahara, Y. Kusumoto and S. Miyamoto, "Solution for Textile Nesting Problems Using Adaptive Meta-Heuristics and Grouping," Soft Computing, Vol. 7, No. 3, 2003, pp. 154-159. doi:10.1007/s00500-002-0203-9
7. E. Burke, R. Hellier, G. Kendall and G. Whitwell, "A New Bottom-Left-Fill Heuristic Algorithm for the Two Dimensional Irregular Packing Problem," Operations Research, Vol. 54, No. 3, 2006, pp. 587-601. doi:10.1287/opre.1060.0293
8. J. A. Bennell and K. A. Dowsland, "Hybridising Tabu Search with Optimisation Techniques for Irregular Stock Cutting," Management Science, Vol. 47, No. 8, 2001, pp. 1160-1172.doi:10.1287/mnsc.47.8.1160.10230
9. A. M. Gomes and J. F. Oliveira, "Solving Irregular Strip Packing Problems by Hybridising Simulated Annealing and Linear Programming," European Journal of Operational Research, Vol. 171, No. 3, 2006, pp. 811-829. doi:10.1016/j.ejor.2004.09.008
10. T. Imamichi, M. Yagiura and H. Nagamochi, "An Iterated Local Search Algorithm Based on Nonlinear Programming for the Irregular Strip Packing Problem," Discrete Optimization, Vol. 6, No. 4, 2009, pp. 345-361. doi:10.1016/j.disopt.2009.04.002
11. S. Umetani, M. Yagiura, S. Imahori, T. Imamichi, K. Nonobe and T. Ibaraki, "Solving the Irregular Strip Packing Problem Via Guided Local Search for Overlap Minimization," International Transactions in Operations Research, Vol. 16, No. 6, 2009, pp. 661-683.doi:10.1111/j.1475-3995.2009.00707.x
12. R. B. Grinde and K. Daniels, "Solving an Apparel Trim Placement Problem Using a Maximum Cover Problem Approach," IIE Transactions, Vol. 31, No. 8, 1999, pp. 763-769.doi:10.1080/07408179908969875
13. M. Fischetti and I. Luzzi, "Mixed-Integer Programming Models for the Nesting Problem," Journal of Heuristic, Vol. 15, No. 3, 2009, pp. 201-226. doi:10.1007/s10732-008-9088-9

14. M. A. Carravilla, C. Ribeiro and J. F. Oliveira, "Solving Nesting Problems with Non-Convex Polygons by Constraint Logic Programming," International Transactions in Operational Research, Vol. 10, No. 6, 2003, pp. 651- 663. doi:10.1111/1475-3995.00434
15. J. F. Oliveira, A. M. Gomes and J. S. Ferreira, "TOPOS— A New Constructive Algorithm for Nesting Problems," OR-Spektrum, Vol. 22, No. 2, 2000, pp. 263-284.doi:10.1007/s002910050105
16. J. A. Bennell and X. Song, "A Beam Search Implementation for the Irregular Shape Packing Problem," Journal of Heuristic, Vol. 16, No. 2, 2010, pp. 167-188.doi:10.1007/s10732-008-9095-x
17. S. Jain and H. C. Gea, "Two-Dimensional Packing Problems Using Genetic Algorithms," Engineering with Computers, Vol. 14, No. 3, 1998, pp. 206-213.doi:10.1007/BF01215974
18. M. Hifi and R. M'Hallah, "A Hybrid Algorithm for the Two-Dimensional Layout Problem: The Cases of Regular and Irregular Shapes," International Transactions in Operational Research, Vol. 10, No. 3, 2003, pp. 195-216. doi:10.1111/1475-3995.00404
19. S. Jakobs, "On Genetic Algorithms for the Packing of Polygons," European Journal of Operational Research, Vol. 88, No. 1, 1996, pp. 165-181. doi:10.1016/0377-2217(94)00166-9
20. H.-Y. Liu and Y.-J. He, "Algorithm for 2D IrregularShaped Nesting Problem Based on the NFP Algorithm and Lowest-Gravity-Center Principle," Journal of Zhejiang University, Science A, Vol. 7, No. 4, 2006, pp. 570- 576. doi:10.1631/jzus.2006.A0570
21. J. F. Oliveira and J. S. Ferreira, "Algorithms for Nesting Problems," In: R. V. V. Vidal, Ed., Applied Simulated Annealing, Lecture Notes in Economics and Mathematical Systems, Vol. 396, Springer-Verlag, Berlin, 1993, pp. 255-274.
22. G. C. Han and S. J. Na, "Two-Stage Approach for Nesting in Two-Dimensional Cutting Problems Using Neural Network and Simulated Annealing," Proceedings of the Institute of Mechanical Engineers, Part B, Journal of Engineering Manufacture, Vol. 210, No. B6, 1996, pp. 509-519.
23. R. Heckmann and T. Lengauer, "Computing Closely Matching Upper and Lower Bounds on Textile Nesting Problems," European Journal of Operational Research, Vol. 108, No. 3, 1998, pp. 473-489. doi:10.1016/S0377-2217(97)00049-0
24. J. Blazewicz, P. Hawryluk and R. Walkowiak, "Using a Tabu Search Approach for Solving the Two-Dimensional Irregular Cutting Problem,"

Annals of Operations Research, Vol. 41, No. 4, 1993, pp. 313-325. doi:10.1007/BF02022998

25. J. Kennedy and R. C. Eberhart, “Particle Swarm Optimization,” Proceedings of IEEE International Conference on Neural Networks, Piscataway, 27 November-1 December 1995, pp. 1942-1948. doi:10.1109/ICNN.1995.488968
26. J. A. Bennell and J. F. Oliveira, “The Geometry of Nesting Problems: A Tutorial,” European Journal of Operational Research, Vol. 184, No. 2, 2008, pp. 397-415.doi:10.1016/j.ejor.2006.11.038
27. M. Konopasek, “Mathematical Treatments of Some Apparel Marking and Cutting Problems,” US Department of Commerce, Washington DC, 1981.
28. E. Hopper, “Two-Dimensional Packing Utilising Evolutionary Algorithms and Other Meta Heuristic Methods,” Ph.D. Thesis, Cardiff University, Cardiff, 2000.
29. C. Blum and D. Merkle, “Swarm Intelligence: Introduction and Applications,” Springer-Verlag, Berlin, 2008. doi:10.1007/978-3-540-74089-6
30. M. F. Tasgetiren, Y. C. Liang, M. Sevkli and G. Gencyilmaz, “Particle Swarm Optimization Algorithm for Makespan and Maximum Lateness Minimization in Permutation Flowshop Sequencing Problem,” Proceedings of the Fourth International Symposium on Intelligent Manufacturing Systems, Sakarya, 6-8 September 2004, pp. 431- 441.
31. J. Egeblad, B. K. Nielsen and A. Odgaard, “Fast Neighborhood Search for Two and Three-Dimensional Nesting Problems,” European Journal of Operational Research, Vol. 183, No. 3, 2007, pp. 1249-1266. doi:10.1016/j.ejor.2005.11.063

Chapter 9

APPLYING NETWORK FLOW OPTIMIZATION TECHNIQUES TO IMPROVE RELIEF GOODS TRANSPORT STRATEGIES UNDER EMERGENCY SITUATION

Novia Budi Parwanto[1,2], Hozumi Morohosi[1], Tatsuo Oyama[1]

[1]National Graduate Institute for Policy Studies (GRIPS), Tokyo, Japan

[2]Institute of Statistics (STIS), Jakarta, Indonesia

ABSTRACT

Given the seriously damaged emergency situation occurring after a large-scale natural disaster, a critical and important problem that needs to be solved urgently is how to distribute the necessary relief goods, such as drinking water, food, and medicine, to the damaged area and how to transport them corresponding to the actual supply and demand situation as quickly as possible. The existing infrastructure, such as traffic roads, bridges, buildings, and other facilities, may suffer from severe damage. Assuming uncertainty related with each road segment's availability, we formulate a transshipment network flow optimization problem under various types of uncertain situations. In order to express the uncertainty regarding the availability of each road segment, we apply the Monte Carlo simulation technique to generate random networks following certain probability distribution conditions. Then, we solve the model to obtain an optimal transport strategy for the relief goods. Thus, we try to implement a necessary and desirable response strategy for managing emergency cases caused by, for example, various natural disasters. Our modeling approach was then applied to the actual road network in Sumatra Island in Indonesia in 2009, when a disastrous earthquake occurred to develop effective and efficient public policies for emergency situations.

INTRODUCTION

Emergencies occur from unexpected, serious, and often dangerous situations, such as natural disasters, human- caused accidents, and major terrorist attacks, and require immediate action. Large-scale emergencies may result in the loss of a large number of lives and/or severe property damage even though such emergencies may not occur so frequently. However, once a large-scale emergency occurred, as we experienced during the Great East Japan Earthquake in March 11, 2011, an enormous amount of confusion and damage to our society resulted.

Regarding disasters caused by nature, according to the International Disaster Database [1] , the incidence of natural disasters has risen dramatically over the last four decades. Natural disasters often cause a large number of fatalities and massive property damage, including those to infrastructure and transportation networks. Large natural disasters, such as earthquakes, hurricanes, and volcanic eruptions, have caused many casualties and/or severe damage to our infrastructure. These disasters include the Haiyuan Earthquake (1920), Tangshan Earthquake (1976), Hanshin-Awaji Great Earthquake (1995), Gujarat Earthquake (2001), Indian Ocean Earthquake and Tsunami (2004), Pakistan Earthquake (2005), Haiti Earthquake (2010), Great East Japan Earthquake and Tsunami (2011), New Zealand Earthquake (2011), the Bhola Cyclone (1970), Nevado del Ruiz Volcano Eruption (1985), Hurricane Andrew (1993), Hurricane Katrina (2005), and Cyclone Nargis (2008). Regarding the Great East Japan Earthquake and Tsunami in 2011, Parwan to and Oyama [2] have discussed details of the occurrence, causes, and damage by investigating the statistical data comparing natural disasters in Japan and Indonesia.

Shortly after a natural disaster occurs, many injured victims need to be immediately taken to hospitals, and survivors need some vital items for survival such as clean drinking water, food, and medicine. Some residents may not have any reserves of such vital items, while others may have them, though only in limited amounts. Unfortunately, the situation sometimes becomes worse, that is, when a large natural disaster occurs, severe damage to the existing infrastructure, such as traffic roads, buildings, bridges and other facilities, may happen. This condition will certainly complicate the process of delivering relief goods to the victims, even though a sufficient amount of goods may be available. Thus, the distribution of relief vital goods under large-scale emergencies, also known as logistics in an emergency, should be undertaken as quickly as possible, so that the victims can have an increased chance of survival. The planning strategy for the distribution of aid by considering all possibilities that occur is very important to implement. As pointed by Jiang et al. [3] , some specific characteristics of large-scale emergencies differ, depending on the type, for logistics in an emergency in the aftermath of a disaster. This condition makes the knowledge of aid logistics and supply chain management immensely important for humanitarian operations. One of the most highly recommended methods and the one often used for humanitarian operations is the operations research (OR) technique, since it has been proven to be beneficial for different planning situations (see Kovacs and Spens [4]). However, given that the application of OR methods in emergency situations under uncertainty is still scarce, this study will investigate this issue, particularly proposing an optimization model regarding how to distribute necessary relief goods to the affected areas as soon as possible using the ground transportation.

We apply our model to the Indonesian case, where a M7.9 earthquake struck West Sumatra Province on September 30, 2009 at 17:16 pm local time and resulted in about 1197 dead and missing, 1798 people injured, and about 249,800 houses damaged. The West Sumatra Earthquake also caused damage to roads and bridges, which in turn disrupted the transportation network. In general, the earthquake caused damage in 13 regions (cities/re- gencies) out of 19 regions. These regions are Padang city, Pariaman city, Solok city, Padang Panjang city, Padang Pariaman regency, Bukittinggi city, Agam regency, Solok regency, Pasaman regency, Pasaman Barat regency, Pesisir Selatan regency, Tanah Datar regency and Mentawai Island regency, in which the last one is an archipelago regency separated from West Sumatra, with the only transportation to Mentawai Islandis by ship or aircraft.

In most cases of large-scale emergencies or large-scale disasters, the amount of relief supplies is limited while the demand is great. Liu et al. [5] argues that the key factor in an effective response under a large-scale emergency is the prompt availability of supplies at emergency sites. Hence, an efficient emergency transportation network becomes important in addressing and optimizing the complex distribution process.

In this paper firstly and generally we formulate a multi-commodity transshipment network flow optimization problem under various types of uncertain situations. Then by solving the network flow optimization problem, we try to find an optimal transport schedule for various kinds of relief goods including cases where all necessary demands are not met. Thus, under two different criteria, i.e. minimizing the total transport distance and minimizing the supply-demand gap, we investigate the characteristics of how the optimal transport schedule would be changed. Also, we take the uncertainty regarding the availability of each traffic road segment into consideration by giving the probability that each traffic road segment would be "blocked". In order to deal with the quantitatively assumed probabilities, we apply the Monte Carlo simulation technique. Thus, we can determine the optimal transport schedule of various relief goods under emergency situations occurring just after natural disasters such as earthquakes and typhoons.

Van Wassen hove [6] describes as follows: disaster relief is about 80% logistics, therefore it would follow then that the only way to achieve this is through slick, efficient and effective logistics operations and more precisely, supply chain management. This condition renders the knowledge of aid logistics and supply chain management immensely important for humanitarian operations. Altay and Green [7] have summarized various applications of OR methods in disaster operations management. Delivery planning in disaster relief operations has also been studied rather extensively. Ozdamar et al. [8] proposed a logistics planning model in emergency situations that involves dispatching commodities to distribution centers in affected areas. Their model addresses the dynamic time-dependent transportation problems that need to be solved repetitively at given time intervals during ongoing aid delivery. Haghani and Oh [9] developed a multi-commodity, multi-modal network flow model for disaster relief operations in which they used penalty costs for unsatisfied demand.

On the other hand, Fiedrich et al. [10] proposed a dynamic optimization model to find the best assignment of available resources to affected areas after an earthquake. In addition, Lin et al. [11] developed a logistics model for disaster relief operations. They argue that the geographic location of the depot is very important as they have shown that they can improve the performance. Also they add that reduction in the number of clusters does not guarantee an improvement in the logistics of humanitarian relief by increasing the number of vehicles.

Meanwhile, Tzeng et al. [12] propose a multi-objective model for the optimal distribution of relief commodities, taking into account cost minimization, travel time minimization, and satisfied demand maximization. Vitoriano et al. also suggest a goal programming approach to support humanitarian organizations in aid distribution decisions [13] . Yi and Ozdamar [14] propose different transportation and network flow models in which they have also taken into account uncertainty, multiple aid items, and multiple time-periods (see also Ozdamar et al. [8] , and Barbarosoglu and Arda [15]).

In most of the above-mentioned studies, the preparation phase or the immediate response of disaster relief has been addressed, in which most of the problems fall into the transportation problem category, namely allowing only shipments that go directly from a supply point to a demand point. However, in many situations, shipments should be allowed between supply points or between demand points, in which there may also be some points (called transshipment points) through which goods can be transshipped on their journey from a supply point to a demand point. Shipping problems with these kinds of characteristics are known as transshipment problems (see e.g. Winston

[16]). Fortunately, the optimal solution to a transshipment problem can be found by solving a transportation problem, although few studies have been conducted in this field. Herer et al. [17] developed a transshipment model in a supply chain, which consists of several retailers and one supplier. Rootkemper et al. [18] developed a mixed-integer programming model to minimize the unsatisfied demand as well as the operational costs by imposing penalty costs for unsatisfied uncertain demand.

We know there are more references dealing with emergency management issues such as the IJISCRAM[1] and the IEEE International Conference on Emergency Management even though we could not include them in this paper. We believe these references could be very useful for our future research.

TRANSSHIPMENT NETWORK FLOW OPTIMIZATION MODEL ANALYSES

Objectives of the Network Flow Optimization Model

We consider a case where an emergency situation, caused by, for example, a relatively large scale natural disaster, has just occurred in a particular area consisting of several regions/cities and has rendered significant damage to property and inflicted human casualties, both deaths and injuries. Some regions have experienced damage and losses more severely than other regions. These regions should immediately get help, in the form of consumable and durable commodities. Some emergency supplies must be transported from several supply centers/depots (airport/harbor or central inventory) to locations with demand (e.g. affected regions). In addition, it is given that, in general, every region has a supply/stockpile of emergency vital needs, albeit for a relatively short time and in limited quantities. Thus, for a relatively short time (i.e. a few days), the need of relief goods vital for survival, such as medicine/drugs, drinking water/clean water, and food, can be supplied by the affected regions of the natural disaster itself; in other words, the affected regions become both demand and supply regions.

We consider the following problem. Suppose we are given a set of regions, e.g. cities, and a set of commodities, e.g. drinking water, food and medicine, and moreover suppose we know that each city has a certain amount of each commodity it can supply, as well as some demand. Then if each city's supply of each commodity is not less than its demand, then we do not need to transport any commodity from one city to another. However, as this case may occur generally, we want to find an optimal transport scheduling strategy so that all the demand for commodities for all cities can be met. Otherwise, namely in the case where all cities' demands for commodities are not necessarily met, we want to find some desirable transport scheduling strategy. We denote regions with excess supply, namely regions where the supply for the commodity is larger than the demand, as RES. Similarly, other affected regions might not be able to satisfy their own demand. We denote such regions with excess demand, namely regions where the demand for the commodity is larger than the supply, as RED. Of course, affected regions that experience excess demand (RED) are regions that should immediately get help in order to minimize the number of casualties as a result of the disaster. The necessary goods for survival must be immediately sent to this region. Thus, logistics in an emergency can be defined as the distribution of relief vital needs commodities under (large-scale) emergencies, which is also known as humanitarian logistics or relief operations, carried out in the second phase of disaster management, i.e. the response phase. The following formulation of a multi-commodity transshipment network flow optimization model provides a solution to the problem described above.

Common practices follow that the available relief aid to be sent immediately should come from the neighboring regions that are less affected (RES) to the regions that are severely affected (RED). However, the immediate relief consignments may not be able to satisfy all the demands. Subsequently, we need to transport the relief logistics from, e.g. the supply center (SC) once they become available. Hence, these problems will be solved by a technique whereby we calculate the maximum demand gap and transport relief aid by using the shortest path technique.

In general, depending on the time when the SC can be established and operated, we consider the stage to deliver relief aid to the affected regions. Firstly, we consider how RES can deliver their surplus to RED, and if there still remains a demand gap in RED, then we consider how the SC can deliver relief aid to meet all the remaining demand "shortage" gap in RED.

General Formulation and Computational Procedures

We denote the network by $G=(N,E)$ where N is a set of vertices corresponding to regions/cities and E is a set of edges corresponding to road segments. We assume that each region is represented by the capital city, thus in order to measure the distance between two regions we define the distance between these capital cities. Set E consists of the edge (i,j) from region i to j defined by $\{(i,j)|i\in N, j\in N\}$. K is the set of relief vital com-

modities such as drinking water, food, medicine/drugs, etc. We formulate a network flow optimization problem, which we call a network flow optimization model (NFOM). We denote input data as follows (i)$\{(i,j)\in E\}$ P_{ij}: probability that the edge (roadsegment) between region i and j is broken; (ii)$\{i\in N,k\in K\}$ S_{ik}: supply for commodity k in region i; (iii)$\{i\in N,k\in K\}$ D_{ik}: demand for commodity k in region i; (iv)$\{(i,j)\in E\}$ T_{ij}: transportation cost (distance or travel time) for road segment(i,j); (v)$\{k\in K\}$ V_k: maximum load capacity of the vehicle for commodity k; (vi)$\{(i,j)\in E\}$ C_{ij}: road capacity for road segment(i,j); (vii)$\{k\in K\}$ Q_k: amount of extra supplies in the SC for commodity k.

We define the decision variables, constraints and objective function of the NFOM as follows:

i) Decision variables

$\{(i,j)\in E,\ k\in K\}$ x_{ijk}: the amount of commodity k traversing road segment(i,j);

$\{j\in N,k\in K\}$ v_{jk}: the demand gap for commodity k in region j;

w_k: the maximum demand gap with respect to commodity k.

ii) Constraints

As the model is based on a network flow formulation, we need to formulate continuity and conservation conditions. The constraints are stated and explained as follows:

$$\sum_{i\in N,(i,j)\in E} x_{ijk} - \sum_{p\in N,(j,p)\in E} x_{jpk} + v_{jk} \geq D_{jk};\quad j\in N,\ k\in K \tag{1}$$

$$\sum_{k\in K} x_{ijk} \leq C_{ij};\quad (i,j)\in E \tag{2}$$

$$\sum_{j\in N,(i,j)\in E} x_{ijk} \leq S_{ik};\quad i\in N,\ k\in K \tag{3}$$

$$v_{jk} \leq w_k;\quad j\in N,\ k\in K \tag{4}$$

Constraint (1) expresses the flow conservation condition satisfying the demand for commodity k in region j. In addition, in constraint (1) the amount of commodity k transported to region j from all other region i may not satisfy the demand needed in region j, then, we assume that the amount of commodity k transported to region j should meet the demand for commodity k in region j by taking a positive value for the variable v_{jk}. Constraint (2) reflects the road capacity constraint. Constraint (3) denotes the supply constraint, in which the amount of commodity k transported from region i should not exceed the supply of commodity k in region i. Constraint (4) indicates the maximum demand gap at region j with respect to commodity k.

iii) Objective function

Our linear programming model has two objectives, namely to minimize the transportation cost (i.e. distance or time travel) and the total demand gap. The objective function of the model is stated in the following:

$$\textbf{Minimizing } z_k = K_1 \sum_{(i,j)\in E} T_{ij} x_{ijk} + K_2 w_k \tag{5}$$

Where $K_1 \geq 0$ and $K_2 \geq 0$

To incorporate the uncertainty condition, we modify one of the constraints namely constraint (2), by adding the probability of broken road segment P_{ij}, that is:

$$\sum_{(i,j)\in E} x_{ijk} \leq \begin{cases} 0 \text{ with probability } P_{ij}; \\ C_{ij} \text{ with probability} \left(1 - P_{ij}\right); \end{cases} \quad k \in K \tag{6}$$

Constraint (6) indicates that a road segment (i,j) is broken with the probability P_{ij}.

We can summarize our computational procedures as follows:

Step 1:		Obtaining an optimal solution for the Reference case $(K_1 = 1,\ K_2 = 0)$
Step 2:		Iterative computation $(K_1 = 1,\ K_2 = 1{,}000{,}000)$
	Step 2-1:	Set $t = 1$
	Step 2-2:	Generate a uniformly distributed random number $\{n_{ij}\}$ on $[0,1]$ for $(i,j) \in E$ Define edge capacities as $C'_{ij} = 0$ if $n_{ij} \leq P_{ij} = C_{ij}$ otherwise
	Step 2-3:	Solve the above optimization problem. Obtain an optimal solution $\{z^t\}$, $\{x^t_{ijk}\}$, $\{v^t_{jk}\}$, $\{w^t_k\}$
	Step 2-4:	If $t > T$, then end; otherwise $t+1 \rightarrow t$ Go to step 2-2
Step 3:		Find the distribution for all solutions $\{z^t\}$, $\{x^t_{ijk}\}$, $\{v^t_{jk}\}$, $\{w^t_k\}$

A flow chart of the Iterative computational procedure is given in Figure 1.

The network flow optimization model, NFOM, described above gives an optimal allocation of supplying relief goods from some region in RES to another region in the RED. Then we consider how we can meet the remaining excess demand of relief aid in RED. We formulate a network flow optimization model to solve this problem for the emergency situation, which we call the demand shortage clearance model (DSCM). Our assumption follows that all the remaining demand of relief commodities in RED will be met by providing commodities from the SC, i.e. a certain designated region in the network, to each region in RED. In order to attain an optimal transportation schedule we formulate the following network flow optimization model.

We define the input data for the network flow optimization model as follows: I: set of destination regions, E_i: set of edges included in the shortest path from the SC to destination i. M: number of simultaneously available vehicles, C: capacity of vehicle, F: velocity of vehicle, T: maximum time horizon of delivery, $T_S = \{1, 2, \cdots, T\}$: set of time horizons, B: maximum number of batches, D_i: demand at destination region $i \in I$, L_i: distance from SC to destination region $i \in I$, N_i: number of vehicles needed for delivering to destination $i \in I$, i.e. $N_i = \lceil D_i / C \rceil$, where $\lceil X \rceil$ indicates the least integer not less than X, T_i: time for delivering supply goods (two-way) to the destination $i \in I$, in which our model assumes an integral value for the sake of simplicity.

Typically $T_i = \lceil 2L_i / F + T_c \rceil$, where T_c is a transaction time, and $D_{kl} = 1$ or 0 corresponding to the case that there exists an edge $(k,l) \in E$ for the computational simulation or not. This model solves a transportation

scheduling problem. Decision variables, constraints and the objective function are stated as follows:

i) Decision variables

x_{ij}: number of vehicles for the delivery to destination i in the j-th batch. $i \in I$, $j = 1, 2, \cdots, B$

ii) Constraints

$$\sum_{j=1:B} x_{ij} \geq N_i \quad \forall i \in I \tag{7}$$

$$\sum_{i \in I} x_{i, \lceil t/T_i \rceil} \leq M, \quad \forall t \in T_S \tag{8}$$

$$x_{il} \leq MC_{kl}, \quad \forall (k,l) \in E_i \tag{9}$$

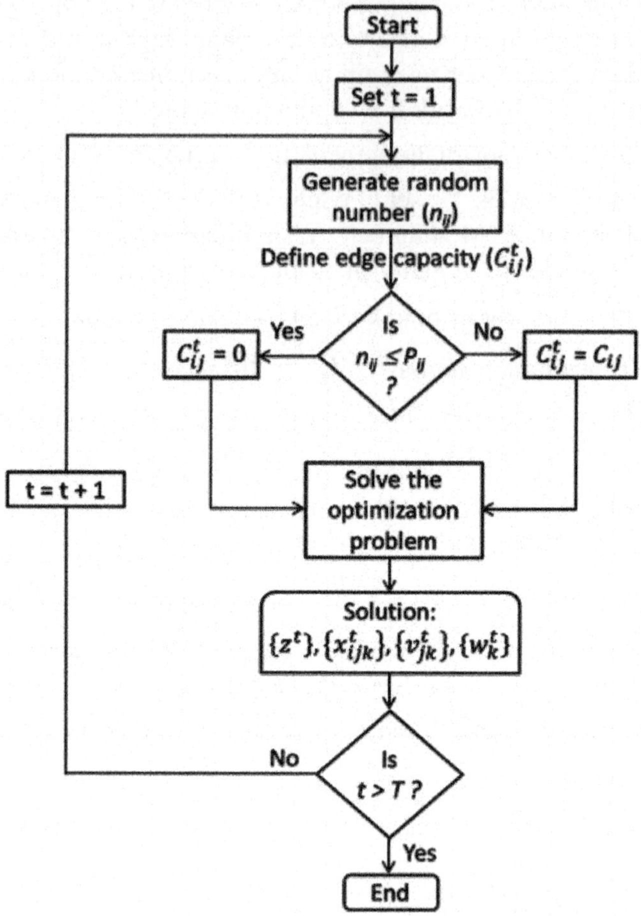

Figure 1: Flow chart of the iterative computational procedure.

Constraint (7) is on the minimum number of vehicles necessary for the delivery to destination i. The summation is taken for all the batches to calculate the total number of vehicles necessary for destination i. Constraint (8) limits the total number of vehicles used at every hour. Incidentally, this constraint (8) becomes redundant for some t values as they give the same constraint, so we need to delete those redundant constraints even though the application software FICO Express 7.4 (Fair Isaac Corporation) ignoredthis redundancy. Constraint (9) is the road capacity constraint.

iii) Objective function

$$\text{Minimize } y = \sum_{i \in I} \sum_{j=1:B} x_{ij} 2^j \qquad (10)$$

We assume the shortest path is used in order to deliver the goods to any destination region. Our objective function aims at minimizing the number of vehicles necessary for the delivery from SC to the destination region in RED in the sense that we try to minimize the criterion given by (10). In the objective Function (10), the coefficient 2^j is to attain as small a number of batches as quickly as possible by giving larger weights to the additional batch of vehicles in order to avoid the late delivery of supplies.

APPLICATION TO THE INDONESIAN SUMATRA EARTH-QUAKE CASE

Data and Assumptions

We apply the network flow optimization model described above to the case of the 2009 West Sumatra earthquake in Indonesia. We know that Sumatra Island is an area included in a zone of high seismic activity, the so- called "Pacific Ring of Fire". Thus, earthquakes are one of the natural hazards that can hit the island at any time. Thus, a disaster management plan that includes emergency response preparations should be done as carefully and quickly as possible, with the principle of "hope for the best but prepare for the worst".

Two earthquakes of 7.6 and 6.2-moment magnitude struck off the coast of West Sumatra, Indonesia on September 30, 2009. The first earthquake occurred at 17:16 p.m. and the second one followed 22 minutes later. The epicenter was 45 km west-northwest of the port city of Padang, the capital of West Sumatra, and it had recorded a depth of 71 km. A third earthquake of 6.8 magnitude struck an inland area, 225 km southeast of Padang, early the next morning (BNPB, 2009). The cumulative impact of these events left a broad swath of destruction. The earthquakes caused serious damage to housing and infrastructure in the communities in 13 regencies/cities[2], destroying livelihoods, and disrupting economic activity and social conditions, causing extensive psychological trauma [19] [20] . Landslides in West Sumatra left scores of houses and several villages buried; i.e., three villages in the path of the disaster in Padang Pariaman regency appeared to have been completely leveled, and most of their inhabitants appear to have been buried due to subsequent landslides. Macro data for the casualties and property damages are given in Table 1, while the corresponding detailed data for the West Sumatra Provinces are shown in Table 2.

Furthermore, to deliver some relief commodities to the affected regions, we need to know the road transportation network system of West Sumatra Province. It should be noted that the entire transportation road network system in West Sumatra province is large as it comprises of state roads, provincial roads, and district roads. Since we only deliver relief aid from SC, located in the provincial capital, Padang city, to the capital of the district/city or to the depot in each district/city, we assume that we only deliver the relief aid up to the SC of each affected region, mostly located in the capital of each regency/city rather than delivering the relief aid directly to the beneficiaries. Thus, we simplify the entire network system by simply taking the main roads linking the capital of the affected district/city, which consists of state roads and provincial roads. Figure 2 shows the road transportation network system with 12 vertices and 22 edges in West Sumatra province as a whole.

In Figure 2, there are 7 regions affected by the disaster, which can be divided into two groups, namely severely affected and less affected regions. Regions that belong to the former group are 1, 2, 5, and 6, while regions 3, 4, and 7 belong to the latter group. Regions in the latter group may be able to satisfy their own demand, and they may have supply surplus for some relief commodities, which then makes them suppliers, and we define these regions as regions with supply surplus or RES. On the other hand, regions in the former group cannot satisfy their own demand, thus creating a demand shortage. We define these regions as regions with demand shortage or RED. In most cases, when there is a disaster, the relief consignments from the supply center, which come from the central government or other parties, usually do not come immediately due to some difficulties.

Table 1: Casualties and property damage caused by the West Sumatra earthquake

	Casualties	
Death	1195	Persons
Severe injured	619	Persons
Slightly injured	1179	Persons
Missing	2	Persons
	Houses and facilities	
Houses (H)	114,797	Units
Houses (M)	67,198	Units
Houses (L)	67,838	Units
Office buildings	442	Units
Educational facilities	4748	Units
Health facilities	153	Units
Places of worship	2851	Units
Markets	58	Units
Bridges	68	Places
Roads	177	km

Source: National Disaster Management Board (BNPB) of Indonesia. H: Heavily; M: Moderately; L: Lightly.

Table 2: Number of population, households, casualties and property damage by affected region

No.	Region	Population	Household	Casualties				Property Damage (Houses)		
				Death	Missing	Severely Injured	Slightly Injured	Severe	Moderate	Slightly
1	Padang city	875,548	197,785	383	2	431	771	33,597	35,815	37,615
2	Solok regency	359,819	77,955	-	-	-	5	11,796	3797	4353
3	South Pesisir regency	448,488	95,747	9	-	7	20	3240	3046	2862
4	Padang Pariaman reg	389,735	81,954	666	-	6	19	57,931	16,291	12,945
5	Pariaman city	70,726	14,023	48	-	74	74	6685	4116	2605
6	Solok city	59,837	13,318	3	-	-	-	2	2	6
7	Padang Panjang city	56,491	12,921	-	-	6	14	17	164	413
8	Tanah Datar regency	336,604	79,615	-	-	-	-	28	115	105
9	Bukittinggi city	107,805	25,381	-	-	-	-	-	-	50
10	Agam regency	431,153	101,557	81	-	90	47	3	-	136
11	West Pasaman reg.	338,567	72,656	5	-	5	25	197	13	931
12	Pasaman regency	261,579	58,918	-	-	-	-	145	243	357
	Total	3,736,352	831,830	1195	2	619	1179	114,797	67,198	67,888

Source: National Disaster Management Board (BNPB) of Indonesia.

Therefore, common practices follow that the available relief aids to be sent immediately should come from the neighboring regions of less affected RES to those of severely affected RED. However, the immediate relief consignments may not be able to satisfy all the demands too. Subsequently, we need to transport the relief logistics from the SC once they become available. Thus, these problems will be solved by calculating the maximum demand gap and transporting relief aid by using the shortest path technique.

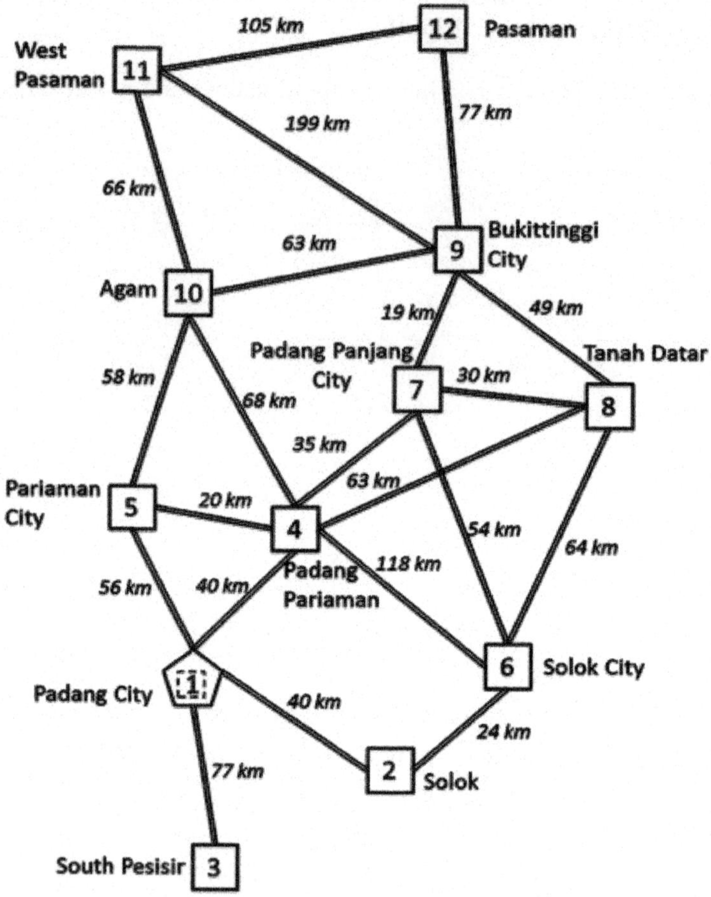

Figure 2: Illustration of road network system and flow commodity to/from a region.

Numerical Results by the Deterministic Approach

Firstly we consider the case that all road segments are available (survive), i.e. $P_{ij} = 0$ for all $(i, j) \in E$. Later, we discuss more general cases such that the road segment $(i, j) \in E$ might be broken with general probability $0 \leq P_{ij} \leq 1$.

Table 3 gives the demand and supply for the commodity, i.e. drinking water for 7 days (one week). The supply and demand columns describe the stockpile/reserve and the need for each commodity in each affected region. In Table 3, the negative value of Gap indicates that the region is in a Region with Excess Demand (RED), while a positive value of Gap indicates that the region is in a Region with Excess Supply (RES). The Total Excess Demand

(TED) from RED is the summation of all the negative values of Gap, which is 823.12 for drinking water from Table 1. The Total Excess Supply (TES) from RES is the summation of all positive values of Gap, which is 71.15 for drinking water from Table 1. As mentioned in Section 2, just after the natural disaster occurs, the affected region should try to satisfy the demands using existing reserves in each region. However, as this can only last for a few days, not long after, the increased demand and the diminishing supplies lead to gaps between supply and demand in some affected regions, especially in regions that have experienced the relatively severe impact of the natural disaster. Of course, this gap should be filled as soon as possible by sending the commodity from the neighboring region and/or the SC. Data sources for the supply for aid commodities are several governmental institutions. Data for drinking water are taken from the regional enterprise of drinking water (PDAM). Furthermore, the estimated relief commodity, drinking water for one week, available in the SC Padang city is 1058.76 (m^3). The number of trucks for transporting food and medicine is estimated to be 56 vehicles with the maximum capacity per vehicle being 14 tons, while the number of tanker trucks for transporting drinking water is estimated to be 39 with the maximum capacity per vehicle being 6 m^3.

Table 3: Demand and supply for drinking water to the affected area for one week by region.

No	Region	Drinking Water (M^3)		
		Supply (S)	Demand (D)	Gap (S-D)
1	Padang city	843.37	984.02	−140.66
2	Solok regency	295.40	323.52	−28.12
3	South Pesisir regency	317.96	403.24	−85.28
4	Padang Pariaman reg.	267.21	438.02	−170.81
5	Pariaman city	37.63	79.49	−41.86
6	Solok city	64.83	53.80	11.03
7	Padang Panjang city	66.52	50.79	15.73
8	Tanah Datar regency	309.10	302.65	6.45
9	Bukittinggi city	134.87	96.93	37.94
10	Agam regency	318.73	387.66	−68.92
11	West Pasaman reg.	21.92	304.41	−282.49
12	Pasaman regency	230.21	235.19	−4.98
	Total	2907.76	3659.71	−751.95

From Table 3, we see that all the commodities have larger total demand than total supply, which leads to "demand gaps" interpreted as "shortages" for each commodity. Furthermore, we can also see that the situation is different for each commodity in each region. Figure 3 illustrates the condition of supply and demand of drinking water by region. For instance, as depicted in Figure 3, regions 1, 2, 3, 4, 5, 10, 11, and 12 encountered excess demand for drinking water denoted by RED, where region 11 has the largest (maximum) demand gap (282.49); while regions 6, 7, 8, and 9 denoted by RES experienced excess supply (71.15). Then by assuming that the main SC takes some times to be established and/or start operating, RES can deliver their excess supply to their neighboring RED.

Our objective function given in (5) consists of two terms in which the first one aims at minimizing the transportation cost (i.e. distance or time travel), while the second one corresponds to minimizing the maximum demand gap. Even though we are aiming at minimizing both the transportation cost and the maximum demand gap simultaneously, the optimal solution obtained from the model depends upon the values of coefficients K_1 and K_2 indicating the relative "importance" of the above two terms. A general computational procedure of the model analysis implies that after estimating the gap, we try to deliver supplying commodities from RES to RED with minimum transportation cost taking the maximum demand gap minimization into consideration.

Table 4 shows the change of optimal solution for water delivery corresponding to the gradual increase of coefficient value K_2 for a fixed value of K_1 (e.g. $K_1 = 1$). From Table 4, we find that when the relative "importance" of the maximum demand gap is rather small, i.e. "gap" cost is small, no delivery from RES to RED is seen from model solution. However, increasing the value of coefficient K_2 larger than 120, the model starts to give a gradual increase with respect to the amount of water supply to be delivered. We know that reducing the maximum demand gap by a certain amount means increasing the delivery by the same amount from some RES to other RED with the minimum transportation cost. Therefore, we can see from Table 4 that for the coefficient value $K_2 = 130$, drinking water has been delivered from RES to RED, that is from region 9 to region 11 $(x_{ijk} = 37.94\ \text{m}^3)$ making the maximum demand gap (w_k) and remaining gap (v_{jk}) at region 11 decrease to 244.55 m^3. Now, regarding the "turning" coefficient value as between $K_2 = 120$ and $K_2 = 130$, this implies that the minimum transportation cost for a unit amount of water supply from some RES to other RED occurs between region 9 and region 11 with a total distance $63 + 69 = 129$. Hence, for a coefficient value less than $K_2 = 129$, there is no supply delivery from RES to RED, while for $K_2 > 129$, the maximum demand gap minimization process starts to provide drinking water from RES region 9 to RED region 11 by the demand gap to be filled as 37.94

m^3 corresponding to the available delivery of drinking water from region 9 given by $134.87 - 96.93 = 37.94$.

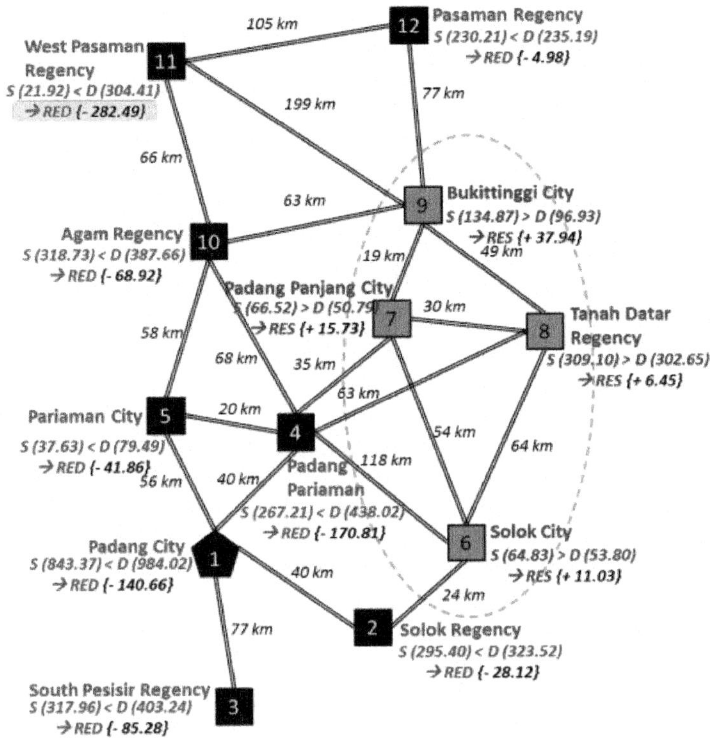

Figure 3: Supply and demand of drinking water (m^3) by region. Note: S: Supply; D: Demand; Number in brackets { } represent the gap; + indicate excess supply; − indicate excess demand; RES: Region with excess supply; RED: Region with excess demand.

Table 4: Optimal solutions for delivering drinking water (m^3).

We can also see that other RES regions (i.e. 6, 7, and 8) have not delivered their supply in order to minimize the sum of the transportation cost and the demand "gap" cost. Subsequently, the next

"turning" coefficient value appears between $K_2 = 140$ and $K_2 = 150$ corresponding to the next shortest distance route from region 7 to region 11 with $19 + 63 + 66 = 148$ as we see in Table 4. Thus, following the increase of the coefficient value K_2, additional deliveries are obtained along the "next" shortest routes.

K_2	Max Gap (m^3)	Total Cost (km)	Total Excess Demand (m^3)		Total Excess Supply (m^3)		Routs	Amount of Relief Aid (Water) to Be Delivered (m^3)
			Original	Improved (Remaining)	Original	Improved (Remaining)		
100	282.49	0	823.12	823.12	71.15	71.15	-	0
110	282.49	0	823.12	823.12	71.15	71.15	-	0
120	282.49	0	823.12	823.12	71.15	71.15	-	0
130	244.55	4894.26	823.12	785.18	71.15	33.21	9 → 10 → **11**	37.94 + 0 = **37.94**
140	244.55	4894.26	823.12	785.18	71.15	33.21	9 → 10 → **11**	37.94 + 0 = **37.94**
150	228.82	7222.30	823.12	769.45	71.15	17.48	7 → 9 → 10 → **11**	15.73 + 37.94 + 0 = **53.67**
160	228.82	7222.30	823.12	769.45	71.15	17.48	7 → 9 → 10 → **11**	15.73 + 37.94 + 0 = **53.67**
170	228.82	7222.30	823.12	769.45	71.15	17.48	7 → 9 → 10 → **11**	15.73 + 37.94 + 0 = **53.67**
180	222.37	8370.40	823.12	763.00	71.15	11.03	7 ↘ 9 → 10 → **11** 8 ↗	15.73+ 37.94 + 0 = **60.12** 6.45+
190	222.37	8370.40	823.12	763.00	71.15	11.03	7 ↘ 9 → 10 → **11** 8 ↗	15.73+ 37.94 + 0 = **60.12** 6.45+
200	222.37	8370.40	823.12	763.00	71.15	11.03	7 ↘ 9 → 10 → **11** 8 ↗	15.73+ 37.94 + 0 = **60.12** 6.45+
210~	**211.34**	**10598.50**	**823.1**	**751.97**	**71.15**	**0.00**	6 → 7 ↘ 9 → 10 → **11** 8 ↗	11.03 + 15.73+ 37.94 + 0 = **71.15** 6.45+

We see no change for our optimal solution after we increase the value of K_2 to over 210, as shown in Table

4. When $K_2 = 210$, we have the final optimal solution where all of the supplies from RES $\left(x_{ijk} = 71.15 \text{ m}^3\right)$

have been delivered to RED (i.e. region 11), resulting in a decrease in w_k and v_{jk} at region 11 to 211.34 m^3. Figure 4 illustrates the routes to be travelled for delivering drinking water from RES to RED in the final optimal solution (i.e. $K_2 = 210$).

The results in Table 4 and Figure 4 show that there still remain demand gaps for the commodity of drinking water in each region with demand shortage (RED). All of these remaining demand gaps in the RED need to be met by delivering additional relief goods from the SC to each destination. The problem then will be how to deliver the relief goods from the SC to the RED as quickly as possible. We try to solve this problem for the emergency situation using what we call DSCM, described in Section 2.2.

From the solutions shown in Figure 4 we see that the maximum excess demand gap in region 11 was reduced from 282.49 to 211.34. However, we know that still there remain excess demand gaps in regions 1, 2, 3, 4, 5, 10 and 12 also with 751.95 in total as shown in Table 3. We try to fill these excess demand gaps by sending vehicles to all regions needed carrying the relief commodity drinking water with the minimum number of batches for sending vehicles. In order to apply the optimization model given in (7)-(10), we use the following data. The number of simultaneously available vehicles, capacity of vehicle, velocity of vehicle, maximum number of batches and maximum time horizon of delivery are given as $M = 39$, $C = 6$ (m³), $F = 30$ (km/hr), $B = 10$ (batches) and $T = 48$ (hours), respectively. For each destination region $i \in I$, demand D_i, distance from SC L_i, time for delivering supply goods (two-way) to the destination T_i, and number of vehicles needed for delivering to destination N_i are given in the left column in Table 5.

Furthermore, as our objective is to deliver the vehicles from the SC, which is at the provincial capital i.e. Padang city, to the RED as quickly as possible, in solving this shortest path problem we exclude region 1. In addition, we also assign the following assumptions: vehicle velocity (F) is 30 km/hour, transaction time for loading (T_c) is 2 hours, maximum number of batches (B) is 5 batches, and maximum time horizon of delivery (T) is 24 hours.

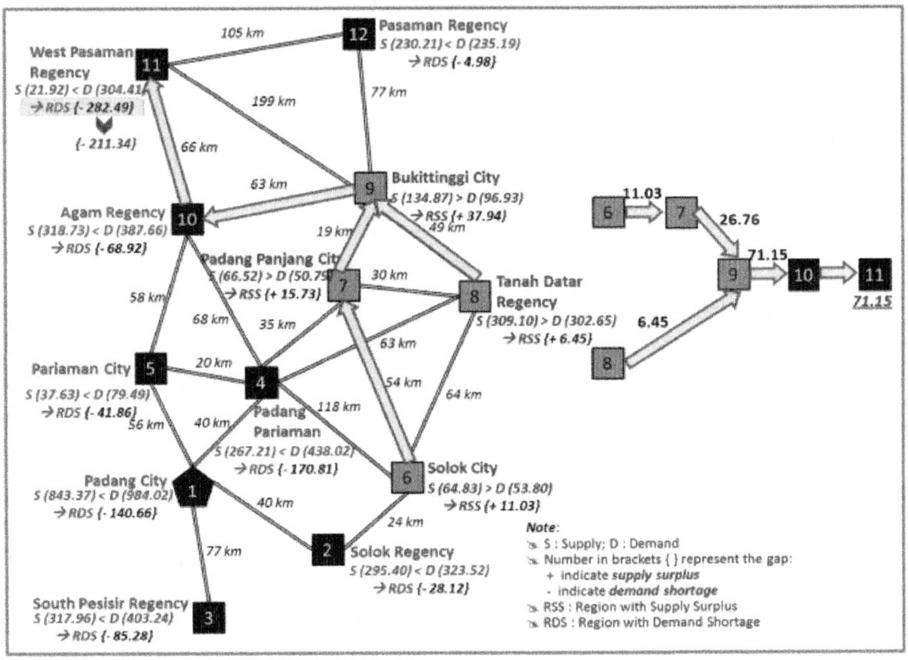

Figure 4: Illustration for delivering drinking water from RES to RED ($K_2 = 210$).

Table 5: Minimum number of vehicles necessary for delivering drinking water.

i (One Batch Time, Cars)	1	2	3	4	5	6	7	8	9	10	11	12	13	14	15	16	17	18	19	20	21
2 (5, 5)	2	2	2	2	2	2	2	2	2	2	1	1	1	1	1	0	0	0	0	0	0
3 (8, 15)	5	5	5	5	5	5	5	5	5	5	5	5	5	5	5	5	5	5	5	5	5
4 (5, 29)	12	12	12	12	12	12	12	12	12	5	5	5	5	5	0	0	0	0	0	0	0
5 (6, 7)	3	3	3	3	3	3	3	3	3	3	3	3	1	1	1	1	1	1	0	0	0
10 (10, 12)	4	4	4	4	4	4	4	4	4	4	4	4	4	4	4	4	4	4	4	4	4
11 (14, 36)	12	12	12	12	12	12	12	12	12	12	12	12	12	12	12	12	12	12	12	12	12
12 (14, 1)	1	1	1	1	1	1	1	1	1	1	1	1	1	1	0	0	0	0	0	0	0
Total	39	39	39	39	39	39	39	39	39	39	31	31	29	29	28	22	22	22	21	21	21

i (One Batch Time, Cars)	22	23	24	25	26	27	28	29	30	31	32	33	34	35	36	37	38	39	40	41	42
2 (5, 5)	0	0	0	0	0	0	0	0	0	0	0	0	0	0	0	0	0	0	0	0	0
3 (8, 15)	5	5	5	0	0	0	0	0	0	0	0	0	0	0	0	0	0	0	0	0	0
4 (5, 29)	0	0	0	0	0	0	0	0	0	0	0	0	0	0	0	0	0	0	0	0	0
5 (6, 7)	0	0	0	0	0	0	0	0	0	0	0	0	0	0	0	0	0	0	0	0	0
10 (10, 12)	4	4	4	4	4	4	4	4	4	0	0	0	0	0	0	0	0	0	0	0	0
11 (14, 36)	12	12	12	12	12	12	12	12	12	12	12	12	12	12	12	12	12	12	12	12	12
12 (14, 1)	0	0	0	0	0	0	0	0	0	0	0	0	0	0	0	0	0	0	0	0	0
Total	21	21	21	16	16	16	16	16	16	12	12	12	12	12	12	12	12	12	12	12	12

We define input data for the model DSCM given by (7)-(10) in order to meet all the demands for the relief

good, i.e. drinking water, in the RED as follows: $I = \{1,2,3,4,5,10,11,12\}$: set of destination regions, $T_S = \{1,2,\cdots,T\}$: set of time horizon, N_i, $i \in I$, and $T_i = \lceil 2L_i/F + T_c \rceil$, where T_c is a transaction time. An optimal solution is given in Table 5. Delivery routes for the relief good of drinking water from SC to RED are shown inFigure 5.

Numerical Results by the Probabilistic Approach

The probabilistic approach considers the probability (P_{ij}) that the road segment is blocked, i.e. becomes unavailable due to a natural disaster, for each road segment (edge) $(i,j) \in E$. This possibility might force the

route for delivering aid commodity as well as the amount of aid commodity to be changed. Following the algorithm shown in Figure 1, we apply conducting $500(=T)$ iterations for the network flow optimization model.

Table 6 presents the computational results giving the mean values of the max-gap, total distance to deliver the relief goods drinking water from RES to RED corresponding to various values of probability of broken road segments. In Table 6, we can see that the available drinking water is delivered from RES to region 11 with the highest demand gap. Figure 6 shows the total distance for delivering the drinking water from SC to RED and the value of the max-gap corresponding to various probability values. From Figure 6, we find that as the probability of broken road segments increases, the total distance for delivering the drinking water from SC to RED region 11 decreases while the max-gap increases. Figure 7 shows the amount of the drinking water from SC to RED corresponding to various probability values. We see that as the probability of broken road segments increases, the amount of the drinking water delivered from SC to region 11 decreases, reaching zero when all road segments were broken with probability 100%.

The higher possibility of broken road segments also suggests alarger remaining demand gap. As we can see from the column of max-gap which represents the amount of remaining maximum demand gap in Table 7, when all the road segments are available$(P_{ij} = 0.00)$, the remaining demand gap of drinking water in region 11 becomes 211.34 m^3, that is after receiving aid of 71.15 m^3 from RES. However, increasing the probability of broken road segments, the remaining demand gap also increases. In addition, there also other consequences; i.e., the routes taken to deliver the drinking water from RES to region 11 also change with the changes of the probability of broken road segments as shown in Table 7. The rows in Table 7 represent the mean of the amount of drinking water delivered to region 11 for certain probabilities of broken road segments (edges), while the columns in Table 7 show the transshipment flows (routes) taken to deliver the drinking water to region 11.

SUMMARY AND CONCLUSIONS

Given that disasters are prominent throughout the world, the importance of disaster management is undeniable, and a large amount of human losses and unnecessary demolition of infrastructure can be avoided with very responsive Disaster Management Action, and thus humanitarian logistics or logistics in an emergency should be well planned in advance.

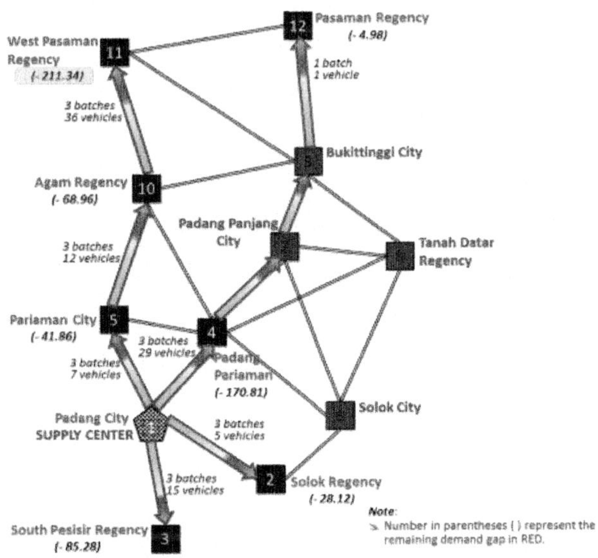

Figure 5: Delivery routes for the drinking water from SC to RED ($K_2 = 210$).

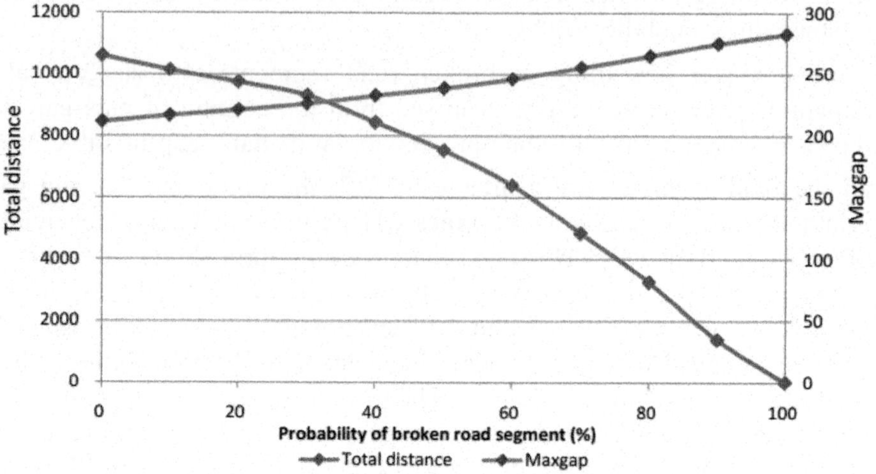

Figure 6: Total distance for delivering drinking water from SC to RED and the max gap.

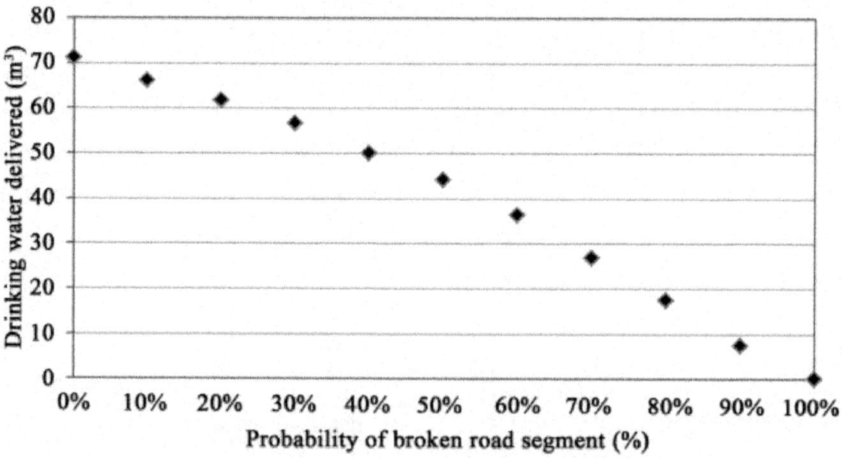

Figure 7: Amount of the drinking water delivered from SC to RED.

Table 6: Mean of max gap, total distance and delivery of drinking water from RES to RED

Probability	Max Gap	Total Distance	Destination Region				
			6	7	8	9	11
0.00	211.34	10598.50	−11.03	−15.73	−6.45	−37.94	71.15
0.10	216.35	10143.16	−7.06	−15.38	−5.75	−37.94	66.14
0.20	220.69	9737.42	−4.54	−14.47	−5.07	−37.71	61.80
0.30	225.84	9309.99	−2.56	−13.09	−4.13	−36.88	56.65
0.40	232.54	8429.27	−1.50	−10.19	−3.20	−35.06	49.95
0.50	238.43	7531.93	−0.90	−8.24	−2.52	−32.40	44.06
0.60	246.04	6397.61	−0.40	−6.01	−1.66	−28.38	36.45
0.70	255.56	4838.31	−0.07	−3.62	−0.94	−22.31	26.93
0.80	264.94	3250.57	−0.02	−1.45	−0.30	−15.78	17.55
0.90	275.07	1384.45	0.00	−0.31	−0.05	−7.06	7.42
1.00	282.49	0.00	0.00	0.00	0.00	0.00	0.00

Note: Transshipment from RES to RED: positive value = imported; negative value = exported.

Table 7: Mean of the transshipment flow of drinking water to region 11.

Probability	Route																	
	4 - 5	4 - 10	5 - 10	6 - 7	7 - 4	7.4	7 - 9	8 - 4	8 - 7	8 - 9	9 - 7	9 - 10	9 - 11	9 - 12	10 - 11	12 - 11	7 - 8	7 - 9
0.0	0.00	0.00	0.00	11.03	0.00	0.00	26.76	0.00	0.00	6.45	0.00	71.15	0.00	0.00	71.15	0.00	26.76	0.00
0.1	0.55	3.19	0.55	7.06	3.19	0.10	20.50	0.55	0.63	4.57	0.61	54.67	0.38	7.35	58.41	7.35	20.50	0.55
0.2	0.77	4.84	0.77	4.54	4.64	0.20	15.69	0.97	0.61	3.62	0.83	42.55	1.44	12.20	48.16	12.20	15.69	0.97
0.3	1.98	6.26	1.98	2.56	6.84	0.30	11.45	1.39	0.71	2.21	2.12	28.78	2.43	17.21	37.01	17.21	11.45	1.39
0.4	0.96	6.89	0.96	1.50	6.50	0.40	8.01	1.34	0.50	1.54	2.50	20.07	4.70	17.33	27.91	17.33	8.01	1.34
0.5	1.01	5.85	1.01	0.90	5.76	0.50	6.25	1.10	0.50	0.98	2.43	15.42	5.84	15.94	22.28	15.94	6.25	1.10
0.6	0.95	4.87	0.95	0.40	4.98	0.60	3.95	0.84	0.35	0.73	2.43	4.37	5.92	11.84	9.17	11.84	1.99	0.64
0.7	0.65	4.16	0.65	0.07	4.16	0.70	1.99	0.64	0.10	0.26	2.43	4.37	5.92	11.84	9.17	11.84	1.99	0.64
0.8	0.12	1.74	0.12	0.02	1.63	0.80	0.90	0.23	0.00	0.06	1.06	1.26	5.77	8.67	3.12	8.67	0.90	0.23
0.9	0.00	0.48	0.00	0.00	0.43	0.90	0.19	0.05	0.00	0.00	0.30	0.38	2.88	3.68	0.86	3.68	0.19	0.05
1.0	0.00	0.00	0.00	0.00	0.00	0.00	0.00	0.00	0.00	0.00	0.00	0.00	0.00	0.00	0.00	0.00	0.00	0.00

In such a situation, an optimal strategy on how to distribute necessary relief commodities to the damaged areas and transport them corresponding to their supply and demand condition as quickly as possible is vitally needed. As part of the action taken in the response phase, our optimization models have tried to address this issue.

In this study, an approach for logistics in emergency under uncertainty has been presented, which can be applied in the occurrence of seriously emergency situations. In such a situation, we try to find an optimal strategy on how to distribute necessary relief commodities to the damaged area and transport them corresponding to their supply and demand condition as quickly as possible. The optimal strategy should take into account two objectives: minimizing the transportation cost (i.e. distance or travel time) and the total demand gap.

In order to solve this problem, we build a multi-commodity transshipment network flow optimization model (NFOM) for making decisions under uncertainty. To evaluate the model and the solution method, we apply the NFOM to the case of the 2009 West Sumatra earthquake in Indonesia. The resulting planning problem consists of the distribution of the relief commodities and the number of vehicles necessary to deliver the relief commodities. Both the NFOM and the demand shortage clearance model (DSCM), including the simulation, and the solution analyses, are applicable to other planning situations such as air transportation and water transportation of various types of relief goods, which have similar characteristics.

The optimal strategy obtained from the NFOM should take into account two objectives, namely, minimizing the transportation cost (i.e. total distance or travel time) and minimizing the maximum demand gap. By assuming a fixed value of the coefficient K_1, if we emphasize more on the importance of the total cost, then we assign a lower value for the coefficient K_2, otherwise, we assign a higher value for K_2. Thus, the general strategy obtained from the NFOM to decrease (minimize) the largest excess demand is to find a set of regions with extra supply (positive value of the gap), and then try to connect those regions to the region with the largest excess demand. Subsequently, depending on the solution obtained from the NFOM, we solve the DSCM to find the shortest path and minimize the number of batches for sending vehicles necessary to deliver relief aid from SC, in order to satisfy all the remaining demand gaps in RED. As a logical consequence which can be seen from the model solution, if we choose to minimize the maximum gap at the NFOM, then at the DSCM, the number of vehicles necessary to deliver relief commodities to the region with the largest gaps can be reduced.

Both NFOM and DSCM are necessary, useful, and important, even though they have their own shortcomings and limitations in the sense that neither can solve the total and comprehensive relief goods transport problem completely. However, we believe that under conditions of uncertainty such as after the occurrence of natural disasters, and given that the amount of damage and casualties may not be known, our model results can give us new insights and findings regarding how we should handle relief goods transport problems under emergency situations.

In this regard, we can say from the viewpoint of public policy that the government must carefully plan the humanitarian logistics in advance. More specifically, the government should always keep an inventory of the supply of relief commodities and update it in terms of both quantity and quality, and keep track of the transportation modes.

ACKNOWLEDGEMENTS

The authors would like to express our sincere thanks to the reviewerswho have given us insightful,invaluable and useful comments. Their comments led us to a significant improvements of the paper.

REFERENCES

1. EM-DAT (2014) The OFDA/CRED International Disaster Database. Universite catholique de Louvain, Brussels. www.emdat.be
2. Parwanto, N.B. and Oyama, T. (2014) A Statistical Analysis and Comparison of Historical Earthquake and Tsunami Disasters in Japan and Indonesia. International Journal of Disaster Risk Reduction, 7, 122-141. http://dx.doi.org/10.1016/j.ijdrr.2013.10.003
3. Jiang, Y., Yuan, Y, Huang, K. and Zhao, L. (2012) Logistics for Large-Scale Disaster Response: Achievements and Challenges. 45th Hawaii International Conference on System Sciences, Maui, 4-7 January 2012, 1277-1285.http://dx.doi.org/10.1109/HICSS.2012.418
4. Kovacs, G. and Spens, K.M. (2007) Humanitarian Logistics in Disaster ReliefOperations.InternationalJournalofPhysicalDistribution&Logistics Management, 37, 99-114.http://dx.doi.org/10.1108/09600030710734820
5. Liu, D., Han, J.Y. and Zhu, J.M. (2007) Vehicle Routing for Medical Supplies in Large-Scale Emergencies. The First International Symposium on Optimization and Systems Biology (OSB2007), Beijing, 8-10 August 2007, 412-419.http://www.aporc.org/LNOR/7/OSB2007F47.pdf
6. Van Wassenhove, L.N. (2006) Humanitarian Aid Logistics: Supply Chain Management in High Gear. Journal of Operational Research Society, 57, 475-489.http://dx.doi.org/10.1057/palgrave.jors.2602125
7. Altay, N. and Green, W.G. (2006) OR/MS Research in Disaster Operations Management. European Journal of Operational Research, 175, 475-493. http://dx.doi.org/10.1016/j.ejor.2005.05.016
8. Ozdamar, L., Ediz, E. and Beste, K. (2004) Emergency Logistics Planning in Natural Disasters. Annals of Operations Research, 129, 217-245.http://dx.doi.org/10.1023/B:ANOR.0000030690.27939.39

9. Haghani, A. and Oh, S.C. (1996) Formulation and Solution of a Multi-Commodity Multi-Modal Network Flow for Disaster Relief Operations. Transportation Research Part A: Policy and Practice, 30, 231-250. http://dx.doi.org/10.1016/0965-8564(95)00020-8
10. Fiedrich, F., Gehbauer, F. and Rickers, U. (2000) Optimized Resource Allocation for Emergency Response after Earthquake Disasters. Safety Science, 35, 41-57.http://dx.doi.org/10.1016/S0925-7535(00)00021-7
11. Lin, Y.H., Batta, R., Rogerson, P.A., Blatt, A. and Flanigan, M. (2009) Application of a Humanitarian Relief Logistics Model to an Earthquake Disaster.http://www.acsu.buffalo.edu/~batta/TRB_Updated.pdf
12. Tzeng, G.H., Cheng, H.J. and Huang, T.D. (2007) Multi-Objective Optimal Planning for Designing Relief Delivery Systems. Transportation Research Part E, 43, 673-686.http://dx.doi.org/10.1016/j.tre.2006.10.012
13. Vitoriano, B., Ortuno, M.T., Tirado, G. and Montero, J. (2011) A Multi-Criteria Optimization Model for Humanitarian Aid Distribution. Journal of Global Optimization, 51, 189-208.http://dx.doi.org/10.1007/s10898-010-9603-z
14. Yi, W. and Ozdamar, L. (2007) A Dynamic Logistics Coordination Model for Evacuation and Support in Disaster Response Activities. European Journal of Operational Research, 179, 1177-1193. http://dx.doi.org/10.1016/j.ejor.2005.03.077
15. Barbarosoglu, G. and Arda, Y. (2004) A Two-Stage Stochastic Programming Framework for Transportation Planning in Disaster Response. Journal of the Operational Research Society, 55, 43-53. http://dx.doi.org/10.1057/palgrave.jors.2601652
16. Winston, W.L. (2003) Operations Research Applications and Algorithms. 4th Edition, Duxbury Press, California.
17. Herer, Y.T., Tzur, M. and Yuecesan, E. (2006) The Multilocation Transshipment Problem. IIE Transactions, 38, 185- 200. http://dx.doi.org/10.1080/07408170500434539
18. Rottkemper, B., Fischer, K. and Blecken, A. (2012) A Transshipment Model for Distribution and Inventory Relocation under Uncertainty in Humanitarian Operations. Socio-Economic Planning Sciences, 46, 98-109.http://dx.doi.org/10.1016/j.seps.2011.09.003
19. West Sumatra and Jambi Natural Disasters: Damage, Loss and Preliminary Needs Assessment. 2009. A Joint Report by the BPNB, Bappenas, and the Provincial and District/City Governments of West Sumatra and Jambi and International Partners.

20. Sugimin, P. (2011) Lesson Learned: Rehabilitation and Reconstruction West Sumatra September 30th, 2009 Earthquake, Building Back Better. Gramedia Ltd. Printing Group, Jakarta.

Chapter 10

OPTIMIZATION OF ENERGY RESOURCE MANAGEMENT FOR ASSEMBLY LINE BALANCING USING ADAPTIVE CURRENT SEARCH

Supaporn Suwannarongsri[1], Tika Bunnag[1], Waraporn Klinbun[2]

[1]Rattanakosin College for Sustainable Energy and Environment (RCSEE), Rajamangala University of Technology Rattanakosin, Nakhon Pathom, Thailand

[2]Faculty of Engineering and Technology, Panyapiwat Institute of Management, Nonthaburi, Thailand

ABSTRACT

This paper aimed to present the optimization of energy resource management in a car factory by the adaptive current search (ACS)—one of the most efficient metaheuristic optimization search techniques. Assembly lines of a specific car factory considered as a case study are balanced by the ACS to optimize their energy resource management. The workload variance of the line is performed as the objective function to be minimized in order to increase the productivity. In this work, the ACS is used to address the number of tasks assigned for each workstation, while the sequence of tasks is assigned by factory. Three real-world assembly line balancing (ALB) problems from a specific car factory are tested. Results obtained by the ACS are compared with those obtained by the genetic algorithm (GA), tabu search (TS) and current search (CS). As results, the ACS outperforms other algorithms. By using the ACS, the productivity can be increased and the energy consumption of the lines can be decreased significantly.

INTRODUCTION

Metaheuristic optimization is one of the most important issues in computer science and operation research. According to the energy resource optimization context, energy resource management is an alternative to increase the productivity or decrease the energy consumption. Especially in car assembly industries, total energy of such industries can be divided into two parts, i.e. main energy used for assembling products and supporting energy used for illumination, ventilation and so on. Minimized workload variance of the assembly line can increase productivity and also increase energy efficiency. Energy resource management can be considered as a class of non-polynomial time hard (NP-hard) combinatorial optimization problem [1,2]. So the problem usually possesses nonlinear and unsymmetrical terms as well as multiple local solutions. These cause the problem complex and difficult to solve by an exact method within a reasonable amount of time. In addition, inefficient algorithms are easily trapped due to its local solutions. To solve the problem, metaheuristic optimization methods are alternatives [3-5]. Over five decades, many metaheuristics such as genetic algorithm (GA), tabu search (TS), particle swarm optimization (PSO) and harmony search (HS) have been developed for combinatorial, continuous and multiobjective optimization problems [5,6]. Metaheuristics has been widely applied to various real-world engineering problems [7,8]. Some metaheuristics such as GA, TS, PSO and HS have been accepted and applied to some energy saving and management applications [9-12].

By literatures, several methods have been launched to solve the assembly line balancing (ALB) problems, for example, heuristic approaches [13,14], metaheuristics such as GA [15] and TS [16,17], and hybrid methods [18,19]. Although GA and TS could provide satisfactory results, they may be trapped by some local optima and spent amount of search time. One of the latest metahueristics is the current search (CS) [20]. The CS has been successfully applied to control engineering [21] and signal processing [22]. However, local entrapments in CS may occur. In order to improve its effectiveness, the CS needs to be modified. The modified version of the CS named adaptive current search (ACS) has been proposed [23]. The ACS shows the superior results to the conventional CS and other metaheuristic algorithms.

This paper proposes the application of the ACS to optimize energy resource management in a car factory. Three real-world assembly line balancing (ALB) problems from a specific car factory are conducted. The workload variance of such lines is minimized in order to increase the productivity and decrease the energy consumption. After an introduction in Section 1, the problem formulation of energy resource management of ALB problems is explained in Section 2. The ACS algorithms are briefly described in Section 3. Results of solving ALB problems by the ACS compared with GA, TS and CS are provided in Section 4, while discussions and conclusions are given in Section 5 and Section 6, respectively.

PROBLEM FORMULATION

The problem formulation of energy resource management of ALB problems of a car factory in sense of energy resource management is presented in this Section. The ALB problem is considered as one of the classical industrial engineering problems [24]. An assembly line is a sequence of workstations connected together by a material handling system. It is used to assemble components or tasks into a final product. The fundamental of the line balancing problems is to assign the tasks to an ordered sequence of workstations in order to minimize the workload variance of the line, whereas satisfying two particular constraints. The first constraint is that the total processing time assigned to each workstation should be less than or equal to the cycle time. The second is that the task assignments should follow the sequential processing order of the tasks or the precedence constraints.

The ALB can be considered as the class of combinatorial optimization problems known as NP-hard [13,25]. Traditionally, the assembly line will be represented by the precedent diagram. Figure 1shows the precedent diagram of a simple assembly line consisting of 29 tasks, where node weight stands for the task time in time units.

In this work, the single-model ALB problem is considered. Balancing of the lines can be measured by the workload variance [14,25]. Therefore, the goals of balancing lines are to minimize the workload variance. Analytical formulations for the ALB problems are stated in (1) - (4), where m is the number of workstations, W is the total processing time, c is the cycle time, T_i is the processing time of the ith workstation, T_{id} is the idle time, w_v is the workload variance and E is the line efficiency.

$$m = W/c \tag{1}$$

$$T_{id} = \sum_{i=1}^{m}(c - T_i) \tag{2}$$

$$w_v = \frac{1}{m}\sum_{i=1}^{m}\left[T_i - (W/m)\right]^2 \tag{3}$$

$$E = \sum_{i=1}^{m} T_i/(m \times c) \tag{4}$$

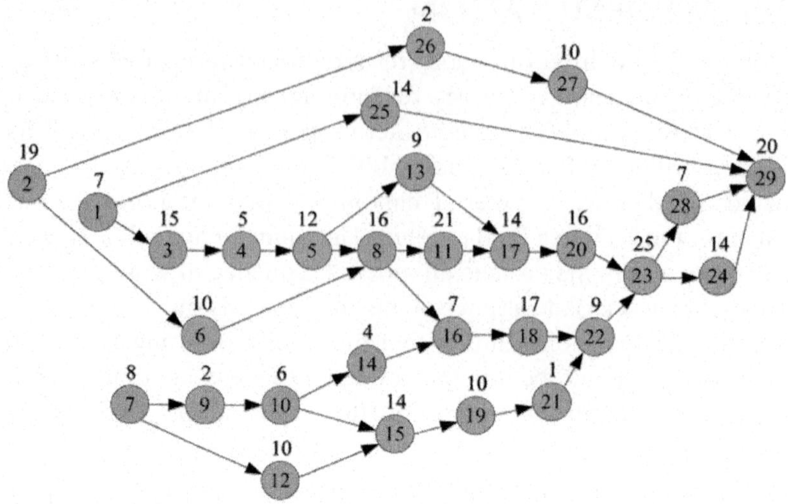

Figure 1: Precedent diagram of simple assembly line.

The ALB problem can be considered as the constrained optimization problem. The workload variance (w_v) is performed as the objective function (J) stated in (5). J will be minimized according to the precedence constraints expressed in (6). Moreover, the constraint in (6) also indicates that the processing time T_i of the ith workstation must equal or less than the cycle time (c).

$$J = w_v = \frac{1}{m}\sum_{i=1}^{m}\left[T_i - (W/m)\right]^2 \tag{5}$$

$$\begin{aligned} &\textbf{minimize} \quad J \\ &\textbf{subject to} \quad T_i \leq c, \\ &\qquad\qquad\quad \textbf{precedent constrains} \end{aligned} \tag{6}$$

Once w_v is reduced, E is then increased. Working time to complete the product can be also reduced. In sense of energy resource management, reduced workload variance can increase productivity and decrease total energy consumption. This approach can be represented by Figure 2. Assume that the total energy of the line consists of main energy used for assembling products and supporting energy used for illumination and ventilation systems. Total energy (Z_{total}) is stated in (7), where Z_{main} is the main energy used for assembly and $Z_{support}$ is the supporting energy used for illumination and ventilation, respectively.

$$Z_{total} = Z_{main} + Z_{support} \tag{7}$$

Referring to Figure 2, a simple assembly line consists of 5 workstations. Time for finishing each product of unbalanced line in Figure 2(a) is equal to cycle time (c_1), while that for finishing each product of balanced line in Figure 2(b) is equal to cycle time (c_2). Once $c_2 < c_1$, this means that time for finishing each product of balanced line is less than unbalanced line does. Total energy per product ($Z_{total/product}$) and supporting energy per product ($Z_{support/product}$) can be calculated by (8) and (9), where N_{item} is numbers of products produced within a certain period. It was found that, although the main energy used for assembling each product is a same value between unbalanced and balanced lines, total energy per product ($Z_{total/product}$) and supporting energy per product ($Z_{support/product}$) can be decreased when cycle time is reduced. By this approach, productivity can be increased and also total energy consumption can be decreased.

$$Z_{total/product} = \frac{Z_{total}}{N_{item}} = \frac{Z_{main} + Z_{support}}{N_{item}} \tag{8}$$

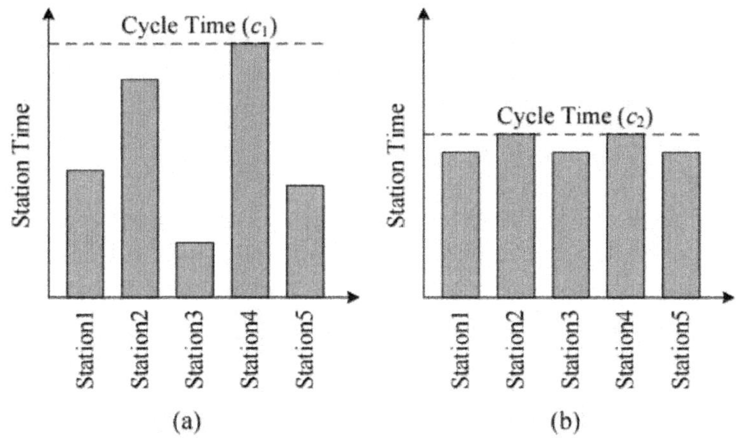

Figure 2: Assembly line consisting of 5 workstations: (a) unbalanced line, (b) balanced line.

$$Z_{support/product} = \frac{Z_{support}}{N_{item}} \tag{9}$$

ACS ALGORITHM

In 2012, the current search (CS) was developed based on the principle of current divider in electric circuit theory [20-22]. The CS possesses both exploration and exploitation properties according to the metaheuristic optimization context. The advantages of the CS are that it can find any local solution in each search direction efficiently. It can perform unlimited search directions defined by users. However, the disadvantages of the CS are that its search process may be trapped or locked by any local solution. In addition, the search time consumed by the CS is depended on the numbers of search directions.

The adaptive current search (ACS), the modified version of the CS, is proposed in 2013 [23]. It provides the memory list (ML) and the adaptive radius (AR) mechanism inserting into the CS algorithms. The ML is used to escape from local entrapment caused by any local solution, while the AR is conducted to speed up the search process. The proposed ML consists of three levels: low, medium and high. The low-level ML is used to store the ranked initial solutions at the beginning of search process, the medium-level ML is conducted to store the solution found along each search direction, and the high-level ML is used to store all local solutions found at the end of each search direction. Performance evaluation of the ACS was performed against benchmark unconstrained and constrained optimization problems [23]. Compared with genetic algorithm (GA), tabu search (TS) and current search (CS), it was found that the ACS shows the superior results to the conventional CS and other algorithms. Some movements of the ACS search process over the search space can be depicted in Figure 3. The ACS algorithm is described as follows:

Step 1 Initialize the search space W, iteration counter $k = j = 1$, maximum allowance of solution cycling j_{max}, number of initial solutions N, number of neighborhood members n, search radius R, and low-level ML Y = Æ, medium-level ML G = Æ, and high-level ML X = Æ.

Step 2 Uniformly random initial solution X_i, $i = 1,\cdots,N$ within W.

Step 3 Evaluate the objective function $f(X_i)$ for “X. Rank X_i that gives $f(X_1) <\cdots< f(X_N)$, then store ranked X_i into the low-level ML Y.

Step 4 Let $x_0 = x_k$ as selected initial solution.

Step 5 Uniformly random neighborhood x_i, $i = 1,\cdots,n$ around x_0 within radius R.

Step 6 Evaluate the objective function $f(x_i)$ for “x. A solution giving the minimum objective function is set as x^*.

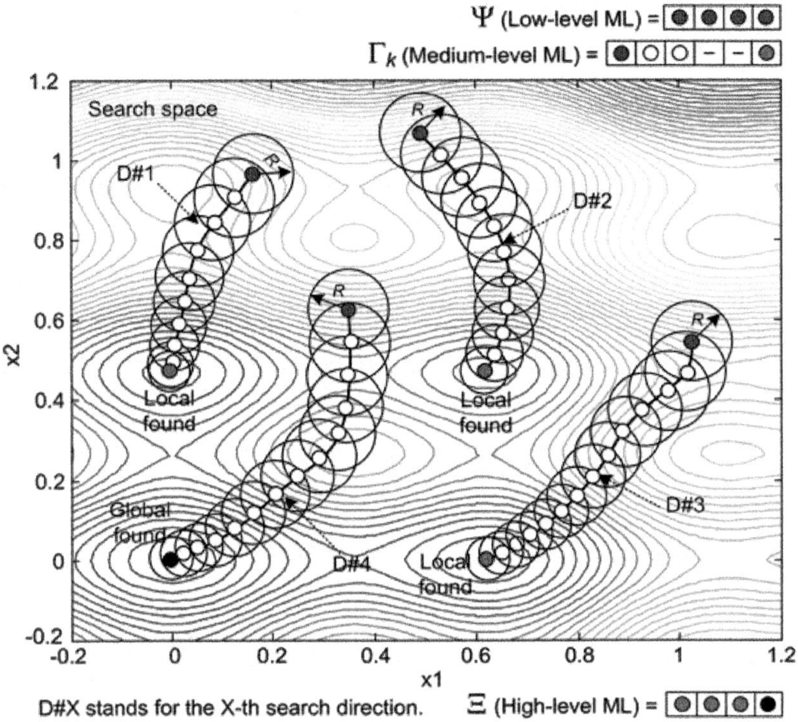

Figure 3: Some movements of ACS algorithm.

Step 7 If $f(x^*) < f(x_0)$, keep x_0 into medium-level ML G_k and set $x_0 = x^*$, set $j = 1$ and return to Step 5. Otherwise keep x^* into medium-level ML G_k and update $j = j + 1$.

Step 8 If the search process moves close to the local solution, activate the AR mechanism by adjusting $R = rR$, $0 < r < 1$.

Step 9 If $j < j_{max}$, return to Step 5. Otherwise keep x_0 into high-level ML X and update $k = k + 1$.

Step 10 Terminate the search process when the termination criteria (TC) are met. The optimum solution found is x_0. Otherwise return to Step 4.

EXPERIMENTAL RESULTS

This Section presents the application of the ACS to optimize energy resource management of assembly line balancing (ALB) problems in a car factory. Sammitr Motors Manufacturing Public Co., Ltd. [26], a car factory located in Samut Sakhon, Thailand, is considered as a case study of this work. It produces car and truck body parts, moulds and fixtures for auto makers such as HINO, ISUZU, MITSUBISHI and NISSAN. Many kinds of car and truck body parts under "Sammitr" trademark are assembled from this factory. For a truck, it comprises of several parts, such as turning disk, dragging arm, frame, rear cap, side cap and so forth. In this factory, each part belongs itself assembly line. For this work, three ALB problems of that car factory, i.e. turning disk, dragging arm and frame assembly lines, are selected. Details of the selected ALB problems are described as follows.

• For the first ALB problem, turning disk assembly line consists of 63 tasks with total working time of 459.63 min. Details of turning disk assembly line are provided in Appendix, for example. Among those tasks, there are 24 tasks for welding (278.20 min.), 9 tasks for grinding and eroding (57.80 min.) and 11 tasks for moving (26.55 min.). Data from factory reveal that there are 5 workstations with the cycle time of 152.33 min. This means that time for finishing each product of this line is equal to 152.33 min.

• For the second ALB problem, dragging arm assembly line consists of 70 tasks with total working time of 307.41 min. Among them, there are 29 tasks for welding (184.67 min.), 7 tasks for grinding and eroding (36.28 min.) and 6 tasks for moving (18.00 min.). Data from factory show that there are 6 workstations with the cycle time of 123.02 min. Time for finishing each product of this line is equal to 123.02 min.

• For the third ALB problem, frame assembly line consists of 75 tasks with total working time of 1,334.27 min. Among them, there are 34 tasks for welding (853.95 min.), 4 tasks for grinding and eroding (55.71 min.) and 9 tasks for moving (165.28 min.). Data from factory indicate that there are 5 workstations with the cycle time of 651.94 min. This means that time for finishing each product of this line is equal to 651.94 min.

All three lines are run 440 min. per day, 25 days per month and 12 months per year. Power used for completing the product consists of one welding machine of 13 kW (a_1 = 13,000 W × 1), one grinding and eroding motor of 1 Hp (a_2 = 746 W × 1) and one crane motor of 2 Hp (a_3 = 1,492 W × 1). Power used for illumination system comes from 18 halogen lamps of 400 W (b_1 = 400 W × 18) and 84 fluorescent lamps with their ballasts of 36 W + 10 W (b_2 = 46 W × 84). Power used for ventilation system comes from 12 motors of 3/4Hp (g = 559.50 W × 12). Therefore, the main energy used for assembly (Z_{main}) and the supporting energy used for illumination and ventilation ($Z_{support}$) of these lines can be yearly calculated in kilowatt-hour (kWh) unit by (10) and (11), where T_1 is time for welding, T_2 is time for grinding and eroding, and T_3 is time for moving. In this work, power for cutting, blowing, drilling and broaching are neglected.

$$Z_{\text{main}} = \frac{N_{\text{item}} \times \sum_{i=1}^{3}(\alpha_i \times T_i)}{60 \times 1{,}000} \quad (10)$$

$$Z_{\text{support}} = \underbrace{\left(\frac{(\beta_1+\beta_2)\times 12\times 25\times 440)}{60\times 1{,}000}\right)}_{\text{illumination}} + \underbrace{\left(\frac{\gamma\times 12\times 25\times 440)}{60\times 1{,}000}\right)}_{\text{ventilation}} \quad (11)$$

Data of all three ALB problems from factory are summarized in Table1 In this work, the ACS is conducted to solve the ALB problems. The ACS is used to address the number of tasks assigned for each workstation, while the sequence of tasks was assigned by factory. The workload variance (w_v) performed as the objective function (J) stated in (5) will be minimized according to the precedence constraints expressed in (6).

In this paper, algorithms of GA and TS are omitted. Readers may refer to [27,28] for GA and [29,30] for TS, respectively. For all three ALB problems, the parameter settings for the GA follow MATLAB-GA Toolbox [28] and for the TS follow [30].

Table 1: Details of selected ALB problems

Data	ALB problems		
	Tuning Disk	**Dragging Arm**	**Frame**
n (tasks)	63	70	75
W (min.)	459.63	307.41	1,334.27
c (min.)	152.33	123.02	651.94
m (stations)	5	6	5
α_1 (W)	13,000 W × 1	13,000 W × 1	13,000 W × 1
α_2 (W)	746 W × 1	746 W × 1	746 W × 1
α_3 (W)	1,492 W × 1	1,492 W × 1	1,492 W × 1
β_1 (W)	400 W × 18	400 W × 18	400 W × 18
β_2 (W)	46 W × 84	46 W × 84	46 W × 84
γ(W)	559.50 W × 12	559.50 W × 12	559.50 W × 12
T_1 (min.)	278.20	184.67	853.95
T_2 (min.)	57.80	36.28	55.71
T_3 (min.)	26.55	18.00	165.28

The common search parameters of the CS and ACS are: n (number of neighborhood members) = 1,000, R (search radius) = 20% of search space and I_{max} (number of iterations) = 1,000. N (number of search directions) = 10 is required to terminate the search. For the ACS, R-adjustment (search radius adjustment of AR mechanism) is set as I_{max} = 500 ® R = 10% of search space and I_{max} = 750 ® R = 5% of search space. The TS, CS and ACS algorithms were coded by MATLAB, while GA was conducted from MATLAB-GA Toolbox. All algorithms were run on Intel Core2 Duo 2.0 GHz 3 Gbytes DDR-RAM computer. Table 2provides the boundaries of number of tasks for each workstation set for the corresponding search spaces. For all algorithms, 50 trials were run to obtain the best solution. Results obtained are summarized in Table 3, while convergent rates are plotted in Figure 4-Figure 6.

DISCUSSIONS

As results of the first ALB problem in Table 3, turning disk assembly line arranged by factory and optimized by GA, TS, CS and ACS can be plotted in Figure 7-Figure 11 and Table 4-Table 8, respectively. The ACS provides superior results to other algorithms and certainly a factory. The workload variance (w_v) obtained by the ACS is less than that obtained by GA, TS and CS. Once w_v is reduced, the line efficiency (E) will be increased. It can be observed that E in Table 3 optimized by the ACS is the highest value.

By those results, percent of decreased total energy per produce (PDTE) and percent of decreased supporting energy per produce (PDSE) can be calculated by (12) and (13), where $Z_{total/product,old}$ is total energy per produce of factory, $Z_{total/product,new}$ is total energy per produce obtained by GA, TS, CS or ACS, $Z_{support/product,old}$ is supporting energy per produce of factory, and $Z_{support/product,new}$ is supporting energy per produce obtained by GA, TS, CS or ACS. Results are summarized in Table9

$$PDTE = 100 \times \left(\frac{Z_{total/product,old} - Z_{total/product,new}}{Z_{total/product,old}} \right) \tag{12}$$

$$PDSE = 100 \times \left(\frac{Z_{support/product,old} - Z_{support/product,new}}{Z_{support/product,old}} \right) \tag{13}$$

As results of the second and third ALB problems in Table 3, dragging arm and frame assembly lines arranged by factory and optimized by GA, TS, CS and ACS are omitted because they have a similar pattern to those of the first ALB problem shown in Figure 7-Figure 11 and Table 4-Table8 Referring to Table 3, the ACS provides superior results to other algorithms and a factory. The workload variance obtained by the ACS is less than that obtained by GA, TS and CS, while E obtained by the ACS is the highest.

Table 2: Lists of search spapces.

ALB problems	Search spaces
Tuning Disk	$S_1 \in [2, 15]$; $S_2 \in [2, 15]$; $S_3 \in [10, 30]$; $S_4 \in [2, 15]$; $S_5 \in [10, 25]$
Dragging Arm	$S_1 \in [2, 10]$; $S_2 \in [10, 30]$; $S_3 \in [2, 10]$; $S_4 \in [2, 20]$; $S_5 \in [10, 20]$; $S_6 \in [2, 15]$
Frame	$S_1 \in [5, 10]$; $S_2 \in [10, 20]$; $S_3 \in [5, 20]$; $S_4 \in [15, 30]$; $S_5 \in [10, 20]$

Table 3: Results of ALB problems

		ALB#1: Turning Disk		
Methods	w_v	T_{id}(min.)	E (%)	Search Time (sec.)
Factory	1,793.10	302.02	60.35	-
GA	457.03	302.02	75.11	510.64
TS	333.03	302.02	76.17	342.58
CS	293.09	302.02	77.02	248.26
ACS	249.28	302.02	81.42	221.45
		ALB#2: Dragging Arm		
Methods	w_v	T_{id}(min.)	E (%)	Search Time (sec.)
Factory	1,286.60	430.71	41.65	-
GA	134.18	430.71	70.82	522.54
TS	122.12	430.71	77.64	397.23
CS	88.59	430.71	79.64	348.67
ACS	72.89	430.71	81.39	291.15
		ALB#3: Frame		
Methods	w_v	T_{id}(min.)	E (%)	Search Time (sec.)
Factory	38,119.43	1,925.43	40.93	-
GA	3,534.78	1,925.43	74.45	566.14
TS	2,430.51	1,925.43	74.45	497.33
CS	2,257.49	1,925.43	77.23	375.36
ACS	1,563.12	1,925.43	78.51	292.48

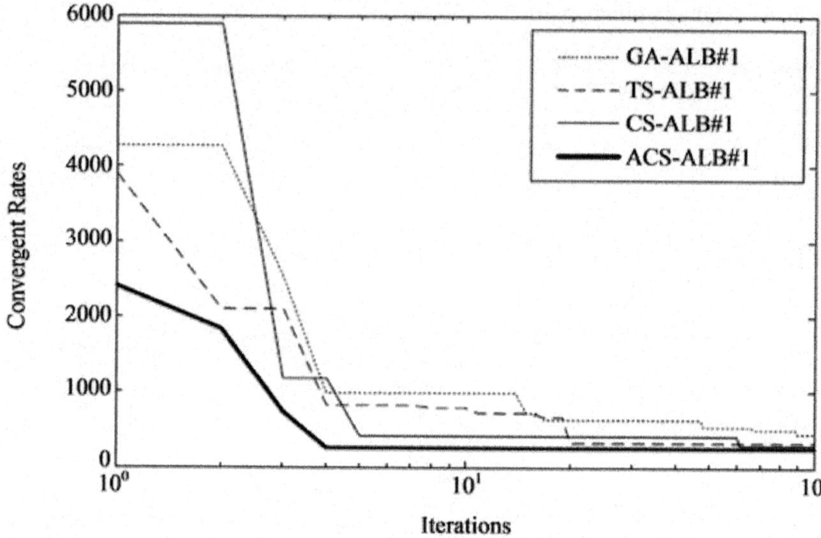

Figure 4: Convergent rate of the first ALB problem.

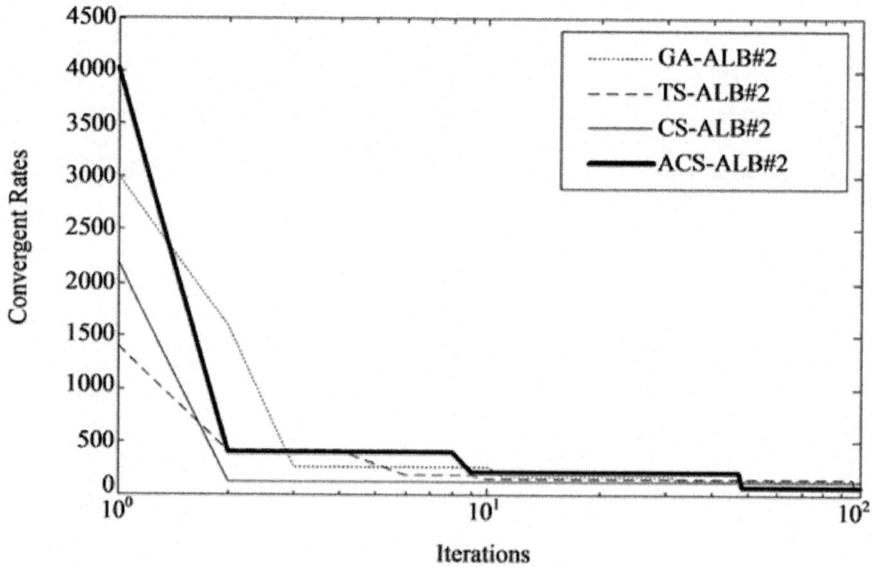

Figure 5: Convergent rate of the second ALB problem.

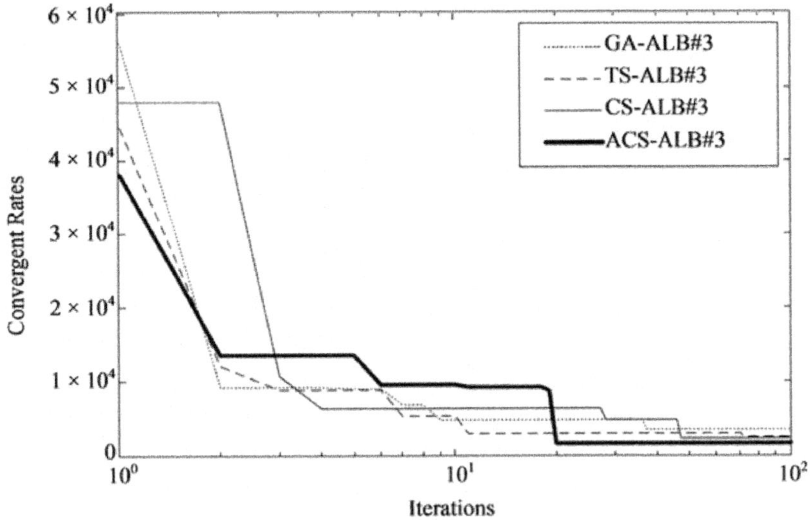

Figure 6: Convergent rate of the third ALB problem.

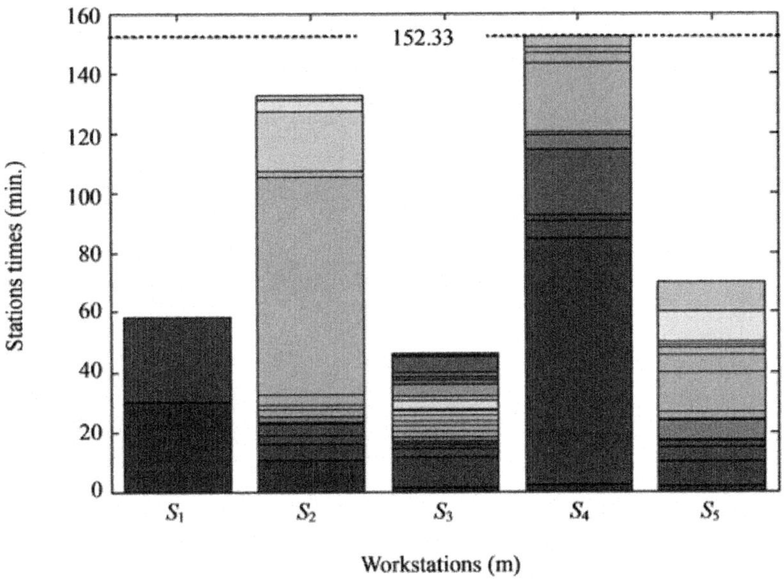

Figure 7: Turning disk assembly line by factory.

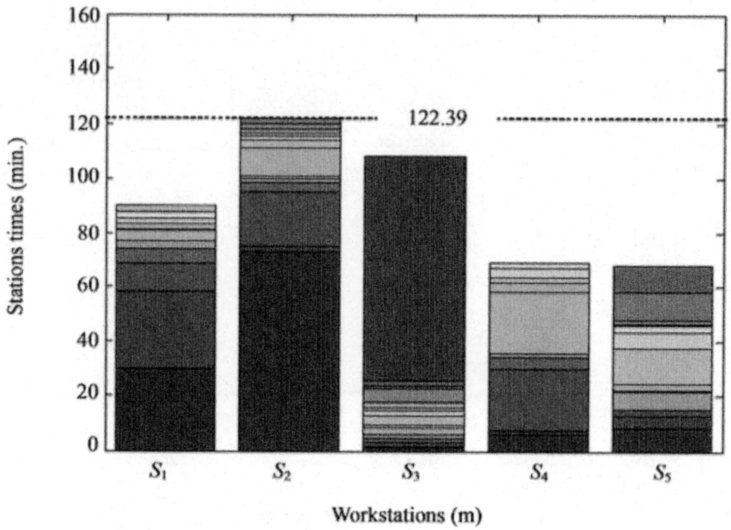

Figure 8: Turning disk assembly line by GA.

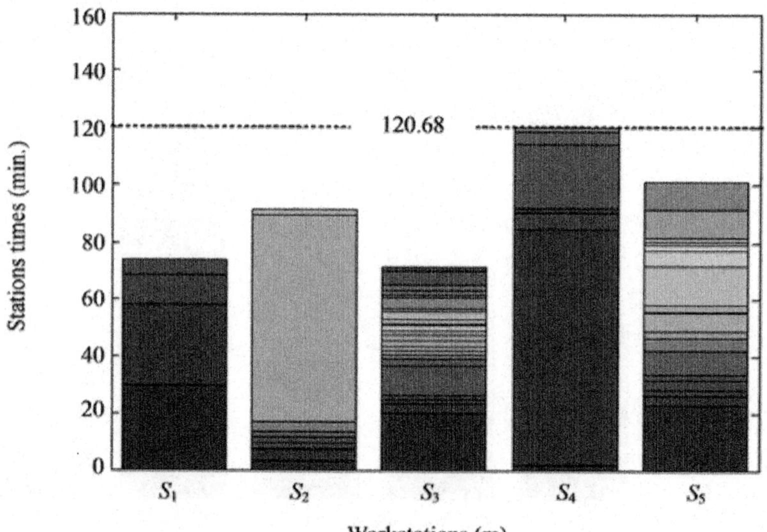

Figure 9: Turning disk assembly line by TS.

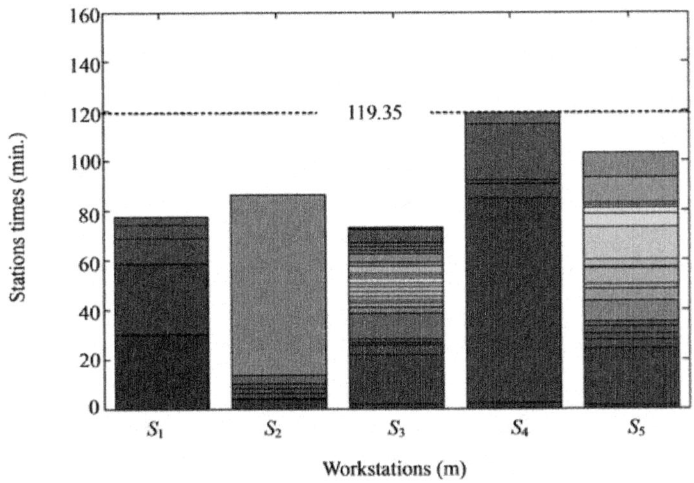

Figure 10: Turning disk assembly line by CS.

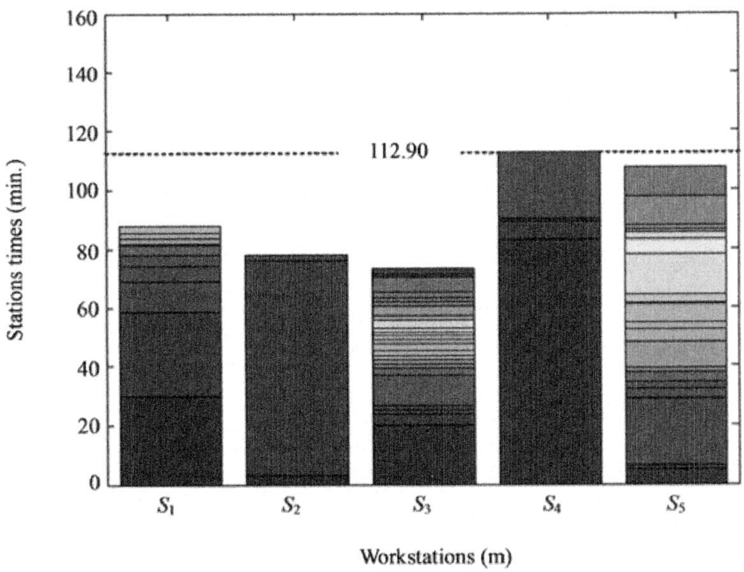

Figure 11: Turning disk assembly line by ACS.

Table 4: Tasks for turning disk assembly line by factory.

Work-Station (m)	Tasks assigned for each workstation	Station Time (min.)	Idle Time (min.)
1	1,2	58.44	93.89
2	3,4,5,6,7,8,9,10,11,12,13,14,15,16	132.42	19.91
3	17,18,19,20,21,22,23,24,25,26,27,28,29,30,31,32,33,34,35,36,37	46.31	106.02
4	38,39,40,41,42,43,44,45,46,47,48	152.33	0.00
5	49,50,51,52,53,54,55,56,57,58,59,60,61,62,63	70.13	82.20

The total idle time (T_{id}) = 303.02 min., The workload variance (w_v) = 1,793.10, The line efficiency (E) = 60.35%.

Table 5: Tasks for turning disk assembly line by GA.

Work-Station (m)	Tasks assigned for each workstation	Station Time (min.)	Idle Time (min.)
1	1,2,3,4,5,6,7,8,9,10,11	90.79	61.54
2	12,13,14,15,16,17,18,19,20,21,22,23,24	122.39	29.94
3	25,26,27,28,29,30,31,32,33,34,35,36,37,38,39	108.84	43.49
4	40,41,42,43,44,45,46,47,48,49	69.33	83.00
5	50,51,52,53,54,55,56,57,58,59,60,61,62,63	68.28	84.05

The total idle time (T_{id}) = 303.02 min., The workload variance (w_v) = 457.03, The line efficiency (E) = 75.11%.

Table 6: Tasks for turning disk assembly line by TS.

Work-Station (m)	Tasks assigned for each workstation	Station Time (min.)	Idle Time (min.)
1	1,2,3,4	74.04	78.29
2	5,6,7,8,9,10,11,12,13	91.73	60.60
3	14,15,16,17,18,19,20,21,22,23,24,25,26,27,28,29,30,31,32,33,34,35,36,37	71.40	80.93
4	38,39,40,41,42,43,44	120.68	31.65
5	45,46,47,48,49,50,51,52,53, 54,55,56,57,58,59,60,61,62, 63	101.78	50.55

The total idle time (T_{id}) = 303.02 min., The workload variance (w_v) = 333.03, The line efficiency (E) = 76.17%.

Table 7: Tasks for turning disk assembly line by CS.

Work-Station (m)	Tasks assigned for each workstation	Station Time (min.)	Idle Time (min.)
1	1,2,3,4,5	77.53	74.80
2	6,7,8,9,10,11,12	86.45	65.88
3	13,14,15,16,17,18,19,20,21,22,23,24,25,26,27, 28,29,30,31,32,33,34,35,36,37	73.19	79.14
4	38,39,40,41,42,43	119.35	32.98
5	44,45,46,47,48,49,50,51,52,53,54,55,56,57,58,59,60,61,62,63	103.11	49.22

The total idle time (T_{id}) = 303.02 min., The workload variance (w_v) = 293.09, The line efficiency (E) = 77.02%.

Table 8: Tasks for turning disk assembly line by ACS.

Work-Station (m)	Tasks assigned for each workstation	Station Time (min.)	Idle Time (min.)
1	1,2,3,4,5,6,7,8,9,10	87.88	64.45
2	11,12,13	77.89	74.44
3	14,15,16,17,18,19,20,21,22,23,24,25,26,27,28,29, 30,31,32,33,34,35,36,37,38	73.32	79.01
4	39,40,41,42	112.90	39.43
5	43,44,45,46,47,48,49,50,51, 52,53,54,55,56, 57,58,59,60, 61,62,63	107.64	44.69
The total idle time (T_{id}) = 303.02 min., The workload variance (w_v) = 249.28, The line efficiency (E) = 81.42%.			

Table 9: Yearly results of energy resource management of ALB problems.

ALB#1: Turning Disk									
Methods	Number of products	Z_{total} (MWh)	$Z_{total/product}$ (kWh)	$PDTE$ (%)	Z_{main} (MWh)	$Z_{main/product}$ (kWh)	$Z_{support}$ (MWh)	$Z_{support/product}$ (kWh)	$PDSE$ (%)
Factory	866.54	92.54	106.79	-	53.43	61.66	39.11	45.14	-
GA	1,078.52	105.61	97.92	8.31	66.50	61.66	39.11	36.26	19.65
TS	1,093.80	106.55	97.41	8.78	67.44	61.66	39.11	35.76	20.78
CS	1,105.99	107.30	97.02	9.15	68.19	61.66	39.11	35.36	21.65
ACS	1,169.18	111.20	95.11	10.94	72.09	61.66	39.11	33.45	25.88
ALB#2: Dragging Arm									
Methods	Number of products	Z_{total} (MWh)	$Z_{total/product}$ (kWh)	$PDTE$ (%)	Z_{main} (MWh)	$Z_{main/product}$ (kWh)	$Z_{support}$ (MWh)	$Z_{support/product}$ (kWh)	$PDSE$ (%)
Factory	1,073.00	83.00	77.36	-	43.90	40.91	39.11	36.45	-
GA	1,824.46	113.75	62.35	19.41	74.64	40.91	39.11	21.44	41.19
TS	2,000.30	120.95	60.46	21.84	81.83	40.91	39.11	19.55	46.36
CS	2,051.92	123.06	59.97	22.49	83.95	40.91	39.11	19.06	47.71
ACS	2,096.90	124.90	59.56	23.01	85.79	40.91	39.11	18.65	48.83
ALB#3: Frame									
Methods	Number of products	Z_{total} (MWh)	$Z_{total/product}$ (kWh)	$PDTE$ (%)	Z_{main} (MWh)	$Z_{main/product}$ (kWh)	$Z_{support}$ (MWh)	$Z_{support/product}$ (kWh)	$PDSE$ (%)
Factory	202.47	77.55	382.99	-	38.43	189.83	39.11	193.17	-
GA	368.25	109.02	296.03	22.71	69.90	189.83	39.11	106.21	45.02
TS	368.25	109.02	296.03	22.71	69.90	189.83	39.11	106.21	45.02
CS	372.33	109.79	294.87	23.01	70.68	189.83	39.11	105.04	45.62
ACS	388.37	112.83	290.53	24.14	73.72	189.83	39.11	100.71	47.87

Productivity, total energy consumption, PDTE and PDSE can be yearly calculated and also summarized in Table9

Referring to Table 9, results of energy resource management of all three ALB problems are yearly summarized. It was found that productivity of all lines is significantly increased. The highest productivity of all lines is achieved due to the least workload variance obtained by the proposed ACS. For the first ALB problem once compared with factory, the productivity of 24.46%, 26.23%, 27.63% and 34.93% can be increased by GA, TS, CS and ACS, respectively. For the second ALB problem once compared with factory, the productivity of 70.03%, 86.42%, 91.23% and 95.42% can be increased by GA, TS, CS and ACS, respectively. For the third ALB problem once compared with factory, the productivity of 81.88%, 81.88%, 83.89% and 91.82% can be increased by GA, TS, CS and ACS, respectively. These cause total and main energies per year are increased proportionally, while supporting energy is unchanged. In contrast, once considering energy per product, it was found that total and supporting energies per product are decreased proportionally, while main energy is unchanged. As results in Table 9, for the first ALB problem once compared with factory, the total energy per product of 8.31%, 8.73%, 9.15% and 10.94% can be decreased by GA, TS, CS and ACS, respectively. For the second ALB problem once compared with factory, the total energy per product of 19.41%, 21.84%, 22.49% and 23.01% can be decreased by GA, TS, CS and ACS, respectively. For the third ALB problem once compared with factory, the total energy per product of 22.71%, 22.71%, 23.01% and 24.14% can be decreased by GA, TS, CS and ACS, respectively. In addition, for the first ALB problem once compared with factory, the supporting energy per product of 19.65%, 20.78%, 21.65% and 25.88% can be decreased by GA, TS, CS and ACS, respectively. For the second ALB problem once compared with factory, the supporting energy per product of 41.19%, 46.36%, 47.71% and 48.83% can be decreased by GA, TS, CS and ACS, respectively. For the third ALB problem once compared with factory, the supporting energy per product of 45.02%, 45.02%, 45.62% and 47.87% can be decreased by GA, TS, CS and ACS, respectively. It can be noticed that for all problems, the maximum PDTE and the maximum PDSE are achieved by the ACS.

CONCLUSION

Optimization of energy resource management for assembly line balancing (ALB) problems in a car factory by the adaptive current search (ACS) has been proposed in this paper. With proposed optimization approach, assembly lines are balanced in order to optimize their energy resources management. The ACS has been used to address the number of tasks assigned for each workstation. Three selected real-world ALB problems, i.e. turning disk, dragging arm and frame assembly lines, from a specific car factory are conducted. By comparison, the ACS outperforms genetic algorithm (GA), tabu search (TS) and current search (CS). Results obtained by the ACS are superior to those obtained by other algorithms. The workload variances of all lines obtained by the ACS are less than those obtained by GA, TS and CS, whereas the line efficiencies of all lines achieved by the ACS are the highest values among those algorithms. In sense of energy resource management optimization, it was found that the least workload variances obtained by the ACS caused the highest productivities of all lines. The maximum percent of decreased total energy per product and the maximum percent of decreased supporting energy per product are also successfully achieved due to the least workload variance obtained by the proposed ACS. It can be concluded that the ACS is one of the most efficient algorithms and can be an alternative to optimize the energy resource management problems.

ACKNOWLEDGEMENTS

The authors wish to thank Sammitr Motors Manufacturing Public Co., Ltd., Samut Sakhon, Thailand for supporting the practical data of assembly lines. Sincere thanks are due to the editor and anonymous reviewers for their constructive comments.

REFERENCES

1. W. C. Turner, "Energy Management Handbook," Fairmont Press, USA, 2004.
2. B. L. Capehart, W. C. Turner and W. J. Kennedy, "Guide to Energy Management," Fairmont Press, USA, 1983.
3. D. T. Pham and D. Karaboga, "Intelligent Optimization Techniques," Springer, London, 2000.http://dx.doi.org/10.1007/978-1-4471-0721-7
4. F. Glover and G. A. Kochenberger, "Handbook of Metaheuristics," Kluwer Academic Publishers, New York, 2003.
5. E. G. Talbi, "Metaheuristics form Design to Implementation," John Wiley & Sons, New Jersey, 2009.
6. X. S. Yang, "Nature-Inspired Metaheuristic Algorithms," Luniver Press, United Kingdom, 2010.
7. X. S. Yang, "Engineering Optimization: An Introduction with Metaheuristic Applications," John Wiley & Sons, New Jersey, 2010. http://dx.doi.org/10.1002/9780470640425
8. S. S. Rao, "Engineering Optimization: Theory and Practice," John Wiley & Sons, New Jersey, 2009. http://dx.doi.org/10.1002/9780470549124
9. K. Sopian, A. Zaharim, Y. Ali and Z. Nopiah, "Optimal Operational

Strategy for Hybrid Renewable Energy System using Genetic Algorithms," WSEAS Transactions on Mathematics, Vol. 7, No. 4,2008, pp. 130-140.

10. Y. A. Katsigiannis and P. S. Georgilakis, "Optimal Sizing of Small Isolated Hybrid Power Systems Using Tabu Search," Journal of Optoelectronics and Advanced Materials, Vol. 10, No. 5, 2008, pp. 1241-1245.
11. V. Aristidis, P. Maria and L. Christos, "Particle Swarm Optimization (PSO) Algorithm for Wind Farm Optimal Design," International Journal of Management Science and Engineering Management, Vol. 5, No. 1, 2010, pp. 53-58.
12. T. Ratniyomchai, A. Oonsivilai, P. Pao-La-Or and T. Kulworawanichpong, "Economic Load Dispatch using Improved Harmony Search," WSEAS Transactions on Systems and Control, Vol. 5, No. 4, 2010, pp. 248-257.
13. M. D. Kilbridge and L. Wester, "A Heuristic Method of Assembly Line Balancing," The Journal of Industrial Engineering, Vol. 12, No. 4, 1961, pp. 292-298.
14. M. Amen, "Heuristic Methods for Cost-Oriented Assembly Line Balancing: A Comparison on Solution Quality and Computing Time," International Journal of Production Economics, Vol. 69, No. 3, 2001, pp. 255-264. http://dx.doi.org/10.1016/S0925-5273(99)00096-1
15. J. Rubinovitz and G. Lavitin, "Genetic Algorithm for Assembly Line Balancing," International Journal of Production Economics, Vol. 41, No. 1-3, 1995, pp. 343-354.http://dx.doi.org/10.1016/0925-5273(95)00059-3
16. W. C. Chiang, "The Application of the Tabu Search Metaheuristic to the Assembly Line Balancing Problem," Journal of Operation Research, Vol. 77, 1998, pp. 209-227.
17. S. D. Lapierre, A. Ruiz and P. Sariano, "Balancing Assembly Lines with Tabu Search," European Journal of Operation Research, Vol. 168, No. 3, 2006, pp. 826-837.http://dx.doi.org/10.1016/j.ejor.2004.07.031
18. J. G. Fernando and J. R. Almeida, "Hybrid Genetic Algorithm for Assembly Line Balancing," Journal of Heuristics, Vol. 8, No. 6, 2002, pp.629-642.http://dx.doi.org/10.1023/A:1020377910258
19. S. Suwannarongsri, S. Limnararat and D. Puangdownreong, "A New Hybrid Intelligent Method for Assembly Line Balancing," Proceeding of the IEEE International Conference on Industrial Engineering and Engineering Management (IEEM 2007), Singapore, 2-4 December 2007, pp.1115-1119. http://dx.doi.org/10.1109/IEEM.2007.4419365
20. Sukulin and D. Puangdownreong, "A Novel Metaheuristic Optimization

Algorithm: Current Search," Proceeding of the 11th WSEAS International Conference on Artificial Intelligence, Knowledge Engineering and Data Bases (AIKED'12), Cambridge, 22-24 February 2012, pp. 125-130.

21. Sukulin and D. Puangdownreong, "Control Synthesis for Unstable Systems via Current Search," Proceeding of the 11th WSEAS International Conference on Artificial Intelligence, Knowledge Engineering and Data Bases (AIKED '12), Cambridge, 22-24 February 2012, pp. 131-136.

22. Sukulin and D. Puangdownreong, "Current Search and Applications in Analog Filter Design Problems," Communication and Computer, Vol. 9, No. 9, 2012, pp. 1083-1096.

23. S. Suwannarongsri, T. Bunnag and W. Klinbun, "Energy Resource Management of Assembly Line Balancing Problem Using Modified Current Search," International Journal of Intelligent Systems and Applications (IJISA), Vol. 6, No. 2, 2013, in Press.

24. L. Gutjahr and G. L. Nemhauser, "An Algorithm for the Balancing Problem," Management Science, Vol. 11, No. 2, 1964, pp. 23-25. http://dx.doi.org/10.1287/mnsc.11.2.308

25. L. Arcus, "COMSOAL: A Computer Method of Sequencing Operations for Assembly Line," International Journal of Production Research, Vol. 4, 1966, pp. 25-32.

26. Sammitr Motors Manufacturing Public Co., Ltd., 2013. http://www.sammitr.com/

27. D. E. Goldberg, "Genetic Algorithms in Search Optimization and Machine Learning," Addison Wesley Publishers, Boston, 1989.

28. MathWorks, "Genetic Algorithm and Direct Search Toolbox: For Use with MATLAB," User's Guide, Version 1, MathWorks, Natick, 2005.

29. F. Glover, "Tabu Search. Part I," ORSA Journal on Computing, Vol. 1, No. 3, 1989, pp. 190-206.

30. F. Glover, "Tabu Search. Part II," ORSA Journal on Computing, Vol. 2, No. 1, 1990, pp. 4-32.http://dx.doi.org/10.1287/ijoc.2.1.4

31.

Chapter 9

RECENT DEVELOPMENTS IN MONITORING OF COMPLEX POPULATION SYSTEMS

Zoltán Varga[1], Manuel Gámez[2], Inmaculada López[2]

[1]Institute of Mathematics and Informatics, Szent István University, Godollo, Hungary

[2]Department of Mathematics, University of Almería, Almería, Spain

ABSTRACT

The paper is an update of two earlier review papers concerning the application of the methodology of mathematical systems theory to population ecology, a research line initiated two decades ago. At the beginning the research was concentrated on basic qualitative properties of ecological models, such as observability and controllability. Observability is closely related to the monitoring problem of ecosystems, while controllability concerns both sustainable harvesting of population systems and equilibrium control of such systems, which is a major concern of conservation biology. For population system, observability means that, e.g. from partial observation of the system (observing only certain indicator species), in principle the whole state process can be recovered. Recently, for different ecosystems, the so-called observer systems (or state estimators) have been constructed that enable us to effectively estimate the whole state process from the observation. This technique offers an efficient methodology for monitoring of complex ecosystems (including spatially and stage-structured population systems). In this way, from the observation of a few indicator species the state of the whole complex system can be monitored, in particular certain abiotic effects such as environmental contamination can be identified. In this review, with simple and transparent examples, three topics illustrate the recent developments in monitoring methodology of ecological systems: stock estimation of a fish population with reserve area; and observer construction for two vertically structured population systems (verticum-type systems): a four-level ecological chain and a stage-structured fishery model with reserve area.

INTRODUCTION AND HISTORICAL OVERVIEW

Mathematical Systems Theory (MST) looks back on several decades of history. In engineering practice, it is a typical situation that an object (e.g. a machine or electronic circuit) is controlled by a human intervention to influence the state of the object, or observing a transform of the state the task is to recover the state process of the object. The corresponding concepts of controllability, observability and the related state space model played an important role in the development of MST. The first comprehensive monograph of this discipline, dealing only with linear systems, was [1], a more recent reference is [2]. Generalizations of controllability and observability theory to nonlinear systems can be found in [3]. Following a successful development of MST for engineering purpose, as a new research line, in [4,5] the application MST to the study of population systems was proposed.

For population system, observability means that, from partial observation of the system, in principle, the whole state process can be recovered. Recently, for different ecosystems, the so-called observer systems (or state estimators) have been constructed that enables us to effectively estimate the whole state process from the observation. This technique offers an efficient methodology for monitoring of complex ecosystems (including spatially and stage-structured population systems). In this way, from the observation of a few indicator species the state of the whole complex system can be monitored, in particular certain abiotic effects such as environmental contamination may be identified.

In fact, the systems-theoretical study of the considered nonlinear frequency-dependent population models required the generalization of general sufficient conditions for controllability and observability to the case of nonlinear systems with invariant manifold (see [4,5]). These results have been applied to a control-theoretical model of artificial selection, phenotypic observation of genetic processes and evolutionary game dynamics, as well as to systems-theoretical models of reaction kinetics, see [6- 15].

Later on, the methodology of MST was used for monitoring of different population systems: observability and system inversion were investigated in density-dependent models of population ecology, ranging from Lotka-Volterra-type ([16-18]) and non Lotka-Volterratype population systems to monitoring of environmental change in a complex ecosystem ([19]). In particular, in [20] the optimal control software developed in [21] and [22], was also used for equilibrium control of a trophic chain. In [23] a new nonlinear system inversion method was applied for the reconstruction of time-dependent abiotic environmental changes, from the observation of certain indicator species. Furthermore, both monitoring and control were studied in a systems-theoretical model of radiotherapy in [24], while in [25-27] tools of MST were applied to biological pest control. Most of these topics and results have been reviewed in the survey papers [28] and [29].

In the present survey we report on recent developments in the methodology and application areas of monitoring in complex ecological systems. Although the presented methodology can be applied for the monitoring of large, but appropriately structured complex population systems, for the sake of simplicity and transparency we illustrate the procedures on observation systems of low dimensions. A particular attention is paid to recent results concerning the so-called verticum-type observation systems. The linear version of such systems have been introduced for modelling certain industrial systems and studied for controllability and observability in [30-38]. Verticum-type systems are composed from several "subsystems" connected sequentially in a particular way: a part of the state variables of each "subsystem" also appears in the next "subsystem" as an "exogenous variable" which can be also interpreted as a control generated by an "exosystem". Therefore, these "subsystems" are not observation systems, but formally can be considered as controlobservation systems. The problem of observability of such systems can be reduced to rank conditions on the "subsystems", which is a kind of decoupling of a complex system into simpler parts. Since most dynamic models of population biology are nonlinear, for the application in this field, it was necessary to extend the basic concepts and theorems of the theory of linear verticumtype systems to the nonlinear case, which has been done recently in [19,39-41].

The paper is organized as follows. In Section 2, based on [42], following a necessary stability analysis according to [43], the stock estimation of a fish population with reserved area is presented, using an appropriate observer design. In Section 3 results from [40] are recalled concerning the monitoring of ecological interaction chains of the type resource-producer-primary user-secondary consumer. The dynamic behaviour of these four-level chains is modelled by a system of differential equations, the linearization of which is a verticum-type system. Section 4 is devoted to the general concept of a nonlinear verticum-type observation system and the corresponding general sufficient condition of observability obtained in [41]. As an application, observer design is also presented for a stage-structured population, decomposing the state estimation according to the verticum structure. In Section 5 further possible application fields of the presented monitoring methodology are summarized. Finally, in the Appendix the theoretical background necessary for the monitoring of nonlinear verticum-type systems is recalled.

STOCK ESTIMATION OF A FISH POPULATION WITH RESERVED AREA

For the basic model of this section, from [43] we recall the dynamics of a fish population moving between two areas, the first, an unreserved one where fishing is allowed, and the second, a reserved one where fishing is prohibited. At time t, let $x_1(t)$ and $x_2(t)$ be the respective biomass densities of the same fish population inside the unreserved and reserved areas, respectively. Assume that the fish subpopulation of the unreserved area migrates into the reserved area at a rate m_{12}, and there is also an inverse migration at rate m_{21}. Let E be a constant fishing effort applied for harvesting in the unreserved area and let us assume that in each area the growth of the fish population follows a logistic model. The dynamics of the fish subpopulations in unreserved and reserved areas are then assumed to be governed by the following system of differential Equations (1) and (2):

$$\dot{x}_1 = r_1 x_1\left(1-\frac{x_1}{K_1}\right) - m_{12}x_1 + m_{21}x_2 - qEx_1 \tag{1}$$

$$\dot{x}_2 = r_2 x_2\left(1-\frac{x_2}{K_2}\right) + m_{12}x_1 - m_{21}x_2 , \tag{2}$$

where r_1 and r_2 are the intrinsic growth rates of the corresponding subpopulations, K_1 and K_2 are the carrying capacities for the fish species in the unreserved and reserved areas, respectively; q is the catchability coefficient in the unreserved area. All parameters r_1, r_2, q, m_{12}, m_{21}, E, K_1 and K_2 are positive constants.

In [43], it was checked that for a unique positive equilibrium $x^* = (x_1^*, x_2^*)$ of the dynamic model (1)-(2) the following set of inequalities are sufficient:

$$\frac{r_2\left(r_1 - m_{12} - qE\right)^2}{K_2 m_{21}} < \frac{\left(r_2 - m_{21}\right)r_1}{K_1} \tag{3a}$$

$$\left(r_2 - m_{21}\right)\left(r_1 - m_{12} - qE\right) < m_{12}m_{21} \tag{3b}$$

$$\frac{r_1 x_1^*}{K_1} > r_1 - m_{12} - qE. \tag{3c}$$

Furthermore, the Lyapunov function

$$V(x) := \left(x_1 - x_1^* - x_1^* \ln \frac{x_1}{x_1^*} \right) + \frac{x_2^* m_{21}}{x_1^* m_{12}} \left(x_2 - x_2^* - x_2^* \ln \frac{x_2}{x_2^*} \right)$$

also implies asymptotic stability of equilibrium x^* for system (1)-(2), globally with respect to the positive orthant of $\mathbf{R}^2$. Throughout the section we shall suppose conditions (3a)-(3c) to guarantee the stable coexistence of the system applying a constant reference fishing effort.

Observability of the Model

From [3], we recall the basic concept of local observability of nonlinear systems and a sufficient condition for a system to have this property, in order to apply it to the considered model and in order to be used in the following sections.

Let

$$m, n \in \mathbf{N}, f \in C^1(\mathbf{R}^n, \mathbf{R}^n), h \in C^1(\mathbf{R}^n, \mathbf{R}^m),$$
$$x^* \in \mathbf{R}^n, f(x^*) = 0, h(x^*) = 0$$

and consider observation system

$$\dot{x} = f(x) \tag{4}$$

$$y = h(x), \tag{5}$$

where function (5) defines the transform of the state, observed instead of the state itself.

Definition 2.1. System (4)-(5) is called locally observable at x^* on $[0,T]$, if for any solutions x^1, x^2 of (4) defined on $[0,T]$, initially close enough to x^*,

$$h \circ x^1 = h \circ x^2 \Rightarrow x^1 = x^2.$$

Linearizing (4)-(5) around x^*, we get

$$A := f'(x^*), C := h'(x^*).$$

Theorem 2.1. ([3]) $\Rightarrow \operatorname{rank}[C|CA|\cdots|CA^{n-1}] = n$

system (4)-(5) is locally observable at x^* on $[0,T]$.

Now, let us consider the problem of stock estimation in the reserve area on the basis of the biomass harvested in the free area. (For technical reason its difference from the equilibrium value is supposed to be observed.) To this end, in addition to dynamics (1)-(2) we introduce an observation equation

$$y = h(x) := qE\left(x_1 - x_1^*\right), \tag{6}$$

representing the observation of the biomass harvested in the free fishing area. Then linearizing observation system (1)-(2), (6), we get the Jacobian of the right-hand side of (1)-(2)

$$A := \begin{bmatrix} r_1 - 2r_1 \dfrac{x_1^*}{K_1} - m_{12} - qE & m_{21} \\ m_{12} & r_2 - 2r_2 \dfrac{x_2^*}{K_2} - m_{21} \end{bmatrix},$$

and the observation matrix

$$C := h'(x^*) = (qE \quad 0).$$

Now, for the linearized system we obviously have $\operatorname{rank}[C|CA]^T = 2.$ Hence, Theorem 2.1 implies local observability of the system near the equilibrium. In other words, in principle the whole system state (in particular the stock of the species in the reserve area) as function of time can be uniquely recovered, observing the biomass harvested per unit time. In the following illustrative example we will see how the state of the system (and hence the total stock) can be effectively calculated from the catch realized in the fishing area, applying the methodology of [44] that we will recall next.

Observer System

Recently, for different ecosystems, the so-called observer system (or state estimators) have been constructed that enables us to effectively estimate the whole state process from the observation. Here we remind the methodology that will be used in this subsection and in the following sections. Consider again observation system (4)-(5)

$$\dot{x} = f(x), y = h(x).$$

Definition 2.2. Given $G \in C^1\left(\mathbf{R}^n \times \mathbf{R}^m, \mathbf{R}^n\right)$, system

$$\dot{z} = G(x, y) \tag{7}$$

iscalledalocal(exponential)observerforsystem(4)-(5)at x^* ,iffor the composite system (4)-(5), (7) we have 1) $x(0) = z(0) \Rightarrow x(t) = z(t), (t \in \mathbf{R}_0^+)$ 2) there exists a neighbourhood V of x^* such that $x(0), z(0) \in V \Rightarrow \lim_{+\infty}(z-x)=0$ (exponentially).

Theorem 2.2. ([44]) Suppose equilibrium x^* of system (4) is Lyapunov stable, and there exists nxn matrix K such that $A-KC$ is stable. Then system

$$\dot{z} = f(z) + K[y - h(z)] \tag{8}$$

is a local exponential observer for observation system (4)-(5).

Example 2.1. For a possible comparison, in this numerical example we use the same parameters as [45]:

$$r_1 = 0.7, r_2 = 0.5, K_1 = 10, K_2 = 2.2,$$
$$m_{12} = 0.2, m_{21} = 0.1, q = 0.25$$

and $E = 0.9$.

$$\dot{x}_1 = 0.7x_1\left(1 - \frac{x_1}{10}\right) - 0.2x_1 + 0.1x_2 - 0.25 \cdot 0.9x_1$$
$$\dot{x}_2 = 0.5x_2\left(1 - \frac{x_2}{2.2}\right) + 0.2x_1 - 0.1x_2 \tag{9}$$

Now the positive equilibrium is $x^* = (4.85, 3.12)$, and with

$$K := \begin{pmatrix} 0 \\ 10 \end{pmatrix},$$

matrix $A-KC$ is Hurwitz; therefore by Theorem 2.2 we have the following observer system

$$\dot{z}_1 = 0.7z_1\left(1 - \frac{z_1}{10}\right) - 0.2z_1 + 0.1z_2 - 0.25 \cdot 0.9z_1$$
$$\dot{z}_2 = 0.5z_2\left(1 - \frac{z_2}{2.2}\right) + 0.2z_1$$
$$-0.1z_2 + 10\left[y - 0.25 \cdot 0.9(z_1 - x_1^*)\right]. \tag{10}$$

If we take an initial condition $x^0 := (30, 120)$ for system (9), and similarly, we consider another nearby initial condition, $z^0 := (35, 100)$ for the observer system (10), then the corresponding solution z of the observer tends to the solution x of the original system, as shown in Figure 1. We note that in this particular case the convergence is much faster than that of the observer constructed in [45].

MONITORING OF A FOUR-LEVEL ECOLOGICAL CHAIN

As a modification of the well-known three-level trophic chain consisting of resource-producer-primary consumer studied in [46], we consider the following four-level ecological interaction chain:

level 0: a resource;

level 1: the producer is a plant, supposed to die out without the resource, and the positive effect of the latter is proportional to the quantity of the resource present in the system;

level 2: the primary user (instead of consumer), i.e. a commensalist animal, making use of the plant as part of its habitat without harming it (e.g. an insect species hosted by the plant), displaying a logistic dynamics in absence of the plant and the secondary consumer;

level 3: the secondary consumer is a monophagous predator of the primary user (e.g. an insectivorous singing bird species), with intraspecific competition.

(For more details on the role of commensalism in ecological communities, we supposed between the producer and the primary user, see e.g. [47]).

For a dynamical model let x_0 be the time-dependent quantity, with a constant supply Q of the resource present in the system, x_1, x_2 and x_3 the time-varying population size (biomass or density) of the producer, the primary user and the secondary consumer, respectively.

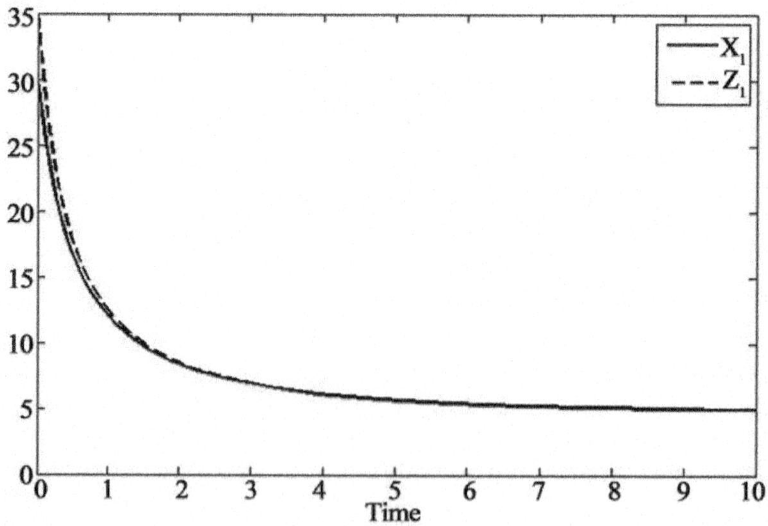

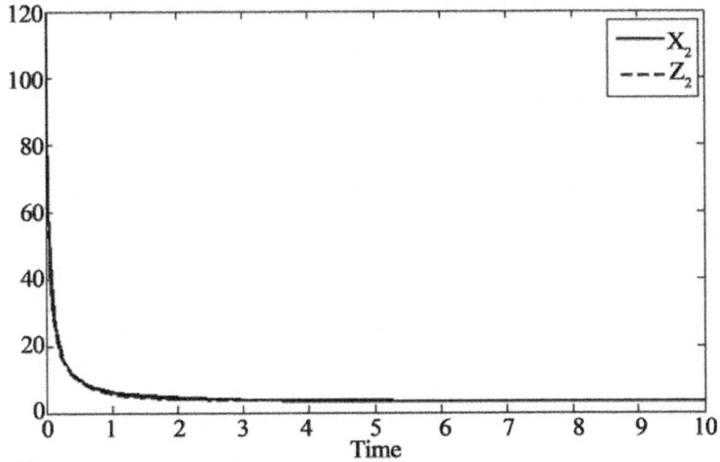

Figure 1: Solution of observer (10), approaching the solution of system (9).

Assume that a unit of biomass of the plant consumes the resource at velocity $\alpha_0 x_0$; however, it increases the biomass of the plant at rate k_1. The relative rate of increase in biomass of the primary user, due to the presence of the plant is $k_2 x_1$. While the plant population is supposed to die out exponentially in the absence of the resource, with Malthus parameter m_1, the primary user displays a logistic growth with Malthus parameter m_2 and is limited by a carrying capacity $\frac{m_2}{\eta_2}$. Furthermore, the secondary consumer would die out at rate m_3, without the presence of the primary user, and there is an intraspecific competition among predators with rate η_3. We will consider a partially closed system, where the dead plants may be recycled into nutrient resource with rate β_1. Then with parameters

$Q, \alpha_0, \alpha_2, m_1, m_2, m_3, \eta_2, \eta_3 > 0;$

$k_1, k_2, k_3 \in]0,1[; \beta_1 \in [0,1[$ we have the following dynamic model for the considered interaction chain:

$$\dot{x}_0 = Q - \alpha_0 x_0 x_1 + \beta_1 m_1 x_1 \tag{11}$$

$$\dot{x}_1 = x_1 \left(-m_1 + k_1 \alpha_0 x_0 \right) \tag{12}$$

$$\dot{x}_2 = x_2 \left(m_2 + k_2 x_1 - \eta_2 x_2 - \alpha_2 x_3 \right) \tag{13}$$

$$\dot{x}_3 = x_3\left(-m_3 + k_3\alpha_2 x_2 - \eta_3 x_3\right) \tag{14}$$

Theorem 3.1. ([40]) Let us suppose that for given biological parameters, the resource supply is high enough,

$$Q > \max\left\{\frac{m_1 m_3 \eta_2}{k_1 k_2 k_3 \alpha_2}, \frac{2m_1^2}{k_1 \alpha_0}\right\}. \tag{15}$$

Then, both the open $(\beta_1 = 0)$ and the partially closed $(\beta_1 > 0)$ ecological chains stably coexist in the sense there exists a positive equilibrium x^* of system calculated in [40], which is asymptotically stable.

Remark 3.1. The conditions of Theorem 3.1 can also be formulated conversely: Given a resource supply Q, biological parameters satisfying condition (15) imply the stable coexistence of the considered ecological chain.

Observability of the Ecological Chain

Let us consider now the following two auxiliary 2-dimension systems

$$\begin{aligned} \dot{x}_0 &= Q - \alpha_0 x_0 x_1 + \beta_1 m_1 x_1 \\ \dot{x}_1 &= x_1\left(-m_1 + k_1 \alpha_0 x_0\right) \end{aligned} \tag{16}$$

and

$$\begin{aligned} \dot{x}_2 &= x_2\left(m_2 + k_2 x_1^* - \eta_2 x_2 - \alpha_2 x_3\right) \\ \dot{x}_3 &= x_3\left(-m_3 + k_3 \alpha_2 x_2 - \eta_3 x_3\right) \end{aligned}. \tag{17}$$

In ecological terms (16) is a subsystem of the original chain (11)-(14), while in (17) the positive effect of the plant on the animal species 2 appears with the equilibrium value x_1^* of the plant. We note that by setting $k_2 = 0$ (i.e. considering the original system without commensalisms), the original ecological chain is split up into two components without interaction.

Remark 3.2. The biological interpretation of system (17) is the following: Suppose that system (11)-(14) is in equilibrium, and the two animal species, by an external disturbance, deviate from their equilibrium densities. Then the resource-primary consumer subsystem can maintain its equilibrium, and the predator-prey subsystem will be governed by system (17).

Continuing the study of systems (16) and (17), we can easily check that they have respective equilibria $w_0^* := (x_0^*, x_1^*)$ and $w_1^* := (x_2^*, x_3^*)$. For system (16) with notation $w_0 := (x_0, x_1)$, let us consider observation function

$$h_0(w_0) = h_0(x_0, x_1) := x_0 - x_0^* . \tag{18}$$

This means that the deviation of the resource from its equilibrium value is observed. In order to check local observability, we calculate the linearization of system (16) at equilibrium w_0^*:

$$A_{00} := \begin{pmatrix} -\alpha_0 x_1^* & -\alpha_0 x_0^* + \beta_1 m_1 \\ k_1 \alpha_0 x_1^* & 0 \end{pmatrix}; C_0 := h_0'(w_0^*) = (1 \quad 0). \tag{19}$$

Hence we easily calculate

$\operatorname{rank}\begin{pmatrix} C_0 \\ C_0 A_{00} \end{pmatrix} = 2$ provided $\beta_1 > 0$. From the classical sufficient condition for the local observability of nonlinear systems, [3], we obtain local observability of system (16) near the equilibrium, with observation (18).

Similarly, suppose that in system (17) the deviation of the density of the prey from its equilibrium value is observed, i.e., with notation $w_1 := (x_2, x_3)$ we consider the observation function

$$h_1(w_1) := x_2 - x_2^* . \tag{20}$$

The linearization of system (17) at equilibrium w_1^* is

$$A_{11} := \begin{pmatrix} -\eta_2 x_2^* & -\alpha_2 x_2^* \\ k_3 \alpha_2 x_3^* & -\eta_3 x_3^* \end{pmatrix}; C_1 := h_1'(w_1^*) = (1 \quad 0) . \tag{21}$$

Checking again the rank condition, by $\alpha_2 > 0$ we get

$\operatorname{rank}\begin{pmatrix} C_1 \\ C_1 A_{11} \end{pmatrix} = 2$ implying local observability of system (17)-(20) near w_1^*. Now, let us observe that with definition

$A_{10} := \begin{pmatrix} 0 & k_2 x_2^* \\ 0 & 0 \end{pmatrix}$ system matrix

$A := \begin{pmatrix} A_{00} & 0 \\ A_{10} & A_{11} \end{pmatrix}$ together with observation matrix

$C := \begin{pmatrix} C_0 & 0 \\ 0 & C_1 \end{pmatrix}$

define a verticum-type linear observation system in the sense defined in the Appendix. Applying Theorem A.2 of the Appendix, we obtain that the linear observation system

$$\dot{w} = Aw \tag{22}$$

$$y = Cw \tag{23}$$

is observable. Since A is just the Jacobian of the righthand side of system (11)-(14), therefore (22) is just the linearization of system (11)-(14). Furthermore, (23) is the linearization of observation function

$$h(x) := col\left(x_0 - x_0^*, x_2 - x_2^*\right) \tag{24}$$

which can be associated with system (11)-(14). Finally, applying again the classical rank condition of [3], we can summarize the reasoning of this subsection in the following theorem.

Theorem 3.2. Let us suppose that ecological chain (11)-(14) is partially closed $(\beta_1 > 0)$. Then with observation function (24), system (11)-(14) is locally observable near equilibrium x^* calculated in [40].

Construction of an Observer System

Following the procedure of [44], let us first determine conditions for the construction of observers for systems (16) and (17), with respective observation functions (18) and (20).

For matrices A_{00} and C_0, figuring in (19), we have to find a matrix $H_0 := col(h_{00}, h_{01})$ such that

$$A_{00} - H_0C_0 = \begin{pmatrix} -\alpha_0 x_1^* - h_{00} & -\alpha_0 x_0^* + \beta_1 m_1 \\ k_1\alpha_0 x_1^* - h_{01} & 0 \end{pmatrix}$$

is a Hurwitz matrix, i.e. all roots of the characteristic polynomial p_0 of matrix $A_{00} - H_0C_0$ have real negative parts. It is easy to check that the latter condition is satisfied if and only if the following inequalities hold:

$$h_{00} > -\alpha_0 x_1^* \tag{25}$$

$$h_{01} < k_1\alpha_0 x_1^*. \tag{26}$$

Simple sufficient conditions for (25) and (26) are $h_{00} > 0$ and $h_{01} < 0$, respectively. By the Theorem of [44], the observer for system (16) with observation function (18) can be determined.

Similarly, for matrices A_{11} and C_1, figuring in (21), we need to find a matrix $H_1 := col(h_{12}, h_{13})$ such that all roots of the characteristic polynomial p_1 of matrix

$$A_{11} - H_1 C_1 = \begin{pmatrix} -\eta_2 x_2^* - h_{12} & -\alpha_2 x_2^* \\ k_3 \alpha_2 x_3^* - h_{13} & -\eta_3 x_3^* \end{pmatrix}$$

have real negative parts. Now a straightforward checking shows that the latter condition is satisfied if and only if h_{12} and h_{13} satisfy the following inequalities:

$$h_{12} > -\eta_2 x_2^* - \eta_3 x_3^*, \tag{27}$$

$$h_{13} < \frac{(\eta_2 x_2^* + h_{21})\eta_3 x_3^* + k_3 \alpha_2^2 x_2^* x_3^*}{\alpha_2 x_2^*}. \tag{28}$$

Similarly to the previous case, in order to satisfy conditions (27) and (28), it is sufficient to set $h_{12} > 0$ and $h_{13} < 0$, and again by the Theorem of [44], the observer for system (17) with observation function (20) can be determined.

Finally, based on the above reasoning, it will be easy to prove the following result:

Theorem 3.3. ([40]) Given $H = \begin{pmatrix} h_{00} & 0 \\ h_{01} & 0 \\ 0 & h_{12} \\ 0 & h_{13} \end{pmatrix}$ with $h_{00}, h_{12} > 0$ and $h_{01}, h_{13} < 0$, and function f defined as the right-hand side of system (11)-(14), system

$$\dot{z} = f(z) + H(y - h(z))$$

is a local exponential observer for system (11)-(14) with observation equation $y = h(x)$, where h is defined in (24).

Example 3.1. We consider the following system

$$\dot{x}_0 = 2.1 - 0.2x_0x_1 + 0.2 \times 0.4x_1$$
$$\dot{x}_1 = x_1(-0.4 + 0.84 \times 0.2x_0)$$
$$\dot{x}_2 = x_2(0.25 + 0.7x_1 - 0.1x_2 - 0.4x_3)$$
$$\dot{x}_3 = x_3(-0.1 + 0.9 \times 0.4x_2 - 0.5x_3). \tag{29}$$

System (29) has a positive equilibrium $x^* = (2.38, 5.3, 10.41, 7.3)$, which is asymptotically stable, because conditions of Theorem 3.1 are satisfied. In Figure 2 it can be seen how, e.g. from initial condition $x(0) := (1,6,9,5)$ near the equilibrium, the solution x of system (29) tends to this positive equilibrium, see **Figure 2**.

Consider now system (29) with observation

$$y = h(x) = (x_0 - x_0^*, x_2 - x_2^*).$$

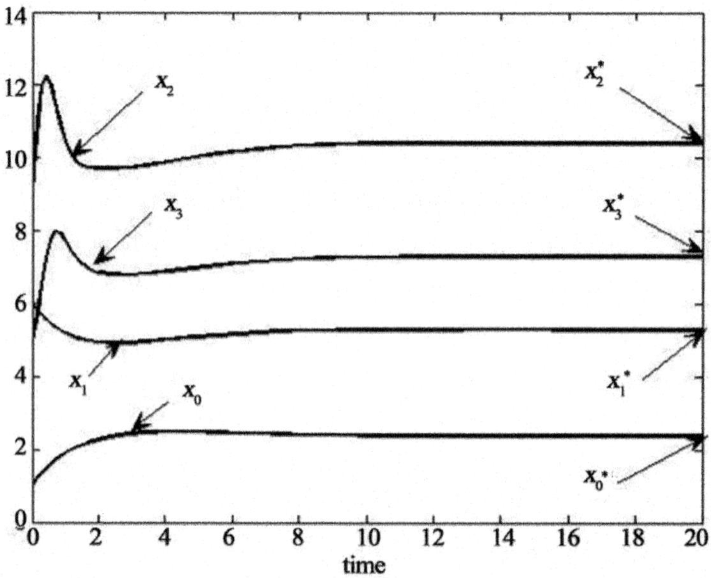

Figure 2: Solution of system (29) with initial condition $x(0) = (1,6,9,5)$.

Since matrix

$$H = \begin{pmatrix} 10 & 0 \\ -20 & 0 \\ 0 & 10 \\ 0 & -0.01 \end{pmatrix}$$ satisfies the conditions of Theorem 3.3, we can construct the following observer

$$\begin{cases} \dot{z}_0 = 2.1 - 0.2z_0z_1 + 0.2\cdot 0.4z_1 \\ \dot{z}_1 = z_1(-0.4 + 0.84\cdot 0.2z_0) \\ \dot{z}_2 = z_2(0.25 + 0.7z_1 - 0.1z_2 - 0.4z_3) \\ \dot{z}_3 = z_3(-0.1 + 0.9\cdot 0.4z_2 - 0.5z_3) \end{cases} + \begin{pmatrix} 10 & 0 \\ -20 & 0 \\ 0 & 10 \\ 0 & -0.01 \end{pmatrix} \cdot (y - h(z)) \quad (30)$$

Solving (30) with initial condition $z(0)=(2,4,7,7)$ near the equilibrium, we can check how this solution tends to recover the corresponding solution of system (29), see **Figure 3**.

A STAGE-STRUCTURED FISHERY MODEL WITH RESERVE AREA

Let us consider a modification of the stage-structured fishery model of [45], supposing that there is reserve area where fishing is not allowed. In what follows, the first index of the biomass density N will indicate the area: $i=1$ for the reserve and $i=2$ for the free area; the second index will refer to the development stage: $j=0$ for the pre-recruits, i.e. the eggs, larvae and the juveniles together, and $j=1$ the exploited stage of the population. The dynamics of the system is modeled by the following autonomous system of differential equations

$$\dot{N}_{10} = -m_{10}N_{10} + f_{11}N_{11} - p_{11}N_{10}N_{11} - p_{10}N_{10}^2 \quad (31a)$$

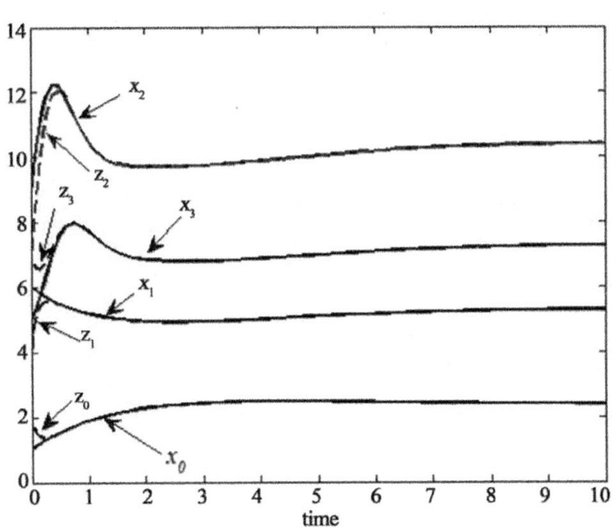

Figure 3: Solutions of systems (29) and (30) with the respective initial conditions $x(0)=(1,6,9,5)$ and $z(0)=(2,4,7,7)$.

$$\dot{N}_{11} = \alpha_{11}N_{10} - m_{11}N_{11} - \beta N_{11} \tag{31b}$$

$$\dot{N}_{20} = -m_{20}N_{20} + f_{21}N_{21} - p_{21}N_{20}N_{21} - p_{20}N_{20}^2 \tag{31c}$$

$$\dot{N}_{21} = \alpha_{21}N_{20} - m_{21}N_{21} + \beta N_{11} - qEN_{21} \tag{31d}$$

where

$m_{ij} =$ natural mortality rate of class ij,

$\alpha_{i1} =$ linear aging coefficient in areas $i = 1, 2,$

$p_{i0} =$ juvenile competition parameter in areas $i = 1, 2,$

$f_{i1} =$ fecundity rate of adult fish in areas $i = 1, 2,$

$p_{i1} =$ predation rate of class 1 on class 0 in areas $i = 1, 2,$

$q =$ catchability coefficient of class 1 in the unreserved area,

$\beta =$ migration rate of the second class from reserved area to unreserved area,

$E =$ constant fishing effort.

From [41] we know that if

$$f_{11}\alpha_{11} - m_{10}\left(m_{11} + \beta\right) > 0, \tag{32}$$

system (31) has a unique positive equilibrium $N^* = \left(N_{10}^*, N_{11}^*, N_{20}^*, N_{21}^*\right)$, which is asymptotically stable under conditions

$$f_{11} - p_{11}N_{10}^* < 0 \text{ and } f_{21} - p_{21}N_{20}^* < 0. \tag{33}$$

Remark 4.1. Since asymptotic stability implies Lyapunov stability, in the next section we can apply Theorem A.3 of the Appendix to the corresponding nonlinear verticum-type observation system.

Observability of the Model

Let $N = (N_1, N_2)$ with $N_1 = (N_{10}, N_{11})$, $N_2 = (N_{20}, N_{21})$ and we consider the observation function $h: R^4 \to R^2$, defined by

$$y = h\left(N_1, N_2\right) = \left(N_{11} - N_{11}^*, qE\left(N_{21} - N_{21}^*\right)\right). \tag{34}$$

Now the observability of observation system (31)-(34) will be analyzed using the results of the Appendix. Consider systems

$$\dot{N}_{10} = -m_{10}N_{10} + f_{11}N_{11} - p_{11}N_{10}N_{11} - p_{10}N_{10}^2$$

$$\dot{N}_{11} = \alpha_{11}N_{10} - m_{11}N_{11} - \beta N_{11} \tag{35}$$

and

$$\dot{N}_{20} = -m_{20}N_{20} + f_{21}N_{21} - p_{21}N_{20}N_{21} - p_{20}N_{20}^2$$

$$\dot{N}_{21} = \alpha_{21}N_{20} - m_{21}N_{21} + \beta N_{11} - qEN_{21} \tag{36}$$

Given observation

$$y_1 = h_1(N_{10}, N_{11}) = N_{11} - N_{11}^*, \tag{37}$$

we calculate its linearization

$$C_1 := h_1'(N_1^*) = (0 \quad 1).$$

It is easy to check that $\operatorname{rank}(C_1|C_1A_1) = 2$, where A_1 is the linearization of (35), therefore by Theorem A.1 of the Appendix we can guarantee local observability of system (35)-(37).

Analogously, for observation

$$y_2 = h_2(N_{20}, N_{21}) = qE(N_{21} - N_{21}^*) \tag{38}$$

of system (36) calculate

$$C_2 := h_2'(N_2^*) = (0 \quad qE).$$

Again we have $\operatorname{rank}(C_2|C_2A_2) = 2$, where A_2 is the linearization of (36), therefore from Theorem A.1 of the Appendix we have local observability of system (36)- (38). Since under the appropriate conditions equilibrium N^* is asymptotically stable and hence also Lyapunov stable, applying Theorem A.3 we obtain Theorem 4.1. Suppose that conditions (32) and (33) hold. Then observation system (31)-(34) is locally observable near the asymptotically stable equilibrium.

Construction of an Observer System

Given the observation system (35)-(37), using the corresponding observer design of [44], it is sufficient to find a matrix K^1 such that $A_1 - K^1C_1$ is Hurwitz. It is easy to check that with $k_2^1 > 0$,

$$K^1 := \begin{pmatrix} 0 \\ k_2^1 \end{pmatrix}$$

is appropriate.

Analogously, for observation systems (36)-(38), with $k_2^2 > 0$,

$$K^2 := \begin{pmatrix} 0 \\ k_2^2 \end{pmatrix},$$

$A_2 - K^2C_2$ is Hurwitz, guaranteeing the construction of the observer system.

From these results, for

$$K := \begin{pmatrix} K^1 \\ K^2 \end{pmatrix}, C := \begin{pmatrix} C_1 & C_2 \end{pmatrix}$$

we can check that $A - KC$ is Hurwitz, which allows the construction of an observer for system (31)-(34) moreover, this observer is composed of the observers constructed for the two subsystems.

Example 4.1. Consider the following model parameters of [48]:

$$m_{10} = 0.4, m_{20} = 0.4, f_{11} = 0.5, f_{21} = 0.5,$$
$$p_{11} = 0.1, p_{21} = 0.1, p_{10} = 0.2, p_{20} = 0.1,$$
$$\alpha_{11} = 0.8, \alpha_{21} = 0.9, m_{11} = 0.05, m_{21} = 0.1,$$
$$\beta = 0.65, q = 0.07, E = 0.5.$$

To construct the observer system for (35)-(37) we can take

$$K^1 = \begin{pmatrix} 0 \\ 10 \end{pmatrix}.$$

Then the observer system is

$$\begin{aligned} \dot{Z}_{10} &= -0.4Z_{10} + 0.5Z_{11} - 0.1Z_{10}Z_{11} - 0.2Z_{10}^2 \\ \dot{Z}_{11} &= 0.8Z_{10} - 0.05Z_{11} - 0.65Z_{11} + 10(N_{11} - Z_{11}) \end{aligned} \tag{39}$$

Considering $N_1(0)=(0.1,0.2)$. as initial value for the system (35), and $Z_1(0)=(0.5,1)$ for the observer (39), in Figure 4 we can see how the solution of the observer system approaches the solution of the original system.

To construct the observer of system (36)-(38) take

$$K^2 = \begin{pmatrix} 0 \\ 10 \end{pmatrix}.$$

Then the observer system is

$$\dot{Z}_{20} = -0.4Z_{20} + 0.5Z_{21} - 0.1Z_{20}Z_{21} - 0.1Z_{20}^2$$
$$\dot{Z}_{21} = 0.9Z_{20} - 0.1Z_{21} + 0.65Z_{11}$$
$$-0.07 \times 0.5Z_{21} + 10 \times 0.07 \times 0.5(N_{21} - Z_{21}). \quad (40)$$

If we consider $N_2(0)=(1,10)$ as initial value for system (36), and $Z_2(0)=(1.5,15)$ for the observer (40)we obtain the result plotted in **Figure 5**.

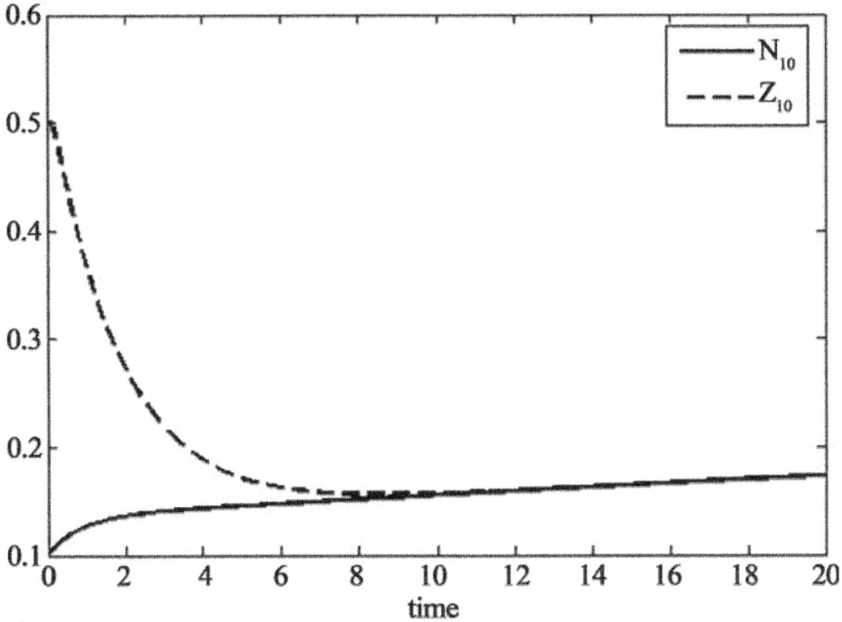

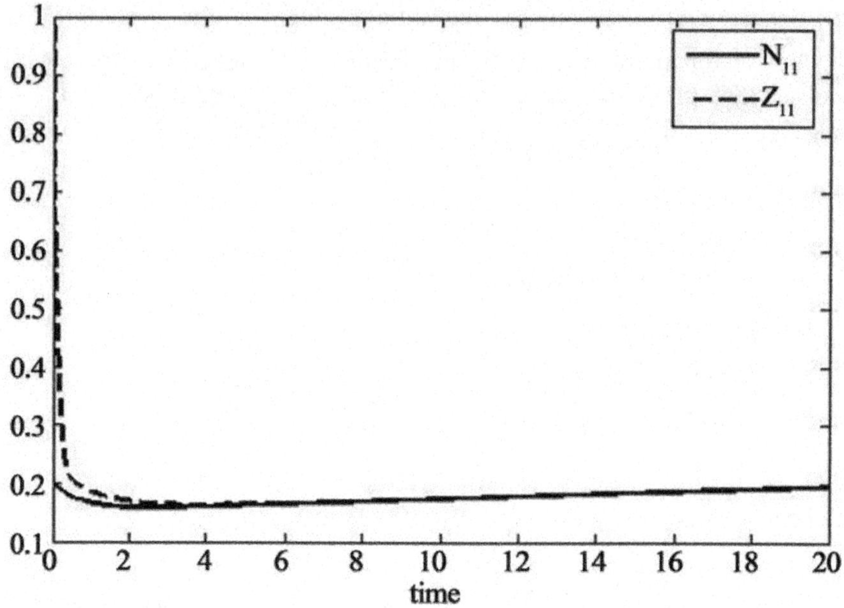

Figure 4: Solution of the observer (39) approaching the solution of the original system (35).

Now the observer for system (31)-(34) can be simply composed from the single observers (39) and (40). In **Figure 6** we can see how the solution of the observer (39)-(40) with initial value $Z(0)=(0.1,0.4,2.5,10)$, estimates the solution of system (31) with initial value $N(0)=(0.3,0.1,0.3,8)$.

DISCUSSION AND OUTLOOK

Observation problems arise in many fields of human activity, when state of an object can be characterized by several numbers (i.e. by a state vector), and it is impossible or too expensive to measure all state variables. Then we may want to recover the whole state vector.

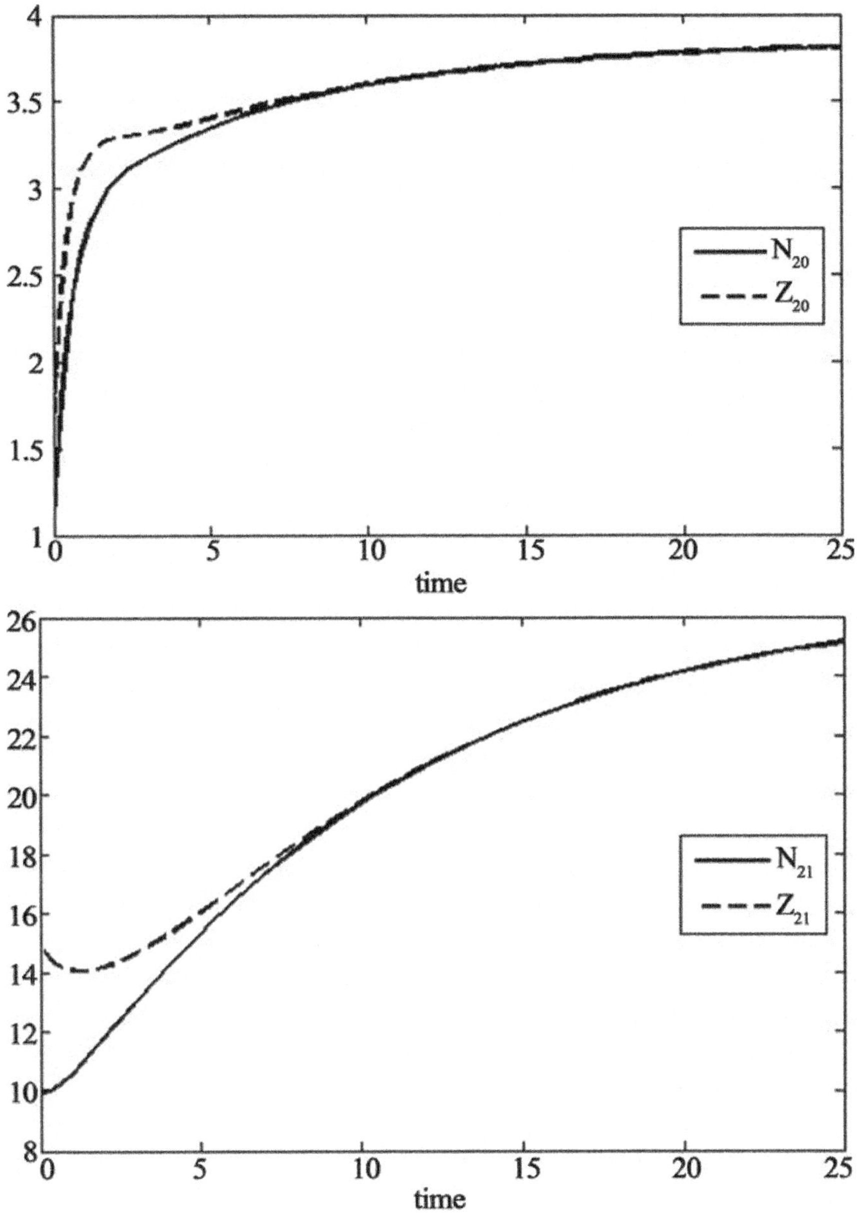

Figure 5: Solution of the observer (40) approaching the solution of the original system (36).

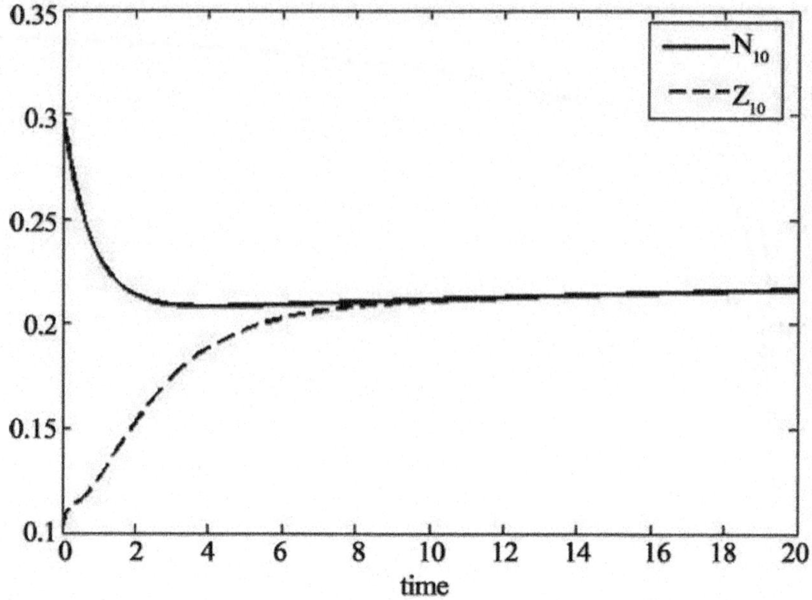
N_{10}
Z_{10}
0.35
0.3
0.25
0.2
0.15
0.1
0 2 4 6 8 10 12 14 16 18 20
time

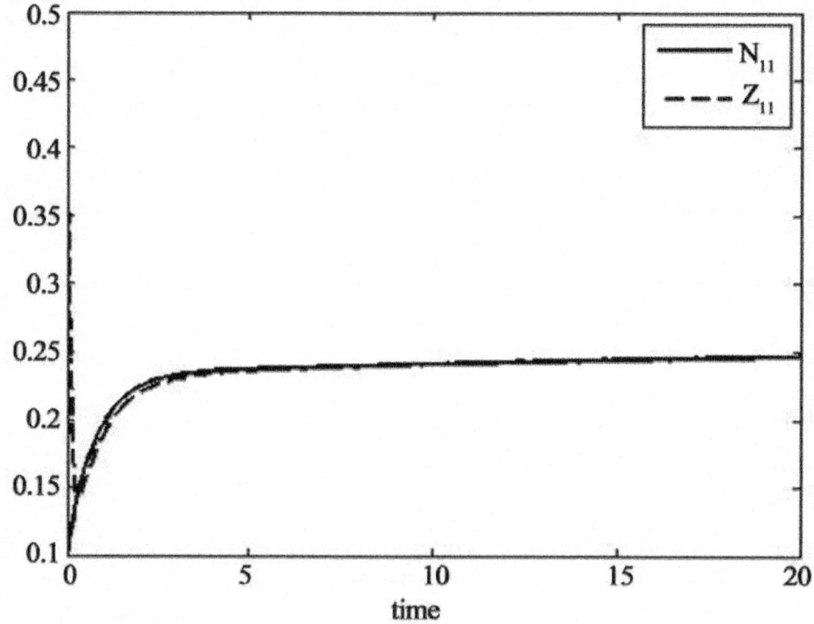
N_{11}
Z_{11}
0.5
0.45
0.4
0.35
0.3
0.25
0.2
0.15
0.1
0 5 10 15 20
time

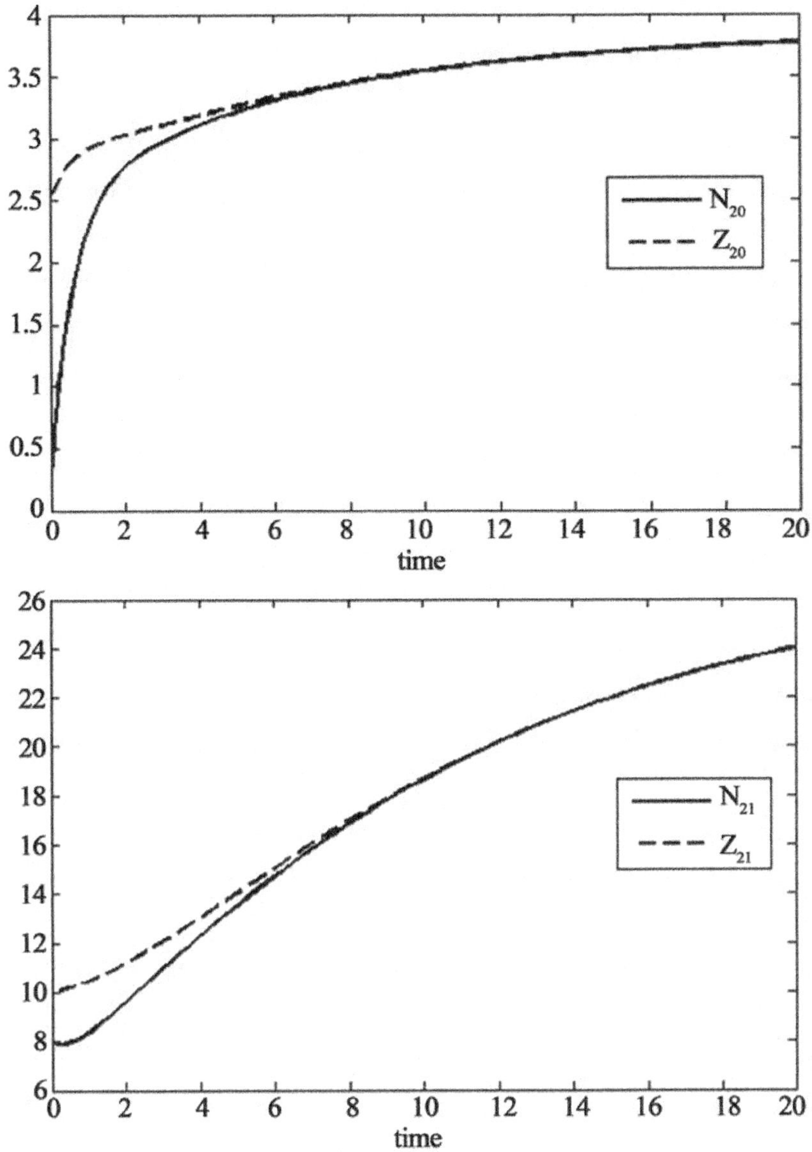

Figure 6: Solution of the observers (39) and (40) approaching the solution of the original system (31) by coordinates.

In a static situation this is clearly impossible, since projection is not invertible. However, in dynamic situation the concepts of observability and observer design of Mathematical Systems Theory turned out to be efficient tools for monitoring of ecological systems, as well. We presented some recent developments in concrete applications to population systems. These systems are not only simple sets of populations, but each of them has a particular structure. In the first case (Section 2) a single species has a spatially structured habitat (with a reserve area, where observation of density by harvesting is not allowed). In the other two cases, verticum-type, i.e. vertically organized dynamic population systems (ecological chains in Section 3, and stage structure of a single species in Section 4), for monitoring purpose "decoupled" observer design may be efficient even in large systems.

These examples anticipate the application of the presented methodology in similar situations. Furthermore, in multispecies models of evolutionary ecology it also opens the way to the monitoring in behaviour-structured population systems. In case of density dependent models, for the monitoring of propagation or extinction of a species we may want to recover the time-dependent density of scarce species, observing a more abundant species of the system. This idea may be applied to the dynamic models of [49-53]. In ecological games the dynamics depends on the behavior types present in the populations, see [54-56]. Then the convergence towards a stable coexistence can be monitored from the observation of certain phenotypes.

Finally, we note that recent papers also show how observer design can be efficiently applied for the monitoring of particular engineering systems. For example, in [57] a real-time local observer was constructed for a linear model of a solar thermal heating system. With different algorithm, a global real-time observer was designed for a more precise nonlinear model for the same solar thermal heating system in [58].

ACKNOWLEDGEMENTS

The present research has been supported in part by the Hungarian Scientific Research Fund OTKA (K81279) and by the Excellence Project Programme of the Ministry of Economy, Innovation and Science of the Andalusian Regional Government, supported by FEDER Funds (P09-AGR-5000).

REFERENCES

1. R. E. Kalman, P. L. Falb and M. A. Arbib, "Topics in Mathematical System Theory," McGraw-Hill, New York, 1969.
2. B. M. Chen, Z. Lin and Y. Shamesh, "Linear Systems Theory. A Structural Decomposition Approach," Birkhauser, Boston, 2004. doi:10.1007/978-1-4612-2046-6
3. E. B. Lee and L. Markus, "Foundations of Optimal Control Theory," Wiley, New York, London, Sydney, 1971.

4. Z. Varga, "On Controllability of Fisher's Model of Selection," In: C. M. Dafermos, G. Ladas and G. Papanicolau, Eds, Differential Equations, Marcel Dekker, New York, 1989, pp. 717-723.
5. Z. Varga, "On Observability of Fisher's Model of Selection," Pure Mathematics and Applications, Series B, Vol. 3, No. 1, 1992, pp. 15-25.
6. G. Farkas, "Local Controllability of Reactions," Journal of Mathematical Chemistry, Vol. 24, No. 1, 1998, pp. 1- 14. doi:10.1023/A:1019150014783
7. G. Farkas, "On Local Observability of Chemical Systems," Journal of Mathematical Chemistry, Vol. 24, No. 1-3, 1998, pp. 15-22. doi:10.1023/A:1019158316600
8. A. Scarelli and Z. Varga, "Controllability of SelectionMutation Systems," BioSystems, Vol. 65, No. 2-3, 2002, pp. 113-121. doi:10.1016/S0303-2647(02)00012-6
9. I. López, "Observabilidad y Controlabilidad en Modelos de Evolución," Ph.D. Dissertation, Universidad de Almería, España, 2003.
10. I. López, M. Gámez, R. Carreño and Z. Varga, "Recovering Genetic Processes from Phenotypic Observation," In: V. Capasso, Ed., Mathematical Modelling & Computing in Biology and Medicine, MIRIAM, Milan, 2003, pp. 356-361.
11. I. López, M. Gámez, R. Carreño and Z. Varga, "Optimization of Mean Fitness of a Population via Artificial Selection," In: R. Bars, Ed., Control Applications of Optimisation, Elsevier, Amsterdam, 2003, pp. 147-150.
12. I. López, M. Gámez and R. Carreño, "Observability in Dynamic Evolutionary Models," BioSystems, Vol. 73, No. 2, 2004, pp. 99-109. doi:10.1016/j.biosystems.2003.10.003
13. I. López, M. Gámez and Z. Varga, "Equilibrium, Observability and Controllability in Selection-Mutation Models," BioSystems, Vol. 81, No. 1, 2005, pp. 65-75.doi:10.1016/j.biosystems.2005.02.006
14. I. López, M. Gámez and Z. Varga, "Observer Design for Phenotypic Observation of Genetic Processes," Nonlinear Analysis: Real World Applications, Vol. 9, No. 2, 2008, pp. 290-302.doi:10.1016/j.nonrwa.2006.10.004
15. M. Gámez, R. Carreño, A. Kósa and Z. Varga, "Observability in Strategic Models of Selection," BioSystems, Vol. 71, No. 3, 2003, pp. 249-255. doi:10.1016/S0303-2647(03)00072-8
16. Z. Varga, A. Scarelli and A. Shamandy, "State Monitoring of a Population System in Changing Environment," Community Ecology, Vol. 4, No. 1, 2003, pp. 73-78.doi:10.1556/ComEc.4.2003.1.11

17. I. López, M. Gámez, J. Garay and Z. Varga, "Monitoring in a Lotka-Volterra Model," BioSystems, Vol. 87, No. 1, 2007, pp. 68-74. doi:10.1016/j.biosystems.2006.03.005

18. M. Gámez, I. López and Z. Varga, "Iterative Scheme for the Observation of a Competitive Lotka-Volterra System," Applied Mathematics and Computation, Vol. 201, No. 1-2, 2008, pp. 811-818. doi:10.1016/j.amc.2007.11.049

19. M. Gámez, I. López and S. Molnár, "Monitoring Environmental Change in an Ecosystem," BioSystems, Vol. 93, No. 3, 2008, pp. 211-217. doi:10.1016/j.biosystems.2008.04.012

20. M. Gámez, I. López and A. Shamandy, "Open-and Closed-Loop Equilibrium Control of Trophic Chains," Ecological Modelling, Vol. 221, No. 16, 2010, pp. 1839- 1846.doi:10.1016/j.ecolmodel.2010.04.011

21. J. R. Banga, E. Balsa-Canto, C. G. Moles and A. A. Alonso, "Dynamic Optimization of Bioprocesses: Efficient and Robust Numerical Strategies," Journal of Biotechnology, Vol. 117, No. 4, 2005, pp. 407-419. doi:10.1016/j.jbiotec.2005.02.013

22. T. Hirmajer, E. Balsa-Canto and J. R. Banga, "DOTcvpSB, a Software Toolbox for Dynamic Optimization in Systems Biology," BMC Bioinformatics, Vol. 10, 2009, p. 199.doi:10.1186/1471-2105-10-199

23. F. Szigeti, C. Vera and Z. Varga, "Nonlinear System Inversion Applied to Ecological Monitoring," Proceedings of the 15th IFAC World Congress on Automatic Control, Barcelona, 21-26 July 2002, pp. 1-5. http://www.nt.ntnu.no/users/skoge/prost/proceedings/ifac2002/data/content/01758/1758.pdf

24. M. Gámez, I. López, J. Garay and Z. Varga, "Observation and Control in a Model of a Cell Population Affected by Radiation," BioSystems, Vol. 96, No. 2, 2009, pp. 172- 177.doi:10.1016/j.biosystems.2009.01.004

25. M. Rafikov, J. M. Balthazar and H. F. von Bremen, "Mathematical Modelling and Control of Population Systems: Applications in Biological Pest Control," Applied Mathematics and Computation, Vol. 200, No. 2, 2008, pp. 557-573. doi:10.1016/j.amc.2007.11.036

26. M. Rafikov and E. H. Limeira, "Mathematical Modelling of the Biological Pest Control of the Sugarcane Borer," International Journal of Computer Mathematics, Vol. 89, No. 3, 2012, pp. 390-401. doi:10.1080/00207160.2011.587873

27. M. Rafikov, A. Del Sole Lordelo and E. Rafikova, "Impulsive Biological Pest Control Strategies of the Sugarcane," Mathematical Problems in Engineering, Vol. 2012, 2012, pp. 1-14. doi:10.1155/2012/726783

28. Z. Varga, "Applications of Mathematical Systems Theory in Population Biology," Periodica Mathematica Hungarica, Vol. 56, No. 1, 2008, pp. 157-168. doi:10.1007/s10998-008-5157-0
29. M. Gámez, "Observation and Control in Densityand Frequency-dependent Population Models," In: W. J. Zhang, Ed., Ecological Modeling, Nova Science Publishers, New York, 2011, pp. 285-306.
30. S. Molnár, "Model Runs for the Definition of the Most Advantageous Integrated Energetical Verticum in the National Economy," Publications of Central Mining Development Institute, Vol. 30, 1887, pp. 121-127.
31. S. Molnár, "Realization of Verticum-Type Systems," Mathematical Analysis and Systems Theory, Vol. 5, 1988, pp. 11-30.
32. S. Molnár, "Optimization of Realization-Independent Cost Functions," Mathematical Analysis and Systems Theory, Vol. 5, 1988, pp. 1-10.
33. S. Molnár, "Observability and Controllability of Decomposed Systems I," Mathematical Analysis and Systems Theory, Vol. 5, 1988, pp. 57-66.
34. S. Molnár, "Observability and Controllability of Decomposed Systems II," Mathematical Analysis and Systems Theory, Vol. 5, 1988, pp. 67-72.
35. S. Molnár, "Observability and Controllability of Decomposed Systems III," Mathematical Analysis and Systems Theory, Vol. 5, 1988, pp. 73-80.
36. S. Molnár, "A Special Decomposition of Linear Systems," Belgian Journal Operations Reseach, Statistics and Computation Science, Vol. 29, No. 4, 1989, pp. 1-19.
37. S. Molnár, "Stabilization of Verticum-Type Systems," Pure Mathematics and Applications, Vol. 4, No. 4, 1993, pp. 493-499.
38. S. Molnár and F. Szigeti, "On Verticum-Type Linear Systems with Time-Dependent Linkage," Applied Mathematics and Computation, Vol. 60, No. 1, 1994, 89-102.doi:10.1016/0096-3003(94)90208-9
39. I. López, M. Gámez and S. Molnár, "Observability and Observers in a Food Web," Applied Mathematics Letters, Vol. 20, No. 8, 2007, pp. 951-957. doi:10.1016/j.aml.2006.09.007
40. M. Gámez, I. López, I. Szabó and Z. Varga, "Verticum- Type Systems Applied to Ecological Monitoring," Applied Mathematics and Computation, Vol. 215, No. 9, 2010, pp. 3230-3238. doi:10.1016/j.amc.2009.10.010
41. S. Molnár, M. Gámez and I. López, "Observation of Nonlinear Verticum-Type Systems Applied to Ecological Monitoring," International Journal of Biomathematics, Vol. 5, No. 6, 2012, pp. 1-15. doi:10.1142/S1793524512500519

42. M. Gámez, I. López, Z. Varga and J. Garay, "Stock Estimation, Environmental Monitoring and Equilibrium Control of a Fish Population with Reserve Area," Reviews in Fish Biology and Fisheries, Vol. 22, No. 3, 2012, pp. 751-766. doi:10.1007/s11160-012-9253-y

43. B. Dubey, P. Chandra and P. Sinha, "A Model for Fishery Resource with Reserve Area," Nonlinear Analysis. Real World Applications, Vol. 4, No. 4, 2003, pp. 625-637.doi:10.1016/S1468-1218(02)00082-2

44. V. Sundarapandian, "Local Observer Design for Nonlinear Systems," Mathematical and Computer Modelling, Vol. 35, No. 1, 2002, pp. 25-36. doi:10.1016/S0895-7177(01)00145-5

45. A. Guiro, A. Iggidr, D. Ngom and H. Touré, "On the Stock Estimation for Some Fishery Systems," Review in Fish Biology and Fisheries, Vol. 19, No. 3, 2009, pp. 313-327.doi:10.1007/s11160-009-9104-7

46. A. Shamandy, "Monitoring of Trophic Chains," Biosystems, Vol. 81, No. 1, 2005, pp. 43-48. doi:10.1016/j.biosystems.2005.02.005

47. P. J. Morin and P. Morin, "Community Ecology," Wiley- Blackwell, Hoboken, 1991.

48. A. Ouahbi, A. Iggidr and M. El Bagdouri, "Stabilization of an Exploited Fish Population," Systems Analysis Modelling Simulation, Vol. 43, No. 4, 2003, pp. 513-524.doi:10.1080/02329290290028543

49. R. Cressman, J. Garay and J. Hofbauer, "Evolutionary Stability Concepts for N-species Frequency-Dependent Interactions," Journal of Theoretical Biology, Vol. 211, No. 1, 2001, pp. 1-10. doi:10.1006/jtbi.2001.2321

50. J. Garay, "Many Species Partial Adaptive Dynamics," BioSystems, Vol. 65, No. 1, 2002, pp. 19-23. doi:10.1016/S0303-2647(01)00196-4

51. R. Cressman and J. Garay, "Evolutionary Stability in Lotka-Volterra Systems," Journal of Theoretical Biology, Vol. 222, No. 2, 2003, pp. 233-245. doi:10.1016/S0022-5193(03)00032-8

52. R. Cressman and J. Garay, "Stablility N-Species Coevolutionary Systems," Theoretical Population Biology, Vol. 64, No. 4, 2003, pp. 519-533. doi:10.1016/S0040-5809(03)00101-1

53. R. Cressman and J. Garay, "A Game-Theoretical Model for Punctuated Equilibrium: Species Invasion and Stasis through Coevolution," BioSystem, Vol. 84, No. 1, 2006, pp. 1-14.doi:10.1016/j.biosystems.2005.09.006

54. R. Cressman, V. Krivan and J. Garay, "Ideal Free Distributions, Evolutionary Games, and Population Dynamics in Multiple-Species Environments," The American Naturalist, Vol. 164, No. 4, 2004, pp. 473-489. doi:10.1086/423827

55. R. Cressman and J. Garay, "A Predator-Prey Refuge System: Evolutionary Stability in Ecological Systems," Theoretical Population Biology, Vol. 76, No. 4, 2009, pp. 248-257.doi:10.1016/j.tpb.2009.08.005
56. R. Cressman and J. Garay, "The Effects of Opportunistic and Intentional Predators on the Herding Behavior of Prey," Ecology, Vol. 92, No. 2, 2011, pp. 432-440. doi:10.1890/10-0199.1
57. R. Kicsiny and Z. Varga, "Real-Time State Observer Design for Solar Thermal Heating Systems," Applied Mathematics and Computation, Vol. 218, No. 23, 2012, pp. 11558-11569. doi:10.1016/j.amc.2012.05.040
58. R. Kicsiny and Z. Varga, "Real-Time Nonlinear Global State Observer Design for Solar Heating Systems," Nonlinear Analysis: Real World Applications, Vol. 14, 2013, pp. 1247-1264. doi:10.1016/j.nonrwa.2012.09.017

APPENDIX

First, we recall the extension of local observability to the case of a control-observation system.

Suppose

$$m,n,r \in \mathbf{N}, F \in C^1\left(\mathbf{R}^n \times \mathbf{R}^m, \mathbf{R}^n\right),$$

$$h \in C^1\left(\mathbf{R}^n, \mathbf{R}^r\right), \left(x^*, u^*\right) \in \mathbf{R}^n \times \mathbf{R}^m$$

such that $F(x^*,u^*)=0$ and $h(x^*)=0$.

Remark A.1. It is known (see e.g. [3]), given a fixed $T>0$, there exists an $\varepsilon_0>0$ such that for all $z_0 \in \mathbf{R}^n$ with $|z_0-x^*|<\varepsilon_0$ there exists a unique continuously differentiable function $x \in C^1\left([0,T], \mathbf{R}^n\right)$ such that

$\dot{x}(t)=F(x(t))$, for all $t \in [0,T]$.

Definition A.1. With the above notation, consider the control-observation system in $\mathbf{R}^n$

$$\dot{x}=F(x,u) \tag{A.1}$$

$$y=h(x) \tag{A.2}$$

System (A.1)-(A.2) is said to be locally observable near the equilibrium if there exists $\varepsilon>0$ such that

$$\left|u(t)-u^*\right|<\varepsilon,\left|x(0)-x^*\right|<\varepsilon,\left|\bar{x}(0)-x^*\right|<\varepsilon,$$

and

$h(x(t))=h(\bar{x}(t))$ for all $t\in[0,T]$

imply that

$$x(0)=\bar{x}(0).$$

Theorem A.1. ([3]) Consider the control-observation system (A.1)-(A.2) in $\mathbf{R}^n$ with

$$A=\frac{\partial}{\partial x}F\left(x^*,u^*\right),C=h'\left(x^*\right).$$

Assume

$$\text{rank}\begin{bmatrix} C \\ CA \\ \vdots \\ CA^{n-1} \end{bmatrix}=n. \tag{A.3}$$

Then system is locally observable near the equilibrium.

Remark A.2. The theorem similar to the previous one is also valid for function F not depending on control, as we have shown in Section 2.

Now, based on [36], we summarize some concepts, notation and a basic sufficient condition for observability of verticum-type systems, in a simplified form used in the present paper.

Let

$$k,n_i,r_i\in\mathbf{N},\mathrm{h}_i:\mathbf{R}^{n_i}\to\mathbf{R}^{r_i},\left(i\in\overline{0,k}\right),f_0\in C^1\left(\mathbf{R}^{n_0},\mathbf{R}^{n_0}\right),$$

$f_i\in C^1\left(\mathbf{R}^{n_i\times n_{i-1}},\mathbf{R}^{n_i}\right)\left(i\in\overline{1,k}\right)$, and consider the nonlinear system

$$\begin{aligned}\dot{x}_0&=f_0\left(x_0\right)\\ y_0&=h_0\left(x_0\right)\end{aligned}, \tag{V_0}$$

and for all $i\in\overline{1,k}$

$$\dot{x}_i = f_i\left(x_i, x_{i-1}\right)$$
$$y_i = h_i\left(x_i\right) \qquad (V_i)$$

Denoting $n := \sum_{i=0}^{k} n_i, r := \sum_{i=0}^{k} r_i;$

let $x = \left(x_0, x_1, \cdots, x_k\right) \in R^n, f : \mathbf{R}^n \to \mathbf{R}^n$ with

$$f\left(x\right) := \left(f_0\left(x_0\right), f_1\left(x_1, x_0\right), \cdots, f_k\left(x_k, x_{k-1}\right)\right);$$

and $h : \mathbf{R}^n \to \mathbf{R}^r$ with

$$h\left(x\right) := \left(h_0\left(x_0\right), h_1\left(x_1\right), \cdots, h_k\left(x_k\right)\right).$$

We shall suppose that there exists

$$x^* = \left(x_0^*, x_1^*, \cdots, x_k^*\right) > 0,$$

such that

$$f\left(x^*\right) = \left(f_0\left(x_0^*\right), f_1\left(x_1^*, x_0^*\right), \cdots, f_k\left(x_k^*, x_{k-1}^*\right)\right) = 0$$

and $h\left(x^*\right) = 0$.

Definition A.2. Observation system

$$\dot{x} = f\left(x\right)$$
$$y = h\left(x\right) \qquad (V)$$

is said to be of verticum type.

Remark A.3. Equations $\left(V_i\right)\left(i \in \overline{1,k}\right)$ do not define a standard observation system in this setting, because of the presence of the "exogenous" variable x_{i-1} connecting it to system $\left(V_{i-1}\right)$.

Remark A.4. It is known that near equilibrium x^* all solutions of system (V) can be defined on the same time interval $[0, T]$. In what follows $T > 0$ will be considered fixed and concerning observability, the reference to T will be suppressed.

For the analysis of observability of system (V), let us linearize systems (V_i), at the respective equilibria $x_i^*, i \in \overline{0,k}$, obtaining the linearized systems

$$\dot{x}_0 = A_{00}x_0$$
$$y_0 = C_0x_0 \quad , \qquad (\mathrm{LV}_0)$$

and for all $i \in \overline{1,k}$

$$\dot{x}_i = A_{ii}x_i + A_{ii-1}x_{i-1}$$
$$y_i = C_ix_i \quad , \qquad (\mathrm{LV}_i)$$

where

$$A_{00} = f_0'(x_0^*), C_0 = h_0'(x_0^*);$$

$$A_{ii} = \frac{\partial f_i}{\partial x_i}(x_i^*, x_{i-1}^*), A_{ii-1} = \frac{\partial f_i}{\partial x_{i-1}}(x_i^*, x_{i-1}^*),$$

$$C_i = h_i'(x_i^*)(i \in \overline{1,k}).$$

Define matrices $A \in \mathbf{R}^{n\times n}, C \in \mathbf{R}^{r\times n}$ as follows:

$$A = \begin{pmatrix} A_{00} & 0 & 0 & \cdots & 0 & \cdots & 0 \\ A_{10} & A_{11} & 0 & \cdots & 0 & \cdots & 0 \\ 0 & A_{21} & A_{22} & \cdots & 0 & \cdots & 0 \\ \vdots & \vdots & \vdots & \ddots & \vdots & \ddots & \vdots \\ 0 & 0 & 0 & \cdots & \vdots & A_{k-1,k-1} & 0 \\ 0 & 0 & 0 & \cdots & \vdots & A_{k,k-1} & A_{k,k} \end{pmatrix},$$

$$C = \begin{pmatrix} C_0 & & & 0 \\ & C_1 & & \\ & & \ddots & \\ & & & \ddots & \\ 0 & & & C_k \end{pmatrix}$$

obtaining linear observation system

$$\dot{x} = Ax$$
$$y = Cx \qquad (\mathrm{LV})$$

of verticum type (see [36]). In the latter paper, a Kalman-type necessary and sufficient condition for observability of linear verticum-type systems was obtained. Here we recall only its "sufficient part" to be applied below.

Theorem A.2. ([36]) Suppose that

$$\operatorname{rank}\begin{bmatrix} C_i \\ C_i A_{ii} \\ \vdots \\ C_i A_{ii}^{n_i-1} \end{bmatrix} = n_i \left(i \in \overline{0,k}\right).$$

Then the linear verticum-type system (LV) is observable.

Remark A.5. If x^* is a Lyapunov stable equilibrium of system

$$\dot{x} = f(x)$$

then $(V_i)\left(i \in \overline{1,k}\right)$ can be considered as a control-observation system with "small" controls in the following sense. By the Lyapunov stability of x^*, for all $\varepsilon > 0$, there exists $\delta > 0$ such that $|x(0)-x^*|<\delta$ implies

$\left|x(t)-x^*\right|<\varepsilon$ (for $t \in [0,T]$). In particular,

$\left|x_{i-1}(t)-x_{i-1}^*\right|<\varepsilon$ for all $t \in [0,T]$.

Considering $u_i(t) := x_{i-1}(t) - x_{i-1}^*$ as a control for system $\dot{x}_i = f_i\left(x_i, x_{i-1}^* + u_i\right)$, (V_i) becomes a controlobservation system in the sense of this Appendix. Suppose that for each $i \in \overline{0,k}$

$$\operatorname{rank}\begin{bmatrix} C_i \\ C_i A_{ii} \\ \vdots \\ C_i A_{ii}^{n_i-1} \end{bmatrix} = n_i$$

then by Theorem A.2 the verticum-type system (LV) is observable.

Hence, the linearization of the observation system (V) is observable. Therefore, by Kalman's theorem on observability of linear systems (see [1]), the rank condition (A.3) is fulfilled, which by Theorem A.1 implies local observability of system (V) near equilibrium x^*.

The above reasoning can be summarized in the following theorem:

Theorem A.3. If equilibrium x^* is Lyapunov stable for system $\dot{x} = f(x)$, and

$$\operatorname{rank}\begin{bmatrix} C_i \\ C_i A_{ii} \\ \vdots \\ C_i A_{ii}^{n_i-1} \end{bmatrix} = n_i \left(i \in \overline{0,k}\right),$$

then observation system (V) is observable near its equilibrium x^*.

Chapter 12

SELECTING THE SIX SIGMA PROJECT: A MULTI DATA ENVELOPMENT ANALYSIS UNIFIED SCORING FRAMEWORK

Mazen Arafah

The Department of Industrial Engineering, Faculty of Engineering & Technology, The University of Jordan, Amman, Jordan

ABSTRACT

The importance of the project selection phase in any six sigma initiative cannot be emphasized enough. The successfulness of the six sigma initiative is affected by successful project selection. Recently, Data Envelopment Analysis (DEA) has been proposed as a six sigma project selection tool. However, there exist a number of different DEA formulations which may affect the selection process and the wining project being selected. This work initially applies nine different DEA formulations to several case studies and concludes that different DEA formulations select different wining projects. Also in this work, a Multi-DEA Unified Scoring Framework is proposed to overcome this problem. This framework is applied to several case studies and proved to successfully select the six sigma project with the best performance. The framework is also successful in filtering out some of the projects that have "selective" excellent performance, i.e. projects with excellent performance in some of the DEA formulations and worse performance in others. It is also successful in selecting stable projects; these are projects that perform well in the majority of the DEA formulations, even if it has not been selected as a wining project by any of the DEA formulations.

INTRODUCTION

Six sigma (SS) is one of a number of quality improvement strategies based on the Shewhart-Deming PDSA cycle [1] . Coronado [2] defines SS as a business improvement strategy used to improve business profitability, to drive out waste and to reduce cost of poor quality and to improve effectiveness and efficiency of all operations so as to meet or even exceed customer's needs and expectations. SS has originated at Motorola Inc. as a long- term quality improvement initiative entitled "The Six Sigma Quality Program". It was launched by the company's chief executive officer (CEO) Bob Galvin [1] .

Antony et al. [3] mention that Juran believes that six sigma improvements must be tackled as projects, which lead to a critical step that precedes the implementation of the SS project, namely, the SS project selection. According to [4] , it has been suggested that perhaps up to 80 percent of all "projects" are not actually projects at all, since they do not include the three project requirements: objectives, budget, and due date. Organizations are faced with a myriad of potential projects to choose from, including six sigma projects. Winning six sigma projects are a major factor in the acceptance of six sigma within the organization [5] .

The project selection for six sigma program is often the most important and difficult priori for the implementation of a six sigma program [6] . Project selection is an important activity that most firms fail to fulfill correctly, eventually resulting in undesirable outcomes. The survey conducted by the Aviation Week magazine identified that 60 percent of the companies selected opportunities for improvement on an ad hoc basis, while only 31 percent relied on a portfolio approach [7] . However, the study shows that companies actually achieve better results when applying the portfolio approach. The main purpose of project selection process is to identify projects that will result in the maximum benefit to the organization from the pool of all available improvement opportunities. As noted in the Aviation Week magazine survey, following a structured approach in project selection will result in better outcomes for the organization and thus a better six sigma experience [6] .

Six sigma projects consume different inputs and are expected to produce multiple outputs, thus the six sigma project selection process is multi criteria-multi objective. In order to manage and optimize the process output, it is important that we identify the key input variables which influence the output [8] . Such factors that play a key role in the success of six sigma initiatives are known as critical success factors (CSFs); close investigation of these factors by the organization leads to higher probability of project success and produces better managerial insights to what factors are more critical than others with respect to the distinct characteristics of the organization.

In this study, we consider a number of CSFs that are most commonly discussed in literature of quality improvement projects which are presented in Table 1. These factors can be considered as resources consumed differently by different projects.

Six sigma project selection can be used to optimize many important objectives. Table 2 presents different objectives for six sigma projects mentioned in the literature.

Many approaches and techniques have been proposed to address the six sigma project selection problem. Table 3 provides a list of the different approaches and techniques used in the selection of six sigma projects.

DEA is one important technique that is used to solve the multi-criteria/multi-objective problem. DEA was first introduced in 1978 [26] . Since that time, a great variety of applications of DEA for use in evaluating the performances of many different kinds of entities have been engaged in many different activities in many different contexts [27] .

Table 1: Critical success factors in six sigma project.

Success Factor	Author
Expected Project Cost	[6] [9]
Level of Leadership and Management Skills	[10]
Training Hours	[11]
Number of Green and Black Belts	[6]
Expected Project Duration	[6] [9]
Level of Management Commitment	[6]
Good Systems and Availability of Information and Resources	[9] [12]
COPQ	[5]
Probability of Implementation	[9]

Table 2: Six sigma project objectives.

Objective	Author
Impact on Business Strategy	[6]
Financial Impact	[6] [9]
Sigma Quality	[6] [9]
Productivity	[6]
Market Share	[13]
Customer Satisfaction	[6] [9] [12]

Table 3: Methods and techniques used for six sigma projects selection.

Proposed Method/Technique	Author
Pareto Analysis	[14]-[16]
Analytic Hierarchy Process (AHP)	[5] [9] [17]-[20]
Project Selection Matrix	[21]
Project Ranking Matrix	[22]
Theory of Constraints (TOC)	[15] [23]
Quality Function Deployment (QFD)	[7] [24]
Pareto Priority Index (PPI)	[15] [16] [23]
Data Envelopment Analysis	[6] [25]

DEA is described as a nonparametric technique that aims at comparing different entities, known as Decision Making Units (DMUs), relying solely on inputs and outputs of the DMUs [28] . The terms entity, inputs, and outputs are very generic. An example of different entities is hospitals, projects and people. Inputs for a hospital could be the number of physicians or nurses and the outputs could be the number of patients treated.

DEA has many different formulations. However, regardless of the major benefits and advantages of the DEA different formulations, it is subject to one major disadvantage; different formulations may lead to selecting different winning projects. The literature rarely discusses or highlights this important DEA shortcoming. For instance, the same project selection problem, when considered under different formulations (benevolent, aggressive, super efficiency, etc.), will produce different wining projects. This work highlights the diverse results of the different DEA formulations for several hypothetical case studies. It also proposes a new framework, Multi- DEA Unified Scoring Framework (Multi-DEA USF), to obtain a final unified score.

DEA FORMULATIONS

DEA is a data oriented approach for evaluating the performance of a set of peer entities called Decision Making Units (DMUs) which convert multiple inputs into multiple outputs [27] . The comparison of the different DMUs is carried out by calculating the relative efficiency score for each DMU while abiding to certain constraints. Basically, DEA provides a categorical classification of the units into efficient and inefficient ones [29] .

The efficiency score in the presence of multiple input and output factors is defined as:

$$\text{Efficiency} = \frac{\text{Weighted sum of Output}}{\text{Weighted sum of Input}} \quad (1)$$

Assuming that there are n DMUs, each with m inputs and s outputs, the relative efficiency score for a test DMU p is given by:

$$\frac{\sum_{k=1}^{s} v_k y_{kp}}{\sum_{j=1}^{m} u_j x_{jp}} \quad (2)$$

where

$k=1$ to s, $j=1$ to m, $i=1$ to n;

$y_{ki} =$ amount of output k produced by DMU i;

$x_{ji} =$ amount of input j utilized by DMU i;

$v_k =$ weight given to output k;

$u_j =$ weight given to output j.

Charnes [30] proposed the following model:

$$\max \frac{\sum_{k=1}^{s} v_k y_{kp}}{\sum_{j=1}^{m} u_j x_{jp}} \quad (3)$$

Subject to

$$\frac{\sum_{k=1}^{s} v_k y_{kp}}{\sum_{j=1}^{m} u_j x_{jp}} \leq 1, \quad \forall i$$

$$u_j, v_k \geq 0; \quad \forall k, j$$

Model (3) is known as the CCR model.

The fractional model presented in (3) is converted to a linear program as shown in (4):

$$\max \sum_{k=1}^{s} v_k y_{kp} \qquad (4)$$

Subject to

$$\sum_{j=1}^{m} u_j x_{jp} = 1$$

$$\sum_{k=1}^{s} v_k y_{kp} - \sum_{j=1}^{m} u_j x_{jp} \leq 0, \quad \forall i$$

$$u_j, v_k \geq 0; \quad \forall k, j$$

The second constraint ensures that the efficiency cannot be greater than one.

The relative efficiency score of DMU k is obtained by maximizing the efficiency score of DMU k by choosing an optimal set of weights that show the DMU at its best. A set of weights is found for each DMU by solving (3) n times. If the relative efficiency (aka the simple score) is 1, then the DMU is said to be efficient. Otherwise, the DMU is inefficient and must increase its output or decrease its input in order to become efficient.

Relying on the simple efficiency score is not enough, mainly because of two deficiencies which are discussed in details in [27]. First, weak discriminating power leads to classifying multiple DMUs as efficient. This is problematic when all DMUs must be ranked or the most efficient DMU must be identified e.g. when the DMUs are projects and one must be selected for implementation. Second, the unrealistic weight-problem where some DMUs may have been classified as efficient by using extreme weights that are not practical.

Researchers have proposed several solutions to overcome these drawbacks. The cross-evaluation method has been proposed. The main idea of cross evaluation is to use DEA in a peer evaluation instead of a self-evaluation mode.

As noted by [31] , there are two principal advantages of cross evaluation: 1) it provides a unique ordering of the DMUs, and 2) it eliminates unrealistic weight schemes without requiring the elicitation of weight restrictions from application area experts [32] .

The optimal weights for the inputs and outputs maximize the efficiency of the DMU being considered. However, we can use the set of weights to calculate the efficiency of other DMUs. This can be thought of as each DMU testing itself with respect to the other DMUs optimal weights. This is called Cross-Efficiency. The result is a Cross-Efficiency Matrix (CEM) with dimensions $n \times n$ where E_{ks} is DMU_s's score using DMU_k's set of weights:

$$E_{ks} = \frac{\sum_{k=1}^{s} v_k y_{ki}}{\sum_{j=1}^{m} u_j x_{ji}} \quad (5)$$

For $k = 1, \cdots, s$, and $j = 1, \cdots, m$.

Note that the diagonal of the CEM shown in Table 4 represents the simple scores of each DMU (E_{kk}).

A DMU with high cross efficiency scores along its column in the CEM is considered a good overall performer.

The column means can be computed to effectively differentiate between good and poor performing DMUs.

$$a_i = \frac{1}{n} \sum_{n=i} E_{si} \quad (6)$$

A problem arises when using the simple CEM. The issue is that there are more than one set of optimal weights that yield the same efficiency score for the DMU being considered i.e. the weights u_r and v_i are not unique. While the simple efficiency score (E_{kk}) will stay the same, the CEM will not. The CEM relies on the sets of weights of each DMU so if they change, so will the CEM. This means that for each problem there are multiple CEMs that describe it. To overcome this problem, a secondary objective is introduced to the linear program giving us the Benevolent and Aggressive formulations. According to

[33] , for the run DMU with the same efficiency score there are two possible cases. Case one, the set of weights leads to a higher cross-efficiency for the other DMUs which is known as the benevolent formulation. Case two, the set of weights reduces the cross- efficiency score for the other DMUs or what is known as the aggressive formulation.

The problem was formulated by [33] :

$$\text{Min}\sum_{k}\left(v_k \sum_{n\neq i} y_{ki}\right) \qquad (7)$$

Subject to

$$\sum_{j}\left(u_j \sum_{n\neq i} x_{ji}\right)=1$$

$$v_k \text{ and } u_j \geq 1$$

$$E_{in} \leq 1 \text{ For } i \neq n$$

$$\sum_{k} y_{ki} v_k - E_{ii} \sum_{j} x_{ji} u_j = 0$$

The Benevolent formulation is the same as (7) but instead of minimizing the objective function we maximize it.

Another way to overcome the lack of discrimination provided in the simple DEA formulation is proposed by calculating the Maverick score. Doyle and Green [33] explained the Maverick score and how it is calculated. The Maverick score measures the deviation between the "self-appraised" efficiency score and the average "peer-appraised" score. It is calculated using Equation (8):

Table 4: Cross efficiency matrix

Rating DMU	Rated DMU			
	1	2	...	N
1	E_{11}	E_{12}	...	E_{1n}
2	E_{21}	E_{22}	...	E_{2n}
.	.	.	.	.
N	E_{n1}	E_{n2}	...	E_{nn}
	a_1	a_2	...	a_n

$$M_i = \frac{(E_{ii} - e_i)}{e_i} \tag{8}$$

$$e_i = \frac{1}{(n-1)} \sum_{n \neq i} E_{ni} \tag{9}$$

Another model that is used for differentiating between efficient projects is the Super Efficiency model. The Super Efficiency model came into prominence as an aid in the sensitivity analysis of classical DEA models [34] . Andersen and Petersen [35] propose the use of super efficiency DEA models in ranking the relative efficiency of each DMU.

The input-oriented super efficiency CCR model is expressed as [36] :

$$\text{Min}\,\theta^{\text{Super}} \tag{10}$$

Subject to

$$\sum_{\substack{j=1 \\ j \neq o}}^{n} u_i x_{ij} \leq \theta^{\text{Super}} x_{io}, \quad i = 1, \cdots, m$$

$$\sum_{\substack{j=1 \\ j \neq o}}^{n} u_j y_{rj} \geq y_{ko}, \quad r = 1, \cdots, s$$

$$\sum_{\substack{j=1 \\ j \neq o}}^{n} u_j = 1$$

$$u_i \geq 0, \quad j \neq o$$

where O is the DMU under evaluation.

METHODOLOGY

Figure 1 shows the methodology followed in this research. which is initiated by project case study generation, (Subsection 3.1) followed by DEA techniques application (Subsection 3.2), then qualitative comparative study (Subsection

3.3) is performed followed by aggregation and winning project selection (Subsection 3.4).

Six Sigma Project Selection Case Studies

The study will be carried out in two parts. For validation purpose, in part one, We start by considering the six sigma case study presented in which included twenty hypothetical six sigma projects. Each Project (DMU) has three inputs and five outputs. In part two, we expand on the previous case by including more factors that are considered imperative factors for decision makers in the implementation of six sigma initiatives.

We added three inputs, namely, "Level of Management Commitment Required", "Required level of leadership and Management Skills", and Training hours, and one output: Percentage increase in Market share. The data for the new inputs and outputs were randomly generated using MATLAB®each according to their possible values. "Level of Management Commitment Required", "Required level of leadership and Management Skills" were obtained by randomly generating numbers between 1 and 10. On this 10 point scale a score of 1 means that not much commitment and skills are required to carry out the project which is more desirable for managers. Based on literature, we found that the training hours for a six sigma initiative are between 40 and 120 hours. Therefore, we randomly generated numbers between 40 and 120 to obtain the data for Training Hours. As for the output "Percentage increase in Market share", we randomly generated numbers between 0% and 35%.

DEA Techniques Application

The different DEA models and formulations applied using MATLAB® are shown in Table 5.

Comparative Study

Since the results of the first seven models are based on the larger the better criterion and the last two (the Mave rick scores) are based on the smaller the better.

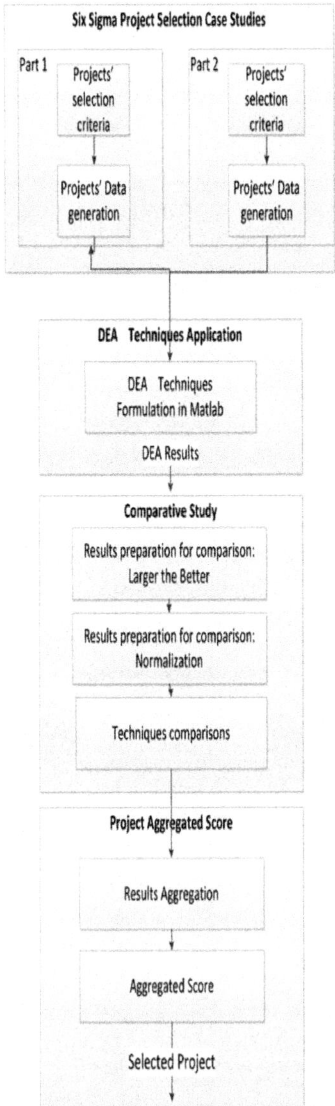

Figure 1: Methodology.

We performed a two-step normalization for the Maverick based scores. First we use Equation (11) to transform the Maverick scores into the larger the better.

$$Y = -10\log_{10}\left(X^2\right) \tag{11}$$

However since some of the Maverick scores are greater than 1 and some are smaller than 1, Y would have both positive and negative values; we used max-min standardization technique as shown in Equation (12).

$$Z = \frac{Y - \text{Min}}{\text{Max} - \text{Min}} \qquad (12)$$

The results of all the formulation are then normalized using Equation (13).

$$Z'_j = \frac{Z_j}{\sum_i Z_i} \qquad (13)$$

A qualitative comparison between the different DEA techniques is performed to explore the diversity in the ranking of projects produced by the different DEA formulation.

Table 5: Summary of the different DEA models used in the study.

Formulation	Brief Description
1. Simple Efficiency Score	• Basic formulation • A self-evaluation mode • Weak discriminating power • Unrealistic weights
2. Aggressive Cross Efficiency Score	• All DMUs are used in the calculation of the efficiency • The set of weights reduces the cross-efficiency score for the other DMUs • Provides a unique ordering of the DMUs • A peer-evaluation mode • Eliminates unrealistic weights
3. Aggressive off Diagonal Cross Efficiency Score	• Diagonal DMUs are not used in the calculation of the efficiency • The set of weights reduces the cross-efficiency score for the other DMUs • Provides a unique ordering of the DMUs • A peer-evaluation mode • Eliminates unrealistic weights
4. Benevolent Cross Efficiency Score	• All DMUs are used in the calculation of the efficiency • The set of weights leads to higher a cross-efficiency scores for the other DMUs • Provides a unique ordering of the DMUs • A peer-evaluation mode • Eliminates unrealistic weights
5. Benevolent off Diagonal Cross Efficiency Score	• Diagonal DMUs are not used in the calculation of the efficiency • The set of weights leads to higher a cross-efficiency scores for the other DMUs • Provides a unique ordering of the DMUs • A peer-evaluation mode • Eliminates unrealistic weights
6. Super Efficiency	• Discrimination is based on using the super efficiency formulation
7. Aggressive Maverick Score	• Discrimination is based on calculating the Maverick score using aggressive efficiencies
8. Benevolent Maverick Score	• Discrimination is based on calculating the Maverick score using benevolent efficiencies

Project Aggregated Score

The normalized scores are summed to obtain a unified score for each project, thus leading to one score to be used for project selection.

RESULTS AND DISCUSSION

In Subsection 4.1, we present the results of applied the Multi-DEA USF to the data provided by [6] . In Subsections 4.2 - 4.4, we present the results of the extended datasets.

Applying the Multi-DEA USF and Validation

For the dataset presented by [6] , we initially applied the simple DEA formulation. Out of the twenty projects, only five projects are efficient. Then, we applied all the other DEA formulations to this dataset. Table 6 present the scores of the five efficient projects. The complete list of scores for the twenty projects is shown in Table 14 (Appendix II) which coincides perfectly with the results provided by [6] .

In Table 6, we notice that the Aggressive and Benevolent scores are less than the simple score for each project. This agrees with the logic of the simple CCR formulation where each project maximizes its own score; while, in the Aggressive and Benevolent formulations a secondary goal constrains the problem and prevents the project from achieving better than its simple score.

Also, note that a lower Maverick score means that the project is less of a Maverick which gives the project a higher rank.

It can be noticed that not all the DEA formulations agreed on the selected projects. For the above case all DEA formulations have selected project 7 except for the aggressive technique which selected project 17. Thus for the data provided by [6] only one of the DEA formulations have disagreed with the rest.

Table 6: Summary of scores for efficient projects Part 1.

Project	Agg. Score	Bnv. Score	X Eff. Agg.	X Eff. Bnv.	OFFD X Eff_Agg	OFFD X Eff_Bnv	Super Efficiency	Mav. Agg.	Mav. Bnv.
2	0.2880	0.7233	0.9542	0.9613	0.9518	0.9593	1.0455	0.0479	0.0403
7	0.2827	**0.7619**	**0.9761**	**0.9884**	**0.9749**	**0.9878**	**1.1365**	**0.0244**	**0.0117**
12	0.3051	0.7435	0.9401	0.9637	0.9370	0.9618	1.0556	0.0637	0.0377
17	**0.5155**	0.7028	0.9204	0.9687	0.9162	0.9670	1.0452	0.0865	0.0323
19	0.2670	0.7356	0.9719	0.9775	0.9704	0.9763	1.1154	0.0289	0.0231

However, this cannot be generalized to other cases as will be shown in the next subsection. The normalized scores for each efficient project were calculated and compared using Figure 2. The normalized scores coincides perfectly with the data provided in Table 5. In that, Project 7 seems to outperform the rest of the projects in all formulations except for the aggressive technique.

Figure 3 presents the aggregated score for the efficient projects. Project 7 is identified as the best six sigma project. This figure shows that project 7 outperforms the rest of the projects and has a clear edge for selection.

Applying the Multi-DEA USF for the Extended Data Sets

We initially applied simple efficiency to select the efficient projects. Then we applied the rest of the DEA formulations. Table 7 and Figure 4 present the scores of the efficient projects. Comparing the efficient projects is more cumbersome in this case because more projects are efficient (14 projects).

The selected project for each DEA formulation is shown in Bold. It's clear from Table 7 and Figure 4that the DEA formulations have diverse decisions. Project 7 has been selected by three DEA formulations, project 19 has been selected by three as well, while project 3 is selected by two and project 10 is selected by one DEA formulation. Thus, the selection process was extremely difficult and requires a rigorous method. The reason of this is the high competitively between the projects. The different DEA techniques are showing high variability in terms of the project expected cost, the best project is project 7. While for expected project duration project 12 is the shortest. Level of management commitment project is 8 the best, etc. For this reason, there was high variability in terms of the selected project using different DEA techniques. For example, aggressive formulation choose project 10. The aggressive cross efficiency choose project 19. The benevolent cross efficiency choose project 7 to be the best project. This diverse decision phenomenon places a lot of doubts on how to pick up the winning one. It is also stresses the need for a unified methodology for choosing the finalized winning project. We suggest in this work to use the Multi-DEA-USF for this purpose.

Figure 5 shows the final aggregate score for the different projects. The suggested technique have successfully selected project 17, although project 7 was a close competing peer project. Figure 6illustrates why this important project should be selected, although it was pick up by only one DEA-formulation (Aggressive) as the winning project. This important project-which might have gone unnoticed through applying only the individual DEA formulations-was always a close competitor in all DEA formulations to the leading projects 7 and 19. It performed better than them in terms of the aggressive scores. The successful selection of project 17-which was shadowed by project 7 and 19 is a major advantage of the suggested technique (multi-DEA-USF) over the individual DEA formulations which allows shadowing of close competitors. The close competitor gives a more stable performance (always performing good enough) in all DEA formulations while the projects picked up by some of the DEA formulations might have worse performance in others (example project 19 in Mav_Bnv).

Applying the Multi-DEA USF to 2nd Dataset

Following to simple DEA application we applied all DEA formulations devised have been applied. Table and Figure 7 show the performance index for each project with respect to each DEA formulation. The selected project by each formulation is in highlighted in bold font. The selection process is highly diverse and it is extremely difficult to pick up a winning project.

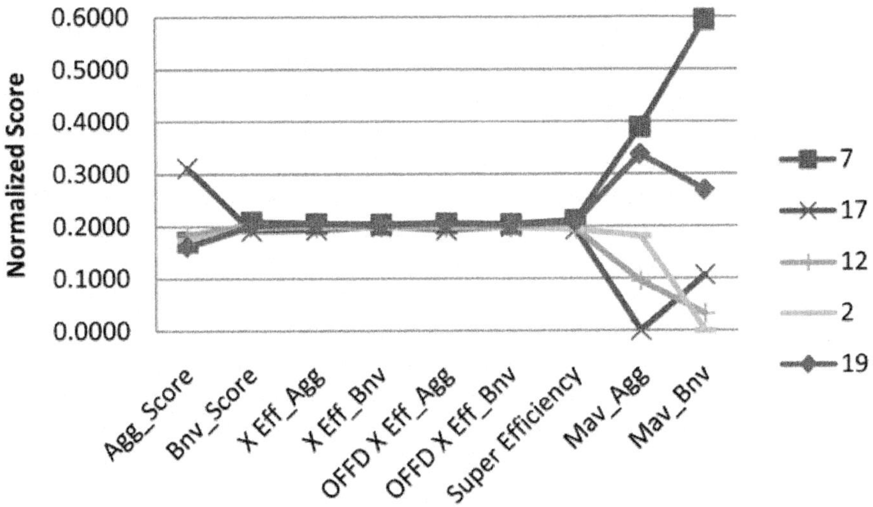

Figure 2: Normalized score for different DEA formulations for efficient projects.

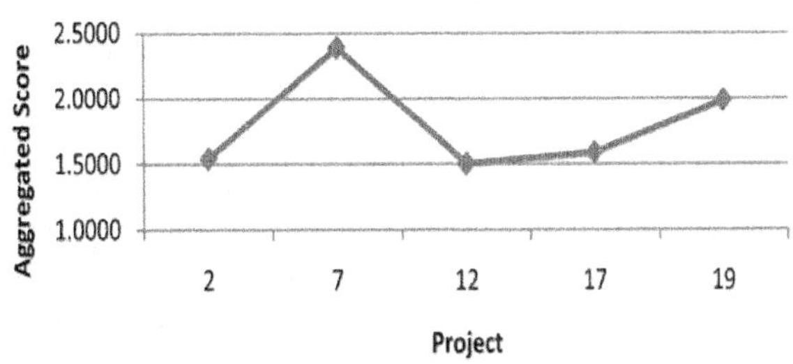

Figure 3: Aggregated scores

Table 7: Summary of scores for efficient projects Part 2 (1st set)

Project	Agg. Score	Bnv. Score	X Eff. Agg.	X Eff. Bnv.	OFFD X Eff_Agg	OFFD X Eff_Bnv	Super Efficiency	Mav_ Agg	Mav. Bnv.
1	0.2632	0.7594	0.4321	0.7332	0.4022	0.7192	1.2238	1.3145	0.3638
2	0.2880	0.7777	0.6731	0.8348	0.6559	0.8261	1.5347	0.4857	0.1979
3	0.0969	**0.7808**	0.5566	0.8394	0.5332	0.8310	**3.3939**	0.7967	0.1913
7	0.2827	0.7804	0.7485	**0.9695**	0.7353	**0.9679**	1.3276	0.3359	**0.0314**
8	0.1760	0.7016	0.4939	0.7050	0.4672	0.6895	2.1999	1.0249	0.4184
9	0.2938	0.4359	0.4223	0.5247	0.3919	0.4997	1.1686	1.3678	0.9060
10	**0.6804**	0.7047	0.3534	0.6784	0.3194	0.6615	1.0087	1.8295	0.4741
12	0.2493	0.7698	0.6694	0.8511	0.6520	0.8433	1.4473	0.4939	0.1749
13	0.5647	0.7153	0.3736	0.7351	0.3406	0.7212	1.0237	1.6768	0.3603
14	0.3979	0.7065	0.3869	0.6726	0.3547	0.6553	1.2400	1.5844	0.4869
15	0.3169	0.7351	0.4360	0.7457	0.4064	0.7323	1.4416	1.2934	0.3410
16	0.3381	0.7136	0.4479	0.6325	0.4188	0.6131	1.3120	1.2328	0.5811
17	0.3585	0.7806	0.7513	0.9660	0.7382	0.9642	1.3115	0.3310	0.0352
19	0.2670	0.7805	**0.7647**	0.9355	**0.7523**	0.9321	1.3145	**0.3077**	0.0690

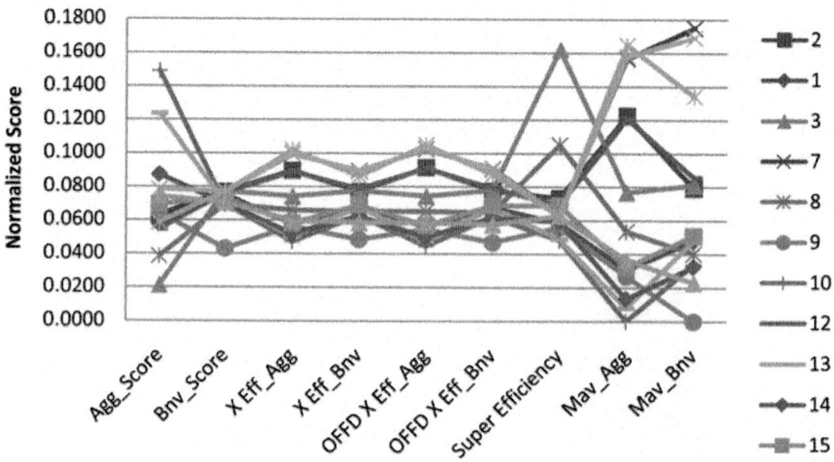

Figure 4: Normalized score for different DEA formulations for efficient projects.

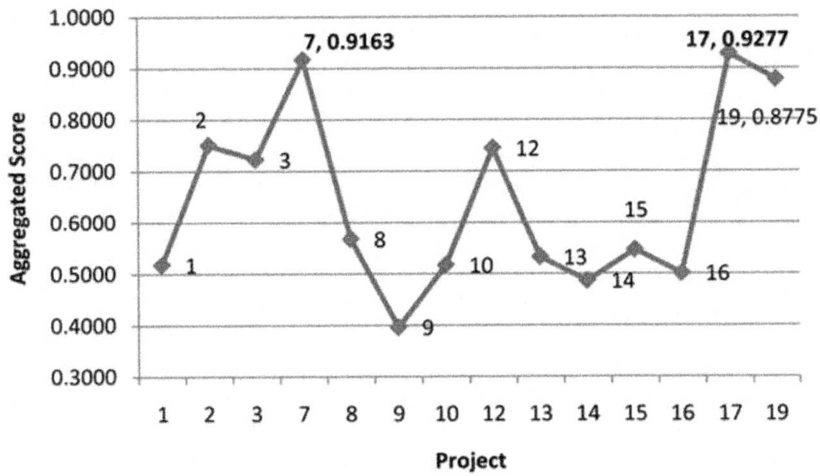

Figure 5: Aggregated score Part 2 (1st set).

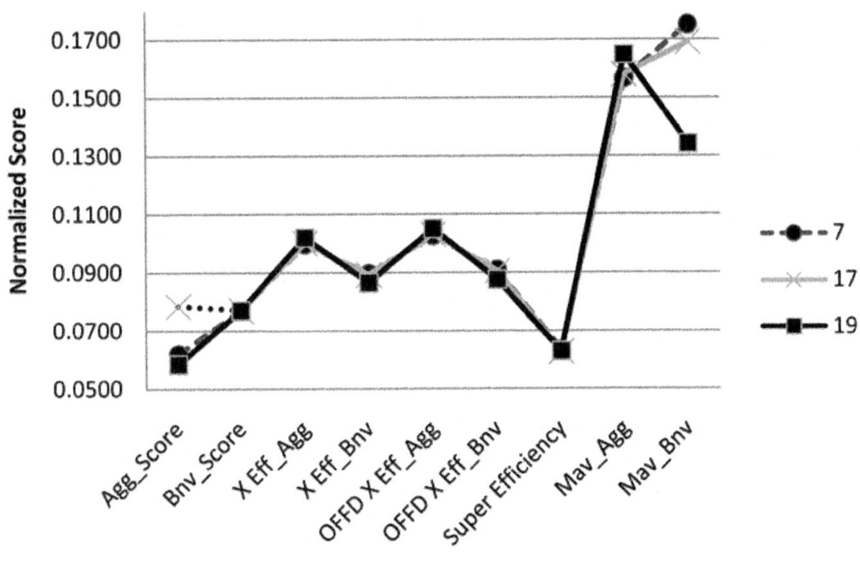

Figure 6: Normalized score for different DEA formulations for competing projects.

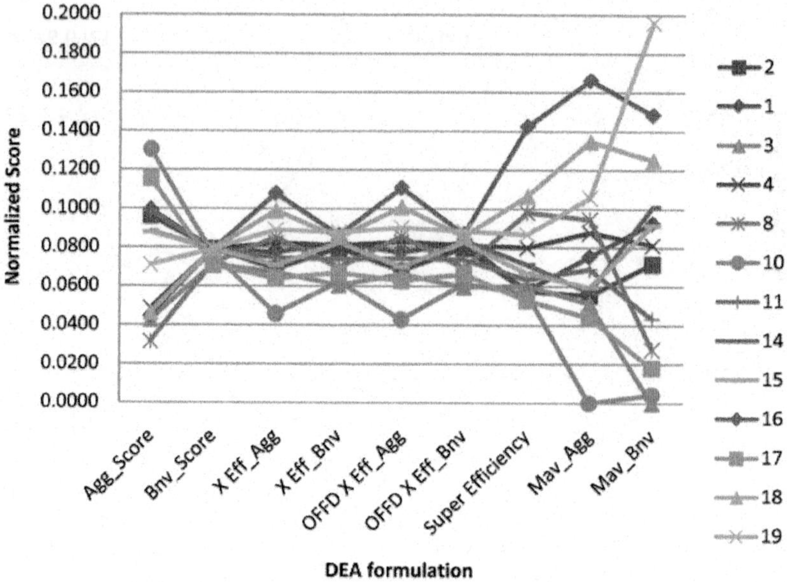

Figure 7: Normalized score for different DEA formulations for efficient projects (2nd set).

Table 8: Summary of scores for efficient projects Part 2 (2nd set).

Project	Agg. Score	Bnv. Score	X Eff. Agg.	X Eff. Bnv.	OFFD X Eff_Agg	OFFD X Eff_Bnv	Super Efficiency	Mav. Agg.	Mav. Bnv.
1	0.3951	0.8546	0.5903	0.9270	0.5688	0.9232	1.2355	0.6939	0.0787
2	0.3817	**0.8549**	0.5258	0.8946	0.5009	0.8891	1.2289	0.9018	0.1178
3	0.1701	0.7673	0.5132	0.6834	0.4876	0.6667	1.2864	0.9486	**0.4633**
4	0.1926	0.8512	0.6282	0.9109	0.6086	0.9062	1.7056	0.5919	0.0978
8	0.1254	0.7783	0.6495	0.7856	0.6311	0.7743	2.0981	0.5396	0.2729
10	**0.5189**	0.8236	0.3507	0.7007	0.3165	0.6850	1.2067	**1.8516**	0.4271
11	0.3905	0.8484	0.5696	0.8325	0.5470	0.8236	1.3568	0.7556	0.2013
14	0.3508	0.8542	0.5317	0.9372	0.5071	0.9339	1.5189	0.8806	0.0670
15	0.3495	**0.8549**	0.5398	0.9254	0.5155	0.9214	1.4298	0.8526	0.0807
16	0.1768	0.8547	**0.8248**	0.9736	**0.8156**	0.9723	**3.0431**	0.2124	0.0271
17	0.4595	0.7632	0.4897	0.7533	0.4629	0.7403	1.1292	1.0419	0.3275
18	0.1841	0.8544	0.7577	0.9595	0.7449	0.9574	2.2811	0.3198	0.0422
19	0.2815	**0.8549**	0.6816	**0.9893**	0.6648	**0.9887**	1.8540	0.4672	0.0109

Figure 7 shows the normalized scores for the different projects, project 16 should be selected while the close competitors are projects 18 and 19.

Figure 8 presents the final aggregate score for the different projects. The suggested technique have successfully selected project 16 although projects 18 and 19 are close competing peer projects.

Figure 9 shows the performance the three competing projects against the different DEA techniques. The figure shows that the project 16 outperforms the other projects in many of the individual DEA techniques. The Multi-DEA-USF was successful in picking up a highly performing project.

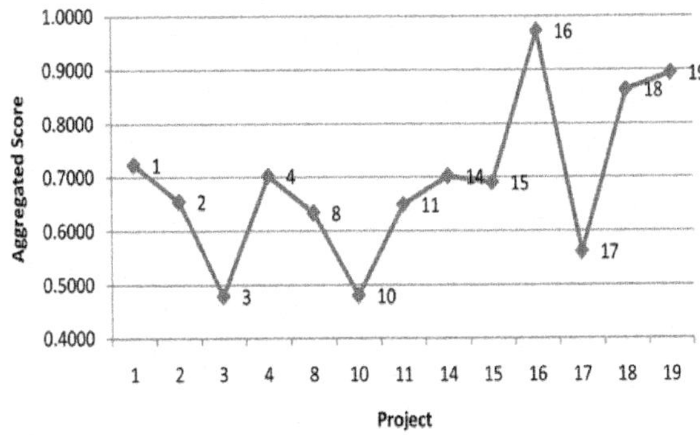

Figure 8. Aggregated score Part 2 (2nd set).

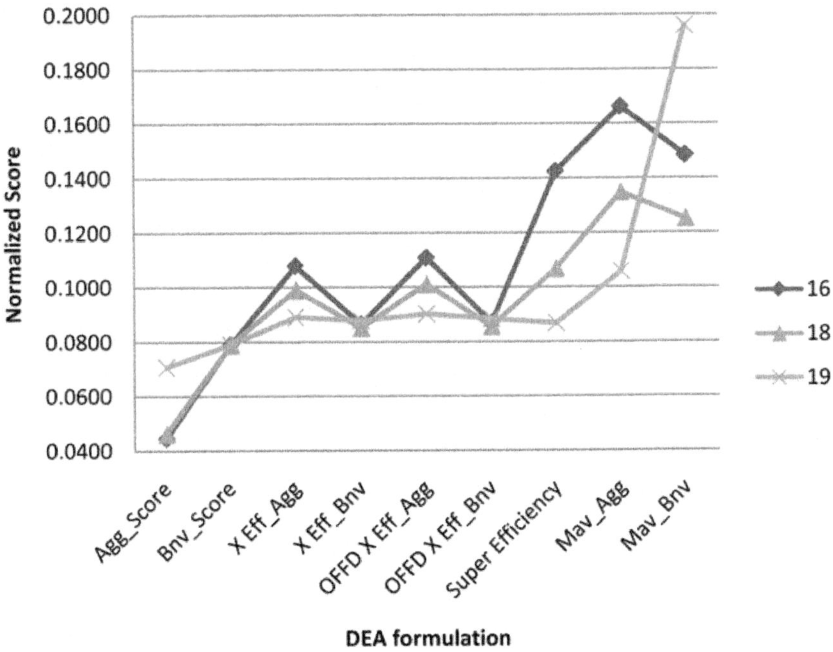

Figure 9. Normalized score for different DEA formulations for competing projects.

Applying the Multi-DEA USF to the 3rd Set

Table 9 and Figure 10 show the performance index for each project with respect to each DEA formulation. It look like that project 1 is highly competitive as it was picked up by some of the DEA techniques. Project 11 shows also some competitive advantage as it has been picked also by some of the individual DEA formulations.

Figure 11 shows the aggregate result for the different projects using the Multi-DEA-USF. Project 1 is the winning project with no close competitors in terms the aggregate index. The Multi-DEA-USF was successful in selecting a highly competitive project.

CONCLUSIONS

The Multi-DEA-USF proposed in this work is used to solve the important six sigma project selection problem which is multi criteria-multi objective. DEA has been used to solve this problem. This work initially solves the six sigma project selection problem using the several DEA formulations proposed in the literature, and concludes that different formulation can give different results in terms of the projects selected. To overcome this diverse DEA result problem, this work proposes using simple normalization and simple weighted score summing as a unified approach to select the winning project.

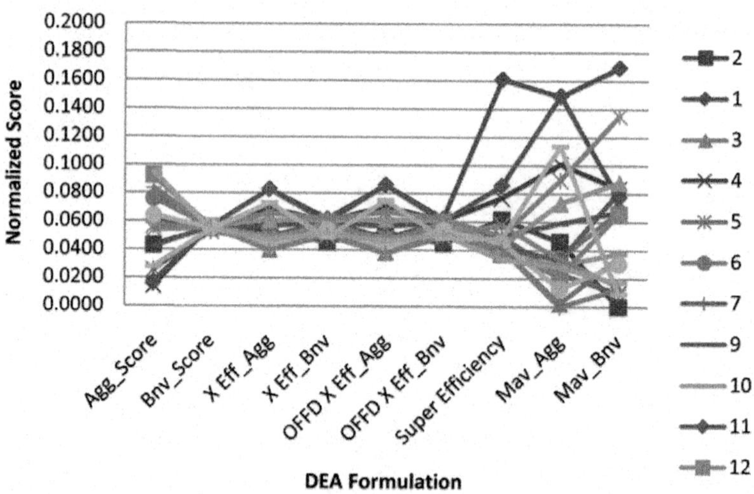

Figure 10: Normalized score for different DEA formulations for efficient projects (3rd set).

Table 9: Summary of scores for efficient projects Part 2 (3rd set).

Project	Agg. Score	Bnv. Score	X Eff. Agg.	X Eff. Bnv.	OFFD X Eff_Agg	OFFD X Eff_Bnv	Super Efficiency	Mav. Agg.	Mav. Bnv.
1	0.1004	0.8951	0.6937	**0.9981**	0.6776	**0.9980**	**5.1648**	0.4416	**0.0019**
2	0.2726	0.8833	0.4435	0.7352	0.4142	0.7213	1.9745	1.2549	0.3601
3	0.5025	0.8907	0.3391	0.8912	0.3043	0.8855	1.3394	1.9490	0.1220
4	0.0911	0.8951	0.5814	0.9735	0.5593	0.9721	2.4629	0.7201	0.0272
5	0.1620	0.8400	0.3880	0.7902	0.3558	0.7792	1.5932	1.5770	0.2655
6	0.4832	0.8945	0.4144	0.9575	0.3836	0.9552	1.2925	1.4132	0.0444
7	0.3875	0.8953	0.4118	0.9716	0.3809	0.9701	1.3662	1.4283	0.0292
9	0.2669	0.8946	0.4764	0.9566	0.4488	0.9543	1.6080	1.0992	0.0454
10	0.5795	0.8898	0.4039	0.8995	0.3725	0.8942	1.1727	1.4762	0.1118
11	0.1074	**0.8958**	**0.6949**	0.9678	**0.6788**	0.9661	2.7466	**0.4391**	0.0333
12	0.5890	0.8943	0.3850	0.9548	0.3526	0.9524	1.1996	1.5974	0.0473
13	0.4042	0.8945	0.5127	0.9769	0.4871	0.9756	1.3690	0.9504	0.0237
15	0.3423	0.8813	0.4128	0.8151	0.3818	0.8054	1.8576	1.4228	0.2268
16	0.3596	0.8952	0.5537	0.9946	0.5302	0.9943	1.6059	0.8062	0.0054
17	0.4064	0.8896	0.3634	0.8770	0.3299	0.8706	1.4556	1.7518	0.1402
18	0.5301	0.8879	0.3335	0.8045	0.2984	0.7942	1.3291	1.9983	0.2430
19	**0.6030**	0.8680	0.3997	0.8257	0.3681	0.8166	1.0734	1.5022	0.2110
20	0.1762	0.8652	0.6136	0.7871	0.5932	0.7759	1.5402	0.6298	0.2704

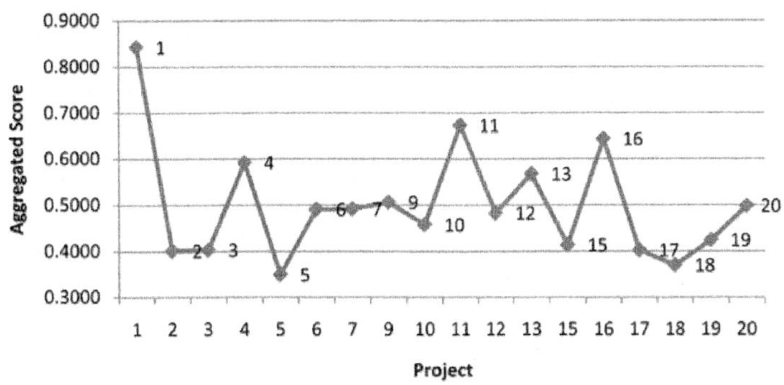

Figure 11: Aggregated score Part 2 (3rd set).

This framework was applied to several case studies and was always successful in picking up "highly competitive" projects. The Multi-DEA-USF was especially successful in picking up stable and well performing project (performing well in all DEA), even though it might have never been selected by any of the DEA formulations (that were excellent projects "shadowed" by slightly better performing projects) and filtering out projects with selective excellent performance; these were projects performing well in some and less well in other DEA formulations.

REFERENCES

1. Henderson, G.R. (2011) Six Sigma Quality Improvement with Minitab. John Wiley & Sons, Hoboken. http://dx.doi.org/10.1002/9781119975328
2. Coronado, R.B. and Antony, J. (2002) Critical Success Factors for the Successful Implementation of Six Sigma Projects in Organisations. The TQM Magazine, 14, 92-99.http://dx.doi.org/10.1108/09544780210416702
3. Antony, J., Antony, F.J., Kumar, M. and Rae Cho, B. (2007) Six Sigma in Service Organisations: Benefits, Challenges and Difficulties, Common Myths, Empirical Observations and Success Factors. International Journal of Quality & Reliability Management, 24, 294-311. http://dx.doi.org/10.1108/02656710710730889
4. Meredith, J.R. and Mantel Jr., S.J. (2011) Project Management: A Managerial Approach. John Wiley & Sons, Hoboken.
5. Saghaei, A. and Didehkhani, H. (2011) Developing an Integrated Model for the Evaluation and Selection of Six Sigma Projects Based on ANFIS and Fuzzy Goal Programming. Expert Systems with Applications, 38, 721-728. http://dx.doi.org/10.1016/j.eswa.2010.07.024
6. Dinesh Kumar, U., Saranga, H., Ramírez-Márquez, J.E. and Nowicki, D. (2007) Six Sigma Project Selection Using Data Envelopment Analysis. The TQM Magazine, 19, 419-441.http://dx.doi.org/10.1108/09544780710817856
7. Zimmerman, J. and Weiss, J. (2005) Six Sigma's Seven Deadly Sins. Quality, 44, 62-66.
8. Deshmukh, S. and Lakhe, R. (2009) Development and Validation of an Instrument for Six Sigma Implementation in Small and Medium Sized Enterprises. 2nd International Conference on Emerging Trends in Engineering and Technology (ICETET).
9. Kazemi, S.M., et al. (2012) Six Sigma Project Selections by Using a Fuzzy Multi Criteria Decision Making Approach: A Case Study in Poly Acryl Corp. CIE42 Proceedings, CIE and SAIIE, 306-1-306-9.
10. Wyman, O. (2007) Keystone of Lean Six Sigma: Strong Middle Management. New York.
11. Antony, J. and Banuelas, R. (2002) Key Ingredients for the Effective Implementation of Six Sigma Program. Measuring Business Excellence,

6, 20-27.http://dx.doi.org/10.1108/13683040210451679

12. Ray, S. and Das, P. (2010) Six Sigma Project Selection Methodology. International Journal of Lean Six Sigma, 1, 293- 309. http://dx.doi.org/10.1108/20401461011096078
13. George, M.O. (2010) The Lean Six Sigma Guide to Doing More with Less: Cut Costs, Reduce Waste, and Lower Your Overhead. John Wiley & Sons, Hoboken.
14. Larson, A. (2003) Demystifying Six Sigma: A Company-Wide Approach to Continuous Improvement. AMACOM, New York.
15. Thomas, P. (2003) The Six Sigma Project Planner. McGraw, New York.
16. Pyzdek, T. and Keller, P.A. (2010) The Six Sigma Handbook: A Complete Guide for Green Belts, Black Belts, and Managers at All Levels. 3rd Edition, McGraw-Hill, New York.
17. Banuelas, R. and Antony, J. (2007) Application of Stochastic Analytic Hierarchy Process within a Domestic Appliance Manufacturer. Journal of the Operational Research Society, 58, 29-38. http://dx.doi.org/10.1057/palgrave.jors.2602060
18. Su, C.T. and Chou, C.J. (2008) A Systematic Methodology for the Creation of Six Sigma Projects: A Case Study of Semiconductor Foundry. Expert Systems with Applications, 34, 2693-2703. http://dx.doi.org/10.1016/j.eswa.2007.05.014
19. Kahraman, C. and Büyüközkan, G. (2008) A Combined Fuzzy AHP and Fuzzy Goal Programming Approach for Effective Six-Sigma Project Selection. Multiple-Valued Logic and Soft Computing, 14, 599-615.
20. Banuelas, R. and Antony, J. (2003) Going from Six Sigma to Design for Six Sigma: An Exploratory Study Using Analytic Hierarchy Process. The TQM Magazine, 15, 334-344.http://dx.doi.org/10.1108/09544780310487730
21. Kelly, W.M. (2002) Three Steps to Project Selection. Six Sigma Forum Magazine, 2, 29-32.
22. Adams, C., Gupta, P. and Wilson, C. (2011) Six Sigma Deployment. Routledge, London.
23. Pyzdek, T. (2000) Selecting Six Sigma Projects.http://www.qualitydigest.com/sept00/html/sixsigma.html
24. Pande, P.S. and Neuman, R.P. (2000) The Six Sigma Way: How GE, Motorola, and Other Top Companies Are Honing Their Performance. 1st Edition, McGraw-Hill, New York.
25. Kumar, M., Antony, J. and Cho, B.R. (2009) Project Selection and Its Impact on the Successful Deployment of Six Sigma.

Business Process Management Journal, 15, 669-686.http://dx.doi.org/10.1108/14637150910987900

26. Charnes, A., Cooper, W.W. and Rhodes, E. (1978) Measuring the Efficiency of Decision Making Units. European Journal of Operational Research, 2, 429-444.http://dx.doi.org/10.1016/0377-2217(78)90138-8
27. Cooper, W.W., Seiford, L.M. and Zhu, J. (2011) Handbook on Data Envelopment Analysis. Springer, Berlin.
28. Lim, S. (2010) A Formulation of Cross-Efficiency in DEA and Its Applications. 40th International Conference on Computers and Industrial Engineering (CIE).
29. Despotis, D.K. (2002) Improving the Discriminating Power of DEA: Focus on Globally Efficient Units. The Journal of the Operational Research Society, 53, 314-323.http://dx.doi.org/10.1057/palgrave.jors.2601253
30. Charnes, A., Cooper, W.W., Lewin, A.Y. and Seiford, L.M. (1994) Data Envelopment Analysis: Theory, Methodology and Applications. Springer, Berlin.
31. Liang, L., Wu, J., Cook, W.D. and Zhu, J. (2008) Alternative Secondary Goals in DEA Cross-Efficiency Evaluation. International Journal of Production Economics, 113, 1025-1030. http://dx.doi.org/10.1016/j.ijpe.2007.12.006
32. Anderson, T.R., Hollingsworth, K. and Inman, L. (2002) The Fixed Weighting Nature of a Cross-Evaluation Model. Journal of Productivity Analysis, 17, 249-255.http://dx.doi.org/10.1023/A:1015012121760
33. Doyle, J. and Green, R. (1994) Efficiency and Cross-Efficiency in DEA: Derivations, Meanings and Uses. The Journal of the Operational Research Society, 45, 567-578.http://dx.doi.org/10.1057/jors.1994.84
34. Charnes, A., Haag, S., Jaska, P. and Semple, J. (1992) Sensitivity of Efficiency Classifications in the Additive Model of Data Envelopment Analysis. International Journal of Systems Science, 23, 789-798. http://dx.doi.org/10.1080/00207729208949248
35. Andersen, P. and Petersen, N.C. (1993) A Procedure for Ranking Efficient Units in Data Envelopment Analysis. Management Science, 39, 1261-1264.http://dx.doi.org/10.1287/mnsc.39.10.1261
36. Zhu, J. (2003) Quantitative Models for Performance Evaluation and Benchmarking: Data Envelopment Analysis with Spreadsheets and DEA Excel Solver. Vol. 51, Springer, Berlin.

APPENDIX I: DATA FOR PROJECTS' INPUT/OUTPUT (TABLES 10-13)

Table 10: The input and output data for the six sigma projects [6] .

Project	Inputs			Outputs				
	Expected Project Cost	Expected Project Duration in Days	Number of Black and Green Belts	Percentage Increase in Customer Satisfaction	Impact on Business Strategy	Financial Impact	Expected Increase in Sigma Quality	Expected Percentage Increase in Productivity
1	212	70	10	11	4	331	0.24	20
2	199	63	3	29	8	342	0.77	23
3	214	88	5	28	4	333	0.33	10
4	280	77	8	29	6	303	0.48	10
5	263	72	11	19	2	240	0.41	11
6	203	70	11	21	5	306	0.52	17
7	196	61	3	31	9	345	0.78	21
8	215	79	5	22	5	264	0.27	6
9	281	71	8	17	3	239	0.66	19
10	233	66	6	10	6	338	0.3	16
11	263	84	10	27	8	310	0.88	19
12	198	60	3	32	8	341	0.51	22
13	220	80	4	15	7	308	0.31	7
14	284	79	6	23	4	325	0.31	7
15	214	87	7	19	7	314	0.54	5
16	235	80	5	27	2	236	0.34	17
17	200	63	4	33	8	339	0.83	23
18	217	75	11	13	5	313	0.74	7
19	198	63	3	31	8	343	0.87	22
20	227	70	5	10	6	317	0.46	17

Table 11: The input and output data for the six sigma projects Part 2 (1st set).

Project Number	Inputs						Outputs					
	Expected Project Cost	Expected Project Duration in Days	Number of Black and Green Belts	Level of Management Commitment Required	Training Hours	Required Level of Leadership and Management Skills	Percentage Increase in Customer Satisfaction	Impact on Business Strategy	Financial Impact	Expected Increase in Sigma Quality	Expected Percentage Increase in Productivity	Percentage Increase in Market Share
1	212	70	10	5	85	2	11	4	331	0.24	20	22
2	199	63	3	5	84	3	29	8	342	0.77	23	22
3	214	88	5	5	74	1	28	4	333	0.33	10	35
4	280	77	8	8	51	6	29	6	303	0.48	10	20
5	263	72	11	4	76	7	19	2	240	0.41	11	16

Continued

6	203	70	11	8	58	6	21	5	306	0.52	17	27
7	196	61	3	5	48	5	31	9	345	0.78	21	30
8	215	79	5	1	52	7	22	5	264	0.27	6	21
9	281	71	8	2	94	7	17	3	239	0.66	19	14
10	233	66	6	8	44	7	10	6	338	0.3	16	6
11	263	84	10	5	54	7	27	8	310	0.88	19	24
12	198	60	3	2	43	10	32	8	341	0.51	22	11
13	220	80	4	4	66	3	15	7	308	0.31	7	2
14	284	79	6	7	40	8	23	4	325	0.31	7	31
15	214	87	7	2	94	3	19	7	314	0.54	5	10
16	235	80	5	8	51	2	27	2	236	0.34	17	18
17	200	63	4	3	45	7	33	8	339	0.83	23	28
18	217	75	11	10	58	5	13	5	313	0.74	7	15
19	198	63	3	3	67	5	31	8	343	0.87	22	30
20	227	70	5	8	46	7	10	6	317	0.46	17	16

Table 12: The input and output data for the six sigma projects Part 2 (2nd set).

	Inputs								Outputs			
Project Number	Expected Project Cost	Expected Project Duration in Days	Number of Black and Green Belts	Level of Management Commitment Required	Training Hours	Required Level of Leadership and Management Skills	Percentage Increase in Customer Satisfaction	Impact on Business Strategy	Financial Impact	Expected Increase in Sigma Quality	Expected Percentage Increase in Productivity	Percentage Increase in Market Share
1	168	81	7	7	55	4	20	10	282	1.31	22	23
2	205	50	4	4	40	10	17	7	247	1.6	20	2
3	224	52	2	9	65	5	34	8	257	0.35	9	30
4	223	61	5	6	95	2	21	8	319	1.75	8	33
5	223	83	9	4	90	10	17	4	326	0.92	12	24
6	214	71	11	10	85	10	12	3	302	0.95	23	27
7	228	71	10	9	75	5	25	1	262	0.82	19	27
8	221	87	1	6	65	2	13	5	297	1.22	18	14
9	168	71	4	7	80	3	11	7	232	0.77	7	23
10	240	45	3	6	100	5	25	3	293	0.89	6	6
11	158	82	4	3	100	6	17	6	301	0.86	23	25
12	209	73	11	4	85	3	19	3	290	1.35	13	2
13	169	67	7	5	100	7	28	3	252	1.41	15	10

Continued

14	172	57	9	3	90	8	32	7	255	1.99	14	2
15	180	56	10	9	50	3	19	9	225	1.9	15	4
16	168	67	2	2	80	2	35	9	320	0.86	10	29
17	196	55	10	3	70	3	34	3	274	0.59	9	25
18	236	47	2	2	45	4	15	4	253	1.86	23	12
19	185	52	9	3	50	5	29	9	317	0.76	24	34
20	203	56	9	5	55	6	26	3	235	1.21	17	2

Table 13: The input and output data for the six sigma projects Part 2 (3rd set).

	Inputs						Outputs						
Project Number	Expected Project Cost	Expected Project Duration in Days	Number of Black and Green Belts	Level of Management Commitment Required	Training Hours	Required Level of Leadership and Management Skills	Percentage Increase in Customer Satisfaction	Impact on Business Strategy	Financial Impact	Expected Increase in Sigma Quality	Expected Percentage Increase in Productivity	Percentage Increase in Market Share
1	205	49	8	1	95	1	34	9	234	1.86	16	16
2	215	89	2	3	75	9	29	7	254	0.41	16	14
3	186	52	5	9	95	10	35	1	249	0.7	6	27
4	228	52	6	1	60	8	16	10	212	1	25	28
5	233	52	6	10	40	1	12	1	277	1.26	10	7
6	218	49	10	8	85	3	19	1	325	1.82	11	18
7	175	65	8	5	60	4	23	10	336	0.63	11	16
8	222	75	9	6	115	7	24	7	309	1.94	5	23
9	162	83	6	3	115	2	35	4	257	1.01	16	25
10	195	81	11	5	70	8	22	6	245	1.62	10	27
11	205	77	1	10	40	2	20	8	304	1.17	13	10
12	214	66	6	6	95	7	35	1	332	0.83	24	24
13	184	77	4	6	105	5	16	8	339	1.01	24	23
14	216	89	6	3	120	8	27	8	212	1.51	7	6
15	230	68	8	5	45	8	35	5	272	1.85	17	5
16	155	59	3	7	75	10	27	6	219	1.81	12	18
17	162	81	7	7	85	9	17	4	238	0.49	20	34
18	228	55	4	4	95	4	23	2	333	0.46	15	12
19	188	55	10	4	100	7	10	4	229	1.81	10	21
20	187	66	2	10	90	2	32	10	218	1.39	15	8

APPENDIX II: SUMMARY OF PROJECTS' SCORES (TABLES 14-17)

Table 14: Summary of project scores Part 1.

Project	Simple	Agg. Score	Bnv. Score	X Eff. Agg.	X Eff. Bnv.	OFFD X Eff_Agg	OFFD X Eff_Bnv	Super Efficiency	Mav. Agg.	Mav. Bnv.
1	0.8870	0.7666	0.7666	0.6159	0.7688	0.6016	0.7626	0.8870	0.4401	0.1537
2	1.0000	0.2880	0.7233	0.9542	0.9613	0.9518	0.9593	1.0455	0.0479	0.0403
3	0.8840	0.7667	0.7667	0.6181	0.7402	0.6041	0.7326	0.8840	0.4303	0.1944
4	0.7070	0.5767	0.5767	0.5191	0.6047	0.5092	0.5993	0.7070	0.3621	0.1693
5	0.5873	0.7560	0.7560	0.4244	0.5111	0.4158	0.5071	0.5873	0.3839	0.1491
6	0.8564	0.7682	0.7682	0.6371	0.7819	0.6255	0.7780	0.8564	0.3442	0.0953
7	1.0000	0.2827	0.7619	0.9761	0.9884	0.9749	0.9878	1.1365	0.0244	0.0117
8	0.6976	0.7759	0.7759	0.5025	0.5979	0.4922	0.5926	0.6976	0.3883	0.1668
9	0.7327	0.5528	0.5528	0.4724	0.5219	0.4587	0.5108	0.7327	0.5509	0.4039
10	0.9011	0.7393	0.7393	0.6226	0.7390	0.6079	0.7304	0.9011	0.4473	0.2194
11	0.7615	0.5131	0.5131	0.5793	0.6652	0.5697	0.6602	0.7615	0.3146	0.1447
12	1.0000	0.3051	0.7435	0.9401	0.9637	0.9370	0.9618	1.0556	0.0637	0.0377
13	0.7954	0.7710	0.7710	0.5603	0.6553	0.5479	0.6480	0.7954	0.4196	0.2136
14	0.7239	0.7480	0.7480	0.5001	0.5974	0.4883	0.5907	0.7239	0.4475	0.2118
15	0.8336	0.7691	0.7691	0.5581	0.6857	0.5435	0.6779	0.8336	0.4938	0.2156
16	0.6963	0.5896	0.5896	0.5073	0.5568	0.4973	0.5495	0.6963	0.3727	0.2505
17	1.0000	0.5155	0.7028	0.9204	0.9687	0.9162	0.9670	1.0452	0.0865	0.0323
18	0.8244	0.7432	0.7432	0.5644	0.7093	0.5507	0.7033	0.8244	0.4608	0.1622
19	1.0000	0.2670	0.7356	0.9719	0.9775	0.9704	0.9763	1.1154	0.0289	0.0231
20	0.7994	0.7519	0.7519	0.6228	0.7113	0.6135	0.7066	0.7994	0.2835	0.1238

Table 15: Summary of project scores Part 2 (1st set).

Project	Simple	Agg. Score	Bnv. Score	X Eff. Agg.	X Eff. Bnv.	OFFD X Eff_Agg	OFFD X Eff_Bnv	Super Efficiency	Mav. Agg.	Mav. Bnv.
1	1.0000	0.2632	0.7594	0.4321	0.7332	0.4022	0.7192	1.2238	1.3145	0.3638
2	1.0000	0.2880	0.7777	0.6731	0.8348	0.6559	0.8261	1.5347	0.4857	0.1979
3	1.0000	0.0969	0.7808	0.5566	0.8394	0.5332	0.8310	3.3939	0.7967	0.1913
4	0.8502	0.5758	0.5758	0.3856	0.6470	0.3612	0.6363	0.8502	1.2049	0.3140
5	0.5998	0.7717	0.7717	0.2727	0.4873	0.2555	0.4814	0.5998	1.1996	0.2308
6	0.8643	0.6357	0.6357	0.4026	0.6536	0.3783	0.6425	0.8643	1.1468	0.3223
7	1.0000	0.2827	0.7804	0.7485	0.9695	0.7353	0.9679	1.3276	0.3359	0.0314
8	1.0000	0.1760	0.7016	0.4939	0.7050	0.4672	0.6895	2.1999	1.0249	0.4184
9	1.0000	0.2938	0.4359	0.4223	0.5247	0.3919	0.4997	1.1686	1.3678	0.9060

Continued

10	1.0000	0.6804	0.7047	0.3534	0.6784	0.3194	0.6615	1.0087	1.8295	0.4741
11	0.9400	0.5316	0.5316	0.5154	0.7344	0.4931	0.7236	0.9400	0.8237	0.2798
12	1.0000	0.2493	0.7698	0.6694	0.8511	0.6520	0.8433	1.4473	0.4939	0.1749
13	1.0000	0.5647	0.7153	0.3736	0.7351	0.3406	0.7212	1.0237	1.6768	0.3603
14	1.0000	0.3979	0.7065	0.3869	0.6726	0.3547	0.6553	1.2400	1.5844	0.4869
15	1.0000	0.3169	0.7351	0.4360	0.7457	0.4064	0.7323	1.4416	1.2934	0.3410
16	1.0000	0.3381	0.7136	0.4479	0.6325	0.4188	0.6131	1.3120	1.2328	0.5811
17	1.0000	0.3585	0.7806	0.7513	0.9660	0.7382	0.9642	1.3115	0.3310	0.0352
18	0.8865	0.5747	0.5747	0.3599	0.6439	0.3322	0.6311	0.8865	1.4631	0.3769
19	1.0000	0.2670	0.7805	0.7647	0.9355	0.7523	0.9321	1.3145	0.3077	0.0690
20	0.9087	0.7022	0.7022	0.4061	0.6587	0.3796	0.6455	0.9087	1.2379	0.3796

Table 16: Summary of project scores Part 2 (2nd set).

Project	Simple	Agg. Score	Bnv. Score	X Eff. Agg.	X Eff. Bnv.	OFFD X Eff_Agg	OFFD X Eff_Bnv	Super Efficiency	Mav. Agg.	Mav. Bnv.
1	1.0000	0.3951	0.8546	0.5903	0.9270	0.5688	0.9232	1.2355	0.6939	0.0787
2	1.0000	0.3817	0.8549	0.5258	0.8946	0.5009	0.8891	1.2289	0.9018	0.1178
3	1.0000	0.1701	0.7673	0.5132	0.6834	0.4876	0.6667	1.2864	0.9486	0.4633
4	1.0000	0.1926	0.8512	0.6282	0.9109	0.6086	0.9062	1.7056	0.5919	0.0978
5	0.8014	0.8465	0.8465	0.3657	0.6657	0.3428	0.6586	0.8014	1.1912	0.2037
6	0.8378	0.7386	0.7386	0.3557	0.7011	0.3303	0.6939	0.8378	1.3553	0.1949
7	0.7486	0.6435	0.6435	0.3845	0.6666	0.3654	0.6623	0.7486	0.9468	0.1230
8	1.0000	0.1254	0.7783	0.6495	0.7856	0.6311	0.7743	2.0981	0.5396	0.2729
9	0.8160	0.6592	0.6592	0.4092	0.6512	0.3878	0.6425	0.8160	0.9940	0.2532
10	1.0000	0.5189	0.8236	0.3507	0.7007	0.3165	0.6850	1.2067	1.8516	0.4271
11	1.0000	0.3905	0.8484	0.5696	0.8325	0.5470	0.8236	1.3568	0.7556	0.2013
12	0.9037	0.7471	0.7471	0.3830	0.7442	0.3556	0.7358	0.9037	1.3596	0.2143
13	0.9513	0.7212	0.7212	0.4455	0.8080	0.4188	0.8005	0.9513	1.1354	0.1773
14	1.0000	0.3508	0.8542	0.5317	0.9372	0.5071	0.9339	1.5189	0.8806	0.0670
15	1.0000	0.3495	0.8549	0.5398	0.9254	0.5155	0.9214	1.4298	0.8526	0.0807
16	1.0000	0.1768	0.8547	0.8248	0.9736	0.8156	0.9723	3.0431	0.2124	0.0271
17	1.0000	0.4595	0.7632	0.4897	0.7533	0.4629	0.7403	1.1292	1.0419	0.3275
18	1.0000	0.1841	0.8544	0.7577	0.9595	0.7449	0.9574	2.2811	0.3198	0.0422
19	1.0000	0.2815	0.8549	0.6816	0.9893	0.6648	0.9887	1.8540	0.4672	0.0109
20	0.9549	0.6852	0.6852	0.4071	0.7653	0.3783	0.7553	0.9549	1.3456	0.2477

Table 17: Summary of project scores Part 2 (3rd set).

Project	Simple	Agg. Score	Bnv. Score	X Eff. Agg.	X Eff. Bnv.	OFFD X Eff_Agg	OFFD X Eff_Bnv	Super Efficiency	Mav. Agg.	Mav. Bnv.
1	1.0000	0.1004	0.8951	0.6937	0.9981	0.6776	0.9980	5.1648	0.4416	0.0019
2	1.0000	0.2726	0.8833	0.4435	0.7352	0.4142	0.7213	1.9745	1.2549	0.3601
3	1.0000	0.5025	0.8907	0.3391	0.8912	0.3043	0.8855	1.3394	1.9490	0.1220
4	1.0000	0.0911	0.8951	0.5814	0.9735	0.5593	0.9721	2.4629	0.7201	0.0272
5	1.0000	0.1620	0.8400	0.3880	0.7902	0.3558	0.7792	1.5932	1.5770	0.2655
6	1.0000	0.4832	0.8945	0.4144	0.9575	0.3836	0.9552	1.2925	1.4132	0.0444
7	1.0000	0.3875	0.8953	0.4118	0.9716	0.3809	0.9701	1.3662	1.4283	0.0292
8	0.9757	0.8372	0.8372	0.4011	0.8842	0.3709	0.8794	0.9757	1.4326	0.1036
9	1.0000	0.2669	0.8946	0.4764	0.9566	0.4488	0.9543	1.6080	1.0992	0.0454
10	1.0000	0.5795	0.8898	0.4039	0.8995	0.3725	0.8942	1.1727	1.4762	0.1118
11	1.0000	0.1074	0.8958	0.6949	0.9678	0.6788	0.9661	2.7466	0.4391	0.0333
12	1.0000	0.5890	0.8943	0.3850	0.9548	0.3526	0.9524	1.1996	1.5974	0.0473
13	1.0000	0.4042	0.8945	0.5127	0.9769	0.4871	0.9756	1.3690	0.9504	0.0237
14	0.9250	0.5656	0.5656	0.3299	0.5811	0.2986	0.5630	0.9250	1.8039	0.5919
15	1.0000	0.3423	0.8813	0.4128	0.8151	0.3818	0.8054	1.8576	1.4228	0.2268
16	1.0000	0.3596	0.8952	0.5537	0.9946	0.5302	0.9943	1.6059	0.8062	0.0054
17	1.0000	0.4064	0.8896	0.3634	0.8770	0.3299	0.8706	1.4556	1.7518	0.1402
18	1.0000	0.5301	0.8879	0.3335	0.8045	0.2984	0.7942	1.3291	1.9983	0.2430
19	1.0000	0.6030	0.8680	0.3997	0.8257	0.3681	0.8166	1.0734	1.5022	0.2110
20	1.0000	0.1762	0.8652	0.6136	0.7871	0.5932	0.7759	1.5402	0.6298	0.2704

CITATION

CHAPTER 1

Q. Zhang, Z. Shen and X. Zhang, "Research on Location Routing Problem (LRP) Based on Chaos Search (CS) and Empirical Analysis," *American Journal of Operations Research*, Vol. 3 No. 1A, 2013, pp. 147-152. doi:10.4236/ajor.2013.31A014.

CHAPTER 2

Rahman, A. , Yella, S. and Dougherty, M. (2014) Simulation and Optimization Techniques for Sawmill Yard Operations—A Literature Review. *Journal of Intelligent Learning Systems and Applications*, **6**, 21-34. doi:10.4236/jilsa.2014.61003.

CHAPTER 3

N. Yan, Z. Chen, Y. Shi, Z. Wang and G. Huang, "Using Non-Additive Measure for Optimization-Based Nonlinear Classification," *American Journal of Operations Research*, Vol. 2 No. 3, 2012, pp. 364-373. doi:10.4236/ajor.2012.23044.

CHAPTER 4

J. Tripto, O. Assaraf and M. Amit, "Mapping What They Know: Concept Maps as an Effective Tool for Assessing Students' Systems Thinking," *American Journal of Operations Research*, Vol. 3 No. 1A, 2013, pp. 245-258. doi:10.4236/ajor.2013.31A022.

CHAPTER 5

Ibe, O. (2015) M/G/1 Vacation Queueing Systems with Server Timeout. *American Journal of Operations Research*,**5**, 77-88. doi: 10.4236/ajor.2015.52007.

CHAPTER 6

E. Kujawski, "Optimization of Critical Systems for Robustness in a Multistate World," *American Journal of Operations Research*, Vol. 3 No. 1A, 2013, pp. 127-137. doi: 10.4236/ajor.2013.31A012

CHAPTER 7

I. Ali and S. Suhaib Hasan, "Bi-Criteria Optimization Technique in Stochastic System Maintenance Allocation Problem," *American Journal of Operations Research*, Vol. 3 No. 1, 2013, pp. 17-29. doi:10.4236/ajor.2013.31002.

CHAPTER 8

M. Shalaby and M. Kashkoush, "A Particle Swarm Optimization Algorithm for a 2-D Irregular Strip Packing Problem," *American Journal of Operations Research*, Vol. 3 No. 2, 2013, pp. 268-278. doi:10.4236/ajor.2013.32024.

CHAPTER 9

Parwanto, N. , Morohosi, H. and Oyama, T. (2015) Applying Network Flow Optimization Techniques to Improve Relief Goods Transport Strategies under Emergency Situation. *American Journal of Operations Research*, **5**, 95-111. doi: 10.4236/ajor.2015.53009.

CHAPTER 10

S. Suwannarongsri, T. Bunnag and W. Klinbun, "Optimization of Energy Resource Management for Assembly Line Balancing Using Adaptive Current Search," *American Journal of Operations Research*, Vol. 4 No. 1, 2014, pp. 8-21. doi: 10.4236/ajor.2014.41002.

CHAPTER 11

Z. Varga, M. Gámez and I. López, "Recent Developments in Monitoring of Complex Population Systems," *American Journal of Operations Research*, Vol. 3 No. 1A, 2013, pp. 167-180. doi: 10.4236/ajor.2013.31A016.

CHAPTER 12

Arafah, M. (2015) Selecting the Six Sigma Project: A Multi Data Envelopment Analysis Unified Scoring Framework.*American Journal of Operations Research*, **5**, 129-150. doi: 10.4236/ajor.2015.53011.

INDEX